NANODISPERSIONS FOR DRUG DELIVERY

NANODISPERSIONS FOR DRUG DELIVERY

Edited by
Raj K. Keservani, MPharm
Anil K. Sharma, MPharm, PhD

Apple Academic Press Inc.
3333 Mistwell Crescent
Oakville, ON L6L 0A2 Canada

Apple Academic Press Inc.
9 Spinnaker Way
Waretown, NJ 08758 USA

Library and Archives Canada Cataloguing in Publication

Nanodispersions for drug delivery / edited by Raj K. Keservani, MPharm, Anil K. Sharma, MPharm, PhD.
Includes bibliographical references and index.
Issued in print and electronic formats.
ISBN 978-1-77188-701-4 (hardcover).--ISBN 978-1-351-04756-2 (PDF)
1. Drug delivery systems. 2. Nanotechnology. I. Keservani, Raj K., 1981-, editor II. Sharma, Anil K., 1980-, editor

RS210.N36 2018	615'.6	C2018-904368-7	C2018-904369-5

Library of Congress Cataloging-in-Publication Data

Names: Keservani, Raj K., 1981- editor. | Sharma, Anil K., 1980- editor.
Title: Nanodispersions for drug delivery / editors, Raj K. Keservani, Anil K. Sharma.
Description: Toronto ; New Jersey : Apple Academic Press, 2019. | Includes bibliographical references and index. Identifiers: LCCN 2018034930 (print) | LCCN 2018036186 (ebook) | ISBN 9781351047562 (ebook) | ISBN 9781771887014 (hardcover)
Subjects: | MESH: Drug Delivery Systems | Nanostructures | Emulsions--administration & dosage
Classification: LCC RS419 (ebook) | LCC RS419 (print) | NLM QV 785 | DDC 615.1/9--dc23
LC record available at https://lccn.loc.gov/2018034930

ABOUT THE EDITORS

Raj K. Keservani

Raj K. Keservani, MPharm, is in the Faculty of B. Pharmacy, CSM Group of Institutions, Allahabad, India. He has more than 10 years of academic experience in various institutes of India imparting pharmaceutical education. He has published 35 peer-reviewed papers in the field of pharmaceutical sciences in national and international reputed journals, 16 book chapters, 2 coauthored books, and 10 edited books. He is also active as a reviewer for several scientific international journals. His research interests encompass nutraceutical and functional foods, novel drug delivery systems (NDDS), transdermal drug delivery/drug delivery, health science/life science, and biology/cancer biology/neurobiology. Mr. Keservani graduated (B. Pharmacy) from the Department of Pharmacy, Kumaun University, Nainital, Uttarakhand, India in 2005. Afterwards, he qualified GATE in same year conducted by IIT Mumbai. He received his Master of Pharmacy (M. Pharmacy) from the School of Pharmaceutical Sciences, Rajiv Gandhi Proudyogiki Vishwavidyalaya, Bhopal, Madhya Pradesh, India, in 2008 with a specialization in pharmaceutics.

Anil K. Sharma

Anil K. Sharma, MPharm, PhD, is currently working as an assistant professor at the Department of Pharmacy, School of Medical and Allied Sciences, GD Goenka University, Gurugram, India. He has more than nine years of academic experience in pharmaceutical sciences. He graduated (B. Pharmacy) from the University of Rajasthan, Jaipur, India, in 2005. Afterwards, he qualified GATE in same year, conducted by IIT Mumbai. He received his Master of Pharmacy (M. Pharmacy) from the School of Pharmaceutical Sciences, Rajiv Gandhi Proudyogiki Vishwavidyalaya, Bhopal, Madhya Pradesh, India, in 2007, with a specialization in pharmaceutics. He has earned his PhD from the University of Delhi. In addition, he has published 29 peer-reviewed papers in the field of pharmaceutical sciences in national and international reputed journals, 15 book chapters, and 10 edited books. His research interests encompass nutraceutical and functional foods, novel drug delivery systems (NDDS), drug delivery, nanotechnology, health science/life science, and biology/cancer biology/neurobiology.

CONTENTS

THE PRESENT BOOK IS DEDICATED TO

OUR BELOVED

AASHNA

ANIKA

ATHARVA

AND

VIHAN

LIST OF CONTRIBUTORS

Jayanthi Abraham
Department of Biomedical Sciences, Microbial Biotechnology Laboratory, School of Bio Sciences and Technology, VIT University, Vellore 632014, Tamil Nadu, India

Javed Ahmad
Department of Pharmaceutics, College of Pharmacy, Najran University, Kingdom of Saudi Arabia (KSA)

Mohammad Zaki Ahmad
Department of Pharmaceutics, College of Pharmacy, Najran University, Kingdom of Saudi Arabia (KSA)

Mohammed Saeed N. Algahtani
Department of Pharmaceutics, College of Pharmacy, Najran University, Kingdom of Saudi Arabia (KSA)

Fatemeh Bahadori
Department of Pharmaceutical Biotechnology, Faculty of Pharmacy, Bezmialem Vakif University, Fatih 34093, Istanbul, Turkey

Fernanda Belincanta Borghi-Pangoni
Postgraduate Program in Pharmaceutical Sciences, Laboratory of Research and Development of Drug Delivery Systems, Department of Pharmacy, State University of Maringá, Maringá, Paraná, Brazil

Marcos Luciano Bruschi
Postgraduate Program in Pharmaceutical Sciences, Laboratory of Research and Development of Drug Delivery Systems, Department of Pharmacy, State University of Maringá, Maringá, Paraná, Brazil

M. M. Chogale
Department of Pharmaceutical Sciences and Technology, Institute of Chemical Technology, Nathalal Parekh Marg, Matunga (E), Mumbai 400019, Maharashtra, India

Jéssica Bassi da Silva
Postgraduate Program in Pharmaceutical Sciences, Laboratory of Research and Development of Drug Delivery Systems, Department of Pharmacy, State University of Maringá, Maringá, Paraná, Brazil

A. A. Date
The Center for Nanomedicine, Wilmer Eye Institute, Johns Hopkins School of Medicine, Baltimore, MD 21205, USA

Sabrina Barbosa de Souza Ferreira
Postgraduate Program in Pharmaceutical Sciences, Laboratory of Research and Development of Drug Delivery Systems, Department of Pharmacy, State University of Maringá, Maringá, Paraná, Brazil

Marcus Vinícius Dias-Souza
Health Sciences Faculty, University Vale do Rio Doce, Minas Gerais, Brazil
Integrated Pharmacology and Drug Interactions Research Group (GPqFAR), Brazil

J. I. Disouza
Department of Pharmaceutics, Tatyasaheb Kore College of Pharmacy, Warananagar, Panhala, Kolhapur, Maharashtra, India

Zahra Eskandari
Department of Chemistry, Biochemistry Division, Faculty of Sciences and Arts, Yildiz Technical University, Istanbul, Turkey
IBSB, Department of Bioengineering, Marmara University, Kadikoy 34722, Istanbul, Turkey

Anudurga Gajendiran
Department of Biomedical Sciences, Microbial Biotechnology Laboratory, School of Bio Sciences and Technology, VIT University, Vellore 632014, Tamil Nadu, India

V. N. Ghodake
Department of Pharmaceutical Sciences and Technology, Institute of Chemical Technology, Nathalal Parekh Marg, Matunga (E), Mumbai 400019, Maharashtra, India

Bey Hing Goh
School of Pharmacy, Monash University, Bandar Sunway, Selangor 45700, Malaysia

Nayan A. Gujarathi
Sandip Institute of Pharmaceutical Sciences, Pune University, Nashik, India

Ashish S. Jain
Department of Pharmaceutics, Shri D. D. Vispute College of Pharmacy & Research Center, Mumbai University, New Panvel, India

Nirmala Rachel James
Department of Chemistry, Indian Institute of Space Science and Technology (IIST), Valiamala, Thiruvananthapuram 695547, Kerala, India

Mariana Volpato Junqueira
Postgraduate Program in Pharmaceutical Sciences, Laboratory of Research and Development of Drug Delivery Systems, Department of Pharmacy, State University of Maringá, Maringá, Paraná, Brazil

P. S. Kakade
Department of Pharmaceutical Sciences and Technology, Institute of Chemical Technology, Nathalal Parekh Marg, Matunga (E), Mumbai 400019, Maharashtra, India

Raj K. Keservani
Faculty of B. Pharmacy, CSM Group of Institutions, Allahabad, India

Tahir Mehmood Khan
School of Pharmacy, Monash University, Bandar Sunway, Selangor 45700, Malaysia
The Institute of Pharmaceutical Sciences, University of Veterinary and Animal Science, Lahore, Pakistan

Learn-Han Lee
School of Pharmacy, Monash University, Bandar Sunway, Selangor 45700, Malaysia

A. S. Manjappa
Department of Pharmaceutics, Tatyasaheb Kore College of Pharmacy, Warananagar, Panhala, Kolhapur, Maharashtra, India

Aristides Ávilo Nascimento
Head, Pharmaceutical School Internship Unit, Institute of Applied Theology (INTA), Ceará, Brazil
Integrated Pharmacology and Drug Interactions Research Group (GPqFAR), Brazil

Ebru Toksoy Oner
IBSB, Department of Bioengineering, Marmara University, Kadikoy 34722, Istanbul, Turkey

Vandana B. Patravale
Department of Pharmaceutical Sciences and Technology, Institute of Chemical Technology, Nathalal Parekh Marg, Matunga (E), Mumbai 400019, Maharashtra, India

S. A. Payghan
Department of Pharmaceutics, Tatyasaheb Kore College of Pharmacy, Warananagar, Panhala, Kolhapur, Maharashtra, India

Bhushan Rajendra Rane
Department of Pharmaceutics, Shri D. D. Vispute College of Pharmacy & Research Center, Mumbai University, New Panvel, India

Kirti Rani
Amity Institute of Biotechnology, Amity University Uttar Pradesh, Noida Sector 125, Noida 201303, Uttar Pradesh, India

Md. Rizwanullah
Department of Pharmaceutics, School of Pharmaceutical Education and Research, Jamia Hamdard, New Delhi 110062, India

P. R. Sarika
Department of Pharmaceutics and Pharmaceutical Chemistry, The University of Utah, Skaggs Research Building, Rm 2760, 30S, 2000E, Salt Lake City, UT 84112, United States of America

Kifayat Ullah Shah
Department of Pharmacy, Quaid-i-Azam University, Islamabad, Pakistan

Thalles Yuri Loiola Vasconcelos
Integrated Pharmacology and Drug Interactions Research Group (GPqFAR), Brazil
Pharmacist of School Internship Unit, Institute of Applied Theology (INTA), Ceará, Brazil

LIST OF ABBREVIATIONS

%EE	% entrapment efficiency
1-H NMR	proton-nuclear magnetic resonance
5-FU	5-fluorouracil
AA	acrylic acid
AFM	atomic force microscopy
AG	alginate sodium
APIs	active pharmaceutical ingredients
ARV	antiretroviral
ATRP	atom transfer radical polymerization
AUC	area under the concentration
AUC	area under the curve
BCS	Biopharmaceutical Classification Systems
BDP	beclomethasone dipropionate
CCD	crosslinkable carbon dots
CD	cyclodextrin
Ce6	chlorin e6
CFC	critical flocculation concentration
CMAs	critical material attributes
CMC	critical micelle concentration
CMCS	carboxymethyl chitosan
CPPs	critical process parameters
CQAs	critical quality attributes
cryo-EM	cryoelectron microscopy
CS	chitosan
CS	chondroitin sulfate
CS-g-PNIPAm	chitosan-g-poly(N-isopropylacrylamide)
CTS	chitosan chains
Cys	cystamine
DDS	drug delivery systems
DEAP	3-diethylaminopropyl isothiocyanate
DEX	dextran
DEX-LA-HEMA	hydroxyl-terminated HEMA-lactate
DHPS	3,5-dihydroxy-4-iso-propylstilbene
DLS	dynamic light scattering

DOX	doxorubicin
DOX-HCL	DOX hydrochloride
DPIs	dry powder inhalers
DSC	differential scanning calorimetry
ECM	extracellular matrix
EDA	ethylenediamine
rPA	recombinant *Bacillus anthracis* protective antigen
ERα	estrogen receptor alpha
FFTEM	freeze-fracture transmission electron microscopy
FITC	fluorescein isothiocyanate
FTIR	Fourier-transform infrared
GAA	gum arabic aldehyde
GAG	glycosaminoglycan
GCS	glycol chitosan
$GdAl_2$	gadolinium–aluminum alloy
GI	gastrointestinal
GIT	gastrointestinal tract
GMP	good manufacturing practice
GPC	gel permeation chromatography
GRAS	generally recognized as safe
GSH	glutathione
GSH	glutathione tripeptide
GSSG	glutathione disulfide
GzmB	granzyme B
HA	hyaluronic acid
HA–EGCG	hyaluronic acid–epigallocatechin gallate conjugates
HIV	human immunodeficiency virus
HLB	hydrophilic–lipophilic balance
HPBCD	hydroxypropyl-β-cyclodextrin
HPC	hydroxypropylcellulose
HPH	high-pressure homogenization
HPMA	methacrylamide
HPMC	hydroxyl propyl methyl cellulose
HSP	heat shock protein
i.v.	intravenous
IFN	interferon
IID	Inactive Ingredients Database
IIST	Indian Institute of Space Science and Technology
INTA	Institute of Applied Theology

IOP	intraocular pressure
KAS	Kingdom of Saudi Arabia
LA	lipoic acid
LBDDS	lipid-based drug delivery systems
LCST	lowest critical solution temperature
LD	laser diffractometry
MALLS	multiangle laser light scattering
MeHA	methacrylated hyaluronic acid
MMP	matrix metalloproteinase
MMs	mixed micelles
MPEG-PCL	methoxy poly(ethylene glycol)-poly(ε-caprolactone)
MPS	mononuclear phagocytic system
MRI	magnetic resonance imaging
MRT	mean residence time
MW	molecular weight
NaTC	sodium taurocholate
NEs	nanoemulsions
NIPAAm	N-isopropylacrylamide
NIR	near infrared
NMR	nuclear magnetic resonance
NPs	nanoparticles
NSAIDs	nonsteroidal anti-inflammatory drugs
O/W	oil-in-water
OA	osteoarthritis
ODN	oligonucleotides
OEG	oligoethyleneglycol
ORI	oridonin
OXA	oxaliplatin
P-gp	P-glycoprotein
PAA	poly(acrylic acid)
PAAc	poly(acrylic acid)
PAE	poly(beta-amino ester)
pAsp	poly(L-aspartic acid)
PBA	poly(N-isopropylacrylamide) functionalized with aminophenylboronic acid
PCL	poly(epsilon-caprolactone)
PCS	photon correlation spectroscopy
PDLLA	poly(D,L-lactide)
PDMS	poly(dimethylsiloxane)

PDS	pyridyl disulfide
PDT	photodynamic therapy
PEG	polyethylene glycol
PEG$_{2000}$-DSPE	1,2-distearoyl-sn-glycero-3-phosphoethanolamine (polyethylene glycol)$_{2000}$
PEG-MA	poly(ethylene glycol)-maleic anhydride
PEI	polyethyleneimine
PEO	poly(ethylene oxide)
PFPE	perfluoropolyether
pGlu	poly(L-glutamic acid)
PHEA	poly(hydroxyethyl acrylate)
PHEA	α,β-poly(N-2-hydroxyethyl)-D,L-aspartamide
pHis	poly(L-histidine)
PI	polydispersity index
PIC	polyion complex
PIT	phase-inversion temperature
PLA	poly(L-lactide)
PLAs	polylactides
PLGA	poly(lactide co-glycolides)
PLP	poly(lactide)-g-pullulan
PMs	polymeric micelles
PNIPAM	poly(N-isopropyl acrylamide)
Poly(ACc)	poly(acrylic acid)
Poly(NIPA)	poly(N-isopropylacrylamide)
PPO	poly(propylene oxide)
PRINT	particle replication in nonwetting templates
PS	average particle size
PS	photosensitizer
PTX	paclitaxel
PVA	polyvinyl alcohol
PVCL	poly(N-vinylcaprolactam)
PVP	poly(N-vinyl pyrrolidone)
PVP	polyvinyl pyrrolidone
QbD	quality by design
QTPP	quality target product profile
RAFT	reversible addition-fragmentation chain transfer
RES	reticuloendothelial system
RESOLV	rapid expansion of supercritical solutions into a liquid solvent

RESS	rapid expansion of supercritical solution
RMs	reverse micelles
RNAi	ribonucleic acid (RNA) interference
SANS	small-angle neutron scattering
SAS	supercritical antisolvent precipitation
SAXS	small-angle-X-ray scattering
SEM	scanning electron microscope
SFRP	stable free radical polymerization
siRNA	small interfering RNA
SNEDDS	self-nanoemulsifying drug delivery system
SOR	surfactant-to-oil ratio
SQUID	superconducting quantum interference device
Tat	cell-penetrating peptide
TEM	transmission electron microscopy
TPGS	D-α-tocopheryl polyethylene-glycol 1000 succinate
TPGS	tocopheryl polyethylene glycol succinate
TPP	target product profile
TPP	tripolyphosphate
TRA	tretinoin
TRF	tocotrienol-rich fraction
UCST	upper critical solution temperature
UV	ultraviolet
VP	N-vinylpirrolidone
VPTT	volume phase transition temperature
W/O	water-in-oil
XRD	X-ray diffraction

PREFACE

The emergence and evolution of nanotechnology has entirely revolution-ized the technical aspects that eventually have paved the way for excellence in technical operations. This has been assisting mankind via a variety of means. The applications pertaining to drug delivery are of prime interest for industry, healthcare personnel, and researchers in academics. There have been consistent efforts to overcome shortcomings of conventional dosage forms by exploiting principles of nanoscience that were in parlance for engi-neering fields.

Nanodispersions possess globules/particles in sizes usually below 1000 nm in which a drug is dispersed in a continuous medium employing surface-active agents as stabilizers. In general, the drug release from such disper-sions occurs in a sustained manner; nevertheless, the polymer used is the key variable governing the release of the drug.

The present book is an endeavor to provide a relevant source of infor-mation to readers. The contents of this book are written by very skilled, experienced, and prominent scientists and researchers across the world with advanced knowledge. They introduce drug delivery capabilities to readers, researchers, academicians, scientists, and industrialists around the globe.

The book, *Nanodispersions for Drug Delivery* consists of eleven chap-ters with three sections, which tell the promise these nanodispersions have demonstrated in the past and shall continue to do so in future as well.

SECTION I: NANOEMULSIONS

Chapter 1, *Nanoemulsions for Drug Delivery*, written by *Anudurga Gajen-diran* and *Jayanthi Abraham* provides an overview of nanoemulsions; it includes an introduction, preparation aspects, characterization techniques, and numerous applications. Further, the challenges in developments of such systems have also been mentioned at the end of chapter.

The details of general principles, classification, and methods of prepara-tion of nanoemulsions have been presented in Chapter 2, *Nanoemulsions: Formulation Insights, Applications, and Recent Advances*, written by *Md. Rizwanullah* and *associates*. The authors have discussed in detail different aspects in the field of nanoemulsion-based drug delivery systems. In addition, the *in vivo* fate of nanoemulsion-based drug delivery systems as well as suitability of such carrier system for the delivery of active therapeutics through oral, parenteral, transdermal, ocular, and intranasal routes are also addressed.

Chapter 3, *Nanoemulsions: Routes of Administration and Applications*, written by *Kifayat Ullah Shah* and *colleagues*, gives focused information relevant to specific routes of applications of nanoemulsions. The chapter begins with an introduction, pros and cons, and fabrication principles of nanoemulsions and with a discuss of detailed applications pertaining to different routes by which nanoemulsions could be delivered.

Specialized applications of nanoemulsions in control of microorganisms are described in Chapter 4, *Immunomodulatory and Antimicrobial Effects of Nanoemulsions: Pharmaceutical Development Aspects and Perspectives on Clinical Treatments* written by *Marcus Vinícius Dias-Souza* and *associates*. The authors have provided details of recent advances on the pharmaceutical development of nanoemulsions for treating inflammatory and infectious diseases. The clinical uses are presented in tabular form as well.

SECTION II: NANOSUSPENSIONS

Chapter 5, *Nanosuspensions: Formulation, Characterization, and Applications*, written by *Chogale* and *colleagues*, deals with different aspects of nanosuspensions. This chapter focuses on the various aspects of nanosuspensions, including their formulation components, preparation methods, unique features, methods of characterization, and applications in various routes of administration. The chapter concludes with a case-study that illustrates the development of a nanosuspension formulation based on the principles of Quality by Design (QbD) and optimization by Design of Experiments (DoE).

The detailed description of nanosuspensions is provided in Chapter 6, *Nanosuspensions for Drug Delivery* written by *Kirti Rani*. The author discusses the concept of preparation and application of nanosuspensions in this chapter by highlighting their effective drug delivery, designing and their administration routes, for example, parenteral, pulmonary, oral, and ocular, which can be considered further for clinical and pharmaceutical purposes.

SECTION III: DIVERSE DISPERSED SYSTEMS

Chapter 7, *Nanomicellar Approaches for Drug Delivery*, written by *Manjappa* and *associates*, gives a detailed description of nanomicelles. This chapter focuses also on the recent developments related to targeted delivery of anti-cancer therapeutics to tumor sites using micellar approaches via passive and active mechanisms. Furthermore, it elaborates on "smart" multifunctional micelles designed for enhanced biological performance by modification of copolymers to contain chemistries that can sense a specific biological environment. This chapter also enumerates the strategies developed to enhance *in vivo* stability of micelles and performances of micelles tested via pulmonary, nasal, and rectal routes. Finally, the clinical toxicity of micelles and strategies to improve clinical translation of micelles or other nanomedicines are pinpointed.

The uses of nanogels for drug delivery have been described in Chapter 8, *Nanogels: General Characteristics, Materials, and Applications in Drug Delivery*, written by *P.R. Sarika* and *Nirmala Rachel James*. The authors have provided information on nanogels, covering a preamble, different polymers used for preparation, and diverse applications pertaining to drug delivery.

Chapter 9, *Nanogels: Preparation, Applications, and Clinical Considerations*, written by *Marcos Luciano Bruschi* and *colleagues*, has assessed and discussed strategies for preparation, applications, characterization, as well as some obstacles to the clinical use of nanogels, a novel approach as a potentially therapeutic modality.

The applications of nanogels derived from carbohydrates are given in Chapter 10, *Polysaccharide-based Nanogels as Drug Delivery Systems*, written by *Zahra Eskandari* and *associates*. The objective of this chapter is to offer an overview of various polysaccharide-based nanogels. After a brief introduction on basic concepts of drug delivery systems, an overview of nanogels is presented, and their advantages and classification are discussed. The succeeding parts focus on polysaccharides, polysaccharide-based nanogel drug delivery systems, and their synthesis methods and applications. Finally, current challenges and future perspectives are discussed.

Chapter 11, *Nanocapsules: Recent Approach in the Field of Targeted Drug Delivery Systems*, written by *Bhushan Rajendra Rane* and *colleagues*, describes the drug delivery approaches based on nanocapsules. The authors have given an elaborated overview of nanocapsules, commencing with an introduction, and covering merits and demerits, materials used for preparation, and lastly uses in drug delivery. The information is also presented as summarized in tabular form.

SECTION I
NANOEMULSIONS

CHAPTER 1

NANOEMULSIONS FOR DRUG DELIVERY

ANUDURGA GAJENDIRAN and JAYANTHI ABRAHAM*

Department of Biomedical Sciences, Microbial Biotechnology Laboratory, School of Bio Sciences and Technology, VIT University, Vellore 632014, Tamil Nadu, India

*Corresponding author. E-mail: jayanthi.abraham@gmail.com

ABSTRACT

In recent years, nanoemulsions (NEs) have gained more attention in the area of drug delivery because of their ease in production, high solubilization capacity, spontaneous formation, thermodynamic stability, enhanced stability of the encapsulated drug, high absorption rates, and ability to encapsulate both hydrophilic as well as hydrophobic drugs. This chapter focuses on structure, formulation of NEs, preparation methods, characterizations, evaluation parameters, patent, and application of NEs in drug delivery.

1.1 INTRODUCTION

An emulsion is a heterogeneous system consisting of two immiscible phases such as oil and water; one phase is the dispersed phase as droplets (internal phase) and the other is the continuous phase (external phase). Depending on the emulsion droplet size it can be classified into three groups such as microemulsions (1–100 nm), nanoemulsions (NEs) (100–1000 nm), and macroemulsions (0.5–100 mm) (Maali and Mosavian, 2013). NEs are characterized by significantly small droplet size typically in the range of 20–200 nm which was developed around 20 year ago (Solans et al., 2005). They are also referred as miniemulsions (El-Aasser and Sudol, 2004; Constantinides et al., 2008), ultrafine emulsions (Guglielmini, 2008), submicron emulsions (Klang and Benita, 1998; Benita, 1999), and translucent emulsion (Fernandez et al., 2004; Tadros et al., 2004). NEs are transparent or translucent systems with blue hue (Salager and Marquez, 2003) and they are kinetically stable and cannot be formed spontaneously; hence, it requires energy as input for their formation. The source of the required energy may be external (dispersion or high-energy methods) or internal (condensation or low-energy methods). Due to the smaller size of NEs, they often possess long-term physical stability and have enhanced properties, when compared to conventional emulsions (Liu et al., 2006).

Many studies have focused on drug delivery systems such as microspheres, nanoparticles, liposomal systems, and traditional emulsions. However, these drug delivery systems have several drawbacks such as low encapsulation efficiency and drug loading, use of organic solvents such as methylene chloride and acetone (which induce toxicity), and low yield and high production cost. Since 1980, many researchers have focused on NEs due to their wide range of application in the field of food science, cosmetics, and pharmaceuticals (Tarver, 2006) and it can act as drug delivery for lipophilic compounds, such as drugs, flavors, nutraceuticals, antioxidants, and antimicrobial agents (McClements et al., 2007; Weiss et al., 2008). The major advantages of NEs as drug delivery carriers include: enhanced solubilization capacity, increased drug-loading capabilities, easy transport, reduced patient variability, dispersibility, controlled drug release, enhanced oral bioavailability, protection from enzymatic degradation, and bioaccessibility of active food and drug components (e.g., carotenoids, α-tocopherol, polyunsaturated fatty acids, hydrophobic vitamins, and aroma compounds). They can also act as excellent encapsulation systems compared to conventional emulsions (Bilbao-Sainz et al., 2010; Kotta et al., 2012). These versatile characteristics make NEs as novel drug delivery systems.

1.2 STRUCTURE OF NANOEMULSION

NEs can be formed by three ways: (1) oil-in-water (o/w), (2) water-in-oil (w/o), and (3) bicontinuous. In o/w NEs, oil droplets are dispersed in continues aqueous phase, whereas in w/o NEs, water droplets are dispersed in continues oil phase. Bicontinuous NEs are systems where the amount of water and oil are similar. All three types of NE interface are stabilized by appropriate combination of surfactants and/or cosurfactants. NEs are made from pharmaceutical surfactants approved for human consumption and common food substances that are "generally recognized as safe" (GRAS) by the Food and Drug Administration (FDA) (Shah et al., 2010). NEs can be divided into two groups based on droplet size, namely, transparent or translucent (50–200 nm) and milky (up to 500 nm) (Izquierdo et al., 2002). Extremely small droplet size of NEs causes a large reduction in the gravity force, and also Brownian motion rate may be sufficient for overcoming gravity, which means no sedimentation or creaming occur on storage (Fernandez et al., 2004).

Although, NE has several advantages in drug delivery system, the major issue is "Ostwald ripening" which destabilizes emulsion systems. The Ostwald ripening process is caused by the polydispersity of emulsion droplets whereby larger droplets are energetically favored over smaller droplets. Many studies have proven that Ostwald ripening can lead to phase separation of emulsions, which may not always be observed because the droplet growing rate gradually decreases with the increase of droplet size. A few strategies have been proposed to reduce Ostwald ripening (Tadros et al., 2004) such as (1) addition of a second disperse phase component with low solubility in continuous phase and (2) utilization of surfactants that are strongly adsorbed at the interface and not easily desorbed during ripening. Nevertheless, detailed research is still needed to overcome the stability problem of NEs in real products.

1.3 FORMULATION OF NANOEMULSIONS

1.3.1 COMPONENTS USED IN PREPARATION OF NANOEMULSIONS

A broad variety of natural and synthetic ingredients has been significantly employed in NEs formulation. Oils, surfactants, cosurfactants, and aqueous phase are used in the preparation of NEs. However, wide variety of oil and surfactants are applicable in formulation; they should be used in limited amount due to their toxicity, irritation capability, and unclear mechanisms of action (Amani et al., 2008; Azeem et al., 2009). Synthesized NE and microemulsion are shown in Figure 1.1.

FIGURE 1.1   **(See color insert.)** (A) Nanoemulsion system and (B) microemulsion system.

1.3.2   OILS/LIPIDS

The most important criterion for selection of oils is the solubility of the drug in the oil phase, whereas, in the case of oral formulation development, this is specifically an important criterion. High solubility of the drug in the oil will reduce the amount of the formulation to be delivered as the thera-peutic dosage of the drug. In NE formulation, oils serve as one of the impor-tant excipients because of the increase in the fraction of lipophilic drugs transported via the intestinal lymphatic systems thereby increasing absorp-tion of drugs from the gastrointestinal tract depending on the molecular nature of the oils (Holm et al., 2002). The w/o NEs are excellent choice for hydrophilic drugs; however, lipophilic drugs have high solubility in oil phase and hence, it can be formulated into o/w NEs. Drug loading in the formulation is a very critical design factor in the development of NEs for poorly soluble drugs, which is dependent on the drug solubility in various formulation components. In NEs formulation, edible oils could offer more logical choice; however, they are not frequently used due to their poor ability to dissolve large amount of lipophilic drugs. Moreover, the formulation of

NEs with oils of low solubility would require more oil to incorporate the target drug dose, which would result in requirement of higher surfactant concentration to achieve oil solubilization. This might finally result in an increase in the toxicity of the system. Modified or hydrolyzed vegetable oils are distinctively used because they form good emulsification system with a large amount of surfactants approved for oral and topical administration and exhibit better drug solubility properties (Taleganokar, 2008). Selection of oil phase is a rational balance between its ability to solubilize the drugs and its capability to develop NEs of desired characteristics. In self-NE formulation, long- and medium chain of triglyceride oils with different degrees of saturations have been widely used in its development. Novel semisynthetic medium-chain derivatives (such as amphiphilic compounds) having surfactant properties are progressively and effectively replacing the regular medium-chain triglyceride oils (Shakeel et al., 2013). Several types of oils and surfactants have been used for formulation of NEs (Table 1.1).

1.3.3   SURFACTANTS

Surfactants are used for stabilization of NEs and they reduce the interfacial tension between two liquids (lipids and aqueous phase) to a very small value to aid dispersion process. It can provide a flexible film that can readily deform around droplets and their lipophilic character serve the correct curvature at the interfacial region for the desired NE type, that is, for w/o, o/w, or bicontinuous (Grigoriev and Miller, 2009; Huang et al., 2011). Based on the nature of the polar group, surfactants can be categorized into four different groups such as nonionic, cationic, anionic, and zwitterionic. Components of NE-based systems are associated with toxicity concerns. Higher amount of surfactants may cause gastrointestinal and skin irritation when administered orally and topically. Thus, appropriate selection of surfactant is crucial and an important factor. Safety and solubility are main determining factors in selection of surfactants. Rational use of the minimal concentration of surfactant in the formulation is recommended. Nonionic surfactants are relatively less toxic than their ionic counterparts and have lower critical micelle concentration. In addition, o/w dosage forms for oral or parenteral use based on nonionic surfactants are likely to offer an *in vivo* stability (Kawakami et al., 2002). Selection of surfactants with proper hydrophilic–lipophilic balance (HLB) value is another important criterion. The appropriate combination of low and high HBL results in formation of stable NEs upon dilution with water. Surfactants with low HLB value (3–6), such as Spans are

generally considered for the development of w/o NEs, whereas surfactants
with high HLB value (8–18) such as Tweens are chosen for o/w system.
Surfactant types and nature are also important factors in NEs formulation.
Nonionic surfactants are generally chosen because they are less affected by
pH, changes in ionic strength, and biocompatibility and are considered as
safe, whereas ionic surfactants are less commonly preferred since they can
cause toxicity to the system. Solubilization of oil with the surfactants is also
an important factor. Surfactant–oil miscibility can give a primary sign on the
possibility of NEs formulation (Giustini et al., 1996; Mehta, 1998).

TABLE 1.1   Oils and Surfactants Used in Nanoemulsion Formulation.

Drug	Oil	Aqueous phase	Surfactant	Preparation method	References
Aspirin	Soybean oil	Water	Polysorbate 80	Microfluidization	Subramanian et al. (2008)
β-carotene	MCT oil	Water	Polysorbate 20, Polysorbate 40, Polysorbate 60, Polysorbate 80	High-pressure homogenization	Yuan et al. (2008a, 2008b)
Flurbiprofen	Isopropyl myristate, soybean oil, coconut oil	Water	Egg lecithin	Ultrasonication	Fang et al. (2004)
Quercetin, methyl quercetin	Octyldodecanol	Water	Lipoid® E-80 (Lipoid KG, Ludwigshafen, Germany)	Spontaneous emulsification	Fasolo et al. (2007)
β-lactamase	Propylene glycol laurate, propylene glycol monocaprylate	–	Polyoxyl 35 castor oil, polyoxyl 40 hydrogenated castor oil, diethylene glycol monoethylene ether (cosurfactant), propylene glycol (cosurfactant)	SNEDDS	Rao et al. (2008a, 2008b)
Celecoxib	Propylene glycol mono-caprylic ester, triacetin	Water	Ceteareth-25, 2-(2-ethyoxyethoxy) ethanol (cosurfactants)	Spontaneous emulsification	Shakeel et al. (2008c)

TABLE 1.1   *(Continued)*

Drug	Oil	Aqueous phase	Surfactant	Preparation method	References
Curcumin	MCT oil	Water	Polysorbate 20	High-pressure homogenization	Wang et al. (2008b)
Primaquine	Capric-caprilic triglyceride	Water + glycerol + sorbitol	Egg lecithin, soybean lecithin, poloxamer 188	High-pressure homogenization	Singh and Vingkar (2008)
Sevoflurane	Perfluorooctyl bromide	NaCl solution (0.9%)	Fluorinated surfactant (F13M5)	Microfluidization	Fast et al. (2008)
Cheliensisin	MCT oil	Water	Soybean lecithin	High-pressure homogenization	Zhao et al. (2008)
Camptothecin	Coconut oil + perfluoro-carbon	Water	phospholipon® 80H (American Lecithin Co., USA), phosphatidyl-ethanolamine, Pluronic F68 (BASF, USA), cholesterol	Ultrasonication	Fang et al. (2008a)
δ-tocopherol	Canola oil	Water	Polysorbate 80	Microfluidization	Kotyla et al. (2008)
Camphor + menthol + methyl salicylate	Soybean oil	Water + propylene glycol	Soybean lecithin, Polysorbate 80, Poloxamer 407	High-pressure homogenization	Mou et al. (2008)
Risperidone	Mono- or diglycerides of caprylic acid	Water	Polysorbate 80, diethylene glycol mono-ethyl ether + propylene glycerol (cosurfactants)	Spontaneous emulsification	Kumar et al. (2008)
Ceramides	Octyldo-decanol	Water + glycerol	Polysorbate 80, lipoid E-80 (Lipoid KG, Ludwigshafen, Germany)	High-pressure homogenization	Yilmaz and Borchert (2005); Yilmaz and Borchert (2006)

TABLE 1.1 *(Continued)*

Drug	Oil	Aqueous phase	Surfactant	Preparation method	References
Paclitaxel + ceramide	Pine nut oil	Water	Lipoid E-80, lipoid PE (Lipoid KG, Ludwigshafen, Germany)	Ultrasonication	Desai et al. (2008)
Cefpodoxime-proxetil	Propylene glycol monocaprylate	–	Polyoxyl 35 castor oil, Akoline® MCM (Aarhus Karlshamn, Sweden; cosurfactants)	SNEDDS	Date and Nagarsenkar (2007)
Atenolol, danazol, metoprolol	Propylene glycol monocaprylate	Water	Pluronic P104, Pluronic® L62, Pluronic L81 (BASF, USA)	Spontaneous emulsification	Brusewitz et al. (2007)
Ramipril	Propylene glycol mono-caprylic ester	Buffer solution (pH 5)	Caprylo caproyl macrogol-8-glyceride, Polysorbate 80, diethylene glycol mono-ethyl ether (cosurfactant), polyglyceryl-6-dioleate (cosurfactant)	Spontaneous emulsification	Shafiq et al. (2007)
Foscan	Capric-caprilic triglyceride	Water	Epikuron® 170 (Cargill Europe BVBA, Belgium), Poloxamer 188	Spontaneous emulsification	Primo et al. (2007)
Paclitaxel	Glycolyzed ethoxylated glycerides	Water	Caprylo caproyl macrogol-8-glyceride	Spontaneous emulsification	Khanda-valli and Panchagnula (2007)
Egg ceramides	Capric-caprilic triglycerides, tripalmitin	Water	Phospholipon® 90 G (American Lecithin Co., USA), Solutol® HS 15 (BASF, USA)	Ultrasonication	Hatzianto-niou et al. (2007)

TABLE 1.1   *(Continued)*

Drug	Oil	Aqueous phase	Surfactant	Preparation method	References
Oligonucle-otides	Capric-caprilic triglycerides	Water + glycerol	Poloxamer 188, Lipoid E-80 (Lipoid KG, Ludwigshafen, Germany)	High-pressure homogenization	Hagigit et al. (2008)

1.3.4   COSURFACTANTS

Cosurfactants are necessary to obtain NEs system at low surfactants concentration (Kreilgaard et al., 2000). Various surfactants are being used in the formulation of NEs such as alcohols, alkanoic acids, alkanediols, and alkyl amines (Bhat et al., 2015). Presence of cosurfactants allows the interfacial film sufficient flexibility to take up various curvatures necessary to form NEs over a broad range of compositions (Warisnoicharoen et al, 2000). Absence of cosurfactants results in the formation of highly rigid films by the surfactants and thus, produces NE over only a very limited range of concentrations. Short to medium chain length alcohols (C3–C8) are commonly used as cosurfactants which have the significant effect to further reduce the interfacial tension and increase the fluidity of the interface and also increase the entropy of the system (Tenjarla, 1999). They also increase the mobility of the hydrocarbon tail and allow greater penetration of the oil into this region. Ethyl esters of fatty acids also act as cosurfactants by penetrating the hydrophobic chain region of the surfactant monolayer.

1.4   PREPARATION OF NANOEMULSION

NEs are nonequilibrated systems and their preparation involves the input of a large amount of either energy or surfactants, and in some cases, a combination of both. As a result, high-energy or low-energy methods can be used in their formulation.

a) High-energy emulsification method is traditionally used for the preparation of NEs formulation. The high-energy emulsification method includes high-energy stirring, ultrasonication, microfluidization, membrane emulsification, and high-pressure homogenization (Tiwari and Amiji, 2006).

b) Low-energy emulsification method has been widely used due to its advantages such as formulation and stability aspects (Anton and Vandamme, 2009; Ravi and Padma, 2011). This method includes phase inversion temperature (PIT), emulsion inversion point, and spontaneous emulsification. Reverse NE is possible to prepare in a highly viscous system using a combined method, which includes the high- and low-energy emulsification.

1.4.1 HIGH-ENERGY EMULSIFICATION TECHNIQUES

A high-energy emulsification technique employs a very high mechanical energy to create NEs with high kinetic energy. This method applies mechanical devices to create intensively disruptive forces which break up the oil and water phases in order to form nanosized droplets (Jaiswal et al., 2015).

1.4.1.1 HIGH-PRESSURE HOMOGENIZATION

Extremely small particle size (up to 1 nm) NE can be prepared in this method using high-pressure homogenizer/piston homogenizer (Anton et al., 2008). In this technique, two liquids, both oily phase and aqueous phase are heated together and dispersed; then, emulsification is carried out using a high shear mixer that results in a course emulsion. The coarse macroemulsion is passed through a small inlet orifice at a high pressure for several times (range of 500–5000 psi); turbulence and hydraulic shear are applied in order to obtain fine particles of emulsion. Hydraulic shear, intense turbulence, and cavitation are the several forces involved during this process to yield NEs with very small droplet size. Parameters such as homogenization pressure, homogenization temperature, intensity of homogenization, as well as the number of homogenization cycles should be optimized carefully to obtain the polydispersity index and desired droplet size (Yuan et al., 2008; Musa at al., 2013). High energy consumption and increase in temperature of emulsion during preparation are the disadvantages of this method.

1.4.1.2 MICROFLUIDIZATION TECHNIQUE

This patented technology is commonly used in large-scale industries to produce extremely small-sized fluid NEs. This technique employs microfluidizer device which applies high-pressure positive displacement pump up to 5000–20,000 psi. This pump forces macroemulsion droplets through the interaction chamber consisting of a series of microchannels. The macroemulsions flow through these microchannels on to an impingement area resulting in formation of very fine particles of submicron range. The NEs with desired size range and dispersity can be obtained by varying the operating pressure and number of passes through interaction chambers. The aqueous and oily phases are combined together and processed in an inline homogenizer to yield a course emulsion. This course emulsion is further introduced into microfluidizer to obtain stable NE.

1.4.1.3   ULTRASONIC EMULSIFICATION

Ultrasonic emulsification has been used in laboratory scale; it is very efficient in reducing droplet size up to 0.2 μm (Fang et al., 2009). In this technique, the energy is provided through sonicator probe that emits ultrasonic waves. Ultrasound can be directly used to produce emulsion. When the sonicator tip contacts the liquid (aqueous and oily phase), it serves mechanical energy and cavitation occurs. Cavitation is the formation of vapor phase of a liquid. NEs with desired properties can be obtained by optimizing the ultrasonic energy input and time.

1.4.1.4   SOLVENT DISPLACEMENT METHOD

This technique is based on the nanoprecipitation method used for polymeric NEs. Formation of NEs can be done by simple stirring at room temperature. The oil phase is dissolved in water-miscible organic solvents followed by the addition of organic phase into an aqueous phase-containing surfactant to yield spontaneous NEs by rapid diffusion of organic solvents. However, the major drawback of this method is it requires additional inputs to remove organic solvent, such as acetone, from the NE after preparation (Lee and McClements, 2010).

1.4.2   LOW-ENERGY METHODS

The high-energy methods cannot be used in certain cases, particularly with thermolabile compounds such as retinoids and marcomolecules such as protein, enzymes, and nucleic acids. In such case, low-energy emulsification techniques such as the spontaneous or self-emulsification method or the PIT method can be preferred. Low-energy emulsification methods require low energy for the fabrication of NEs.

1.4.2.1   PIT TECHNIQUE

This method involves temperature-dependent solubility of nonionic surfactants, such as polyethoxylated surfactants, to modify affinities for water and oil. Polyoxyethylene surfactant tends to become lipophilic on heating. At low temperature, the surfactant monolayer causes o/w emulsion, and at high temperature it forms w/o emulsion. During cooling of the emulsion, the system crosses a point of zero spontaneous curvature and minimal surface tension, promoting the formation of finely dispersed droplet (Solans et al., 2005). It has been observed that the characteristics of the NEs are mainly dependent on the structure of the surfactant at HLB temperature and also on the addition of surfactant and oil ratio. The process of changing temperature can be repeated to obtain the reduction in size of oil droplet. Initially, PIT method was believed to be suitable for the fabrication of o/w NEs. However, in recent years, the application of the PIT method has been established for the fabrication of w/o emulsions and NE (Wang et al. 2007).

1.4.2.2   SPONTANEOUS EMULSIFICATION OR SOLVENT DISPLACEMENT METHOD

The solvent displacement method for spontaneous fabrication of NEs has been adopted from the nanoprecipitation method used for polymeric nanoparticles. In this method, oily phase is dissolved in water-miscible organic solvents, such as acetone, ethanol, methyl alcohol, and ketone. This method involves three steps: (1) preparation of homogenous organic solution containing oil and lipophilic surfactant in water-miscible solvent and hydrophilic surfactant, (2) the organic phase is introduced into the aqueous phase containing surfactant to yield o/w emulsion by rapid diffusion of organic solvents, and (3) the aqueous phase (organic solvents) is removed by evaporation under reduced pressure (Porras et al., 2008).

1.4.2.3 PHASE-INVERSION COMPOSITION METHOD (SELF-NANOEMULSIFICATION METHOD)

This method produces NEs at room temperature without the use of organic solvents, high shear, or heat, thus, being the most preferred method. Kinetically stable NEs with small droplet size (~50 nm) can be achieved by gradual addition of water into the phase containing surfactant mix and oil, with gentle stirring at constant temperature. The spontaneous nanoemulsification can be related to the phase transitions during the emulsification process and involves lamellar liquid-crystalline phases or D-type bicontinuous microemulsion during the process. NEs obtained from the spontaneous nanoemulsification process are not thermodynamically stable; however, they possess high kinetic energy.

1.5 FACTORS TO BE CONSIDERED DURING THE PREPARATION OF STABLE NANOEMULSIONS

Stability is the major parameter to be considered in the development of NEs. It should strengthen both the physical and chemical nature of the drugs. Stability analysis is carried out by storing NEs at different temperature (room temperatures and refrigerator) for few months (Tenjarla, 1999). During storage time, the impact of viscosity, droplet size, and refractive index of the NEs are determined. To examine the accelerated stability, the NE formulation is placed at accelerated temperatures and the samples are periodically collected at regular intervals of time. The collected samples are analyzed for drug content by stability-indicating assay technique. The amount of drug degraded and remaining in NE formulation is determined at regular interval of time. NE formulation stability can be enhanced by controlling few aspects such as class and concentration of surfactant and cosurfactant, type of oil phase, addition of additives, preparation methods, and process variables (Shakeel et al., 2008a; 2008b). Other important factors to be considered during NE preparations are as follows (Haritha et al., 2013).

(i) Surfactant must be conscientiously chosen such that an ultralow interfacial tension may be attained at the oil–water interface.

(ii) Surfactant concentration must be high to stabilize the microdroplets to produce nanosized droplets.

(iii) The interface must be flexible to promote the formation of NEs.

1.6 CHARACTERIZATION OF NANOEMULSIONS

Physical and chemical characterizations of NEs are important to understand the greater depth of this system. They are determined by evaluating the parameters such as compatibility of the components involved in preparation of NE, uniformity of the content, method, isotropicity of the formulation, color, pH viscosity, density, conductivity, surface tension, size, and zeta potential of the dispersed phase.

1.6.1 *PHYSICAL CHARACTERIZATION TECHNIQUES*

1.6.1.1 *DROPLET SIZE STUDIES*

Droplet size of the NE is an important factor in self-nanoemulsification performance because it determines the rate and extent of drug release as well as absorption. It can be analyzed by dynamic light scattering; it also called as photon correlation spectroscopy or quasielastic light scattering, and static light scattering. Dynamic light scattering/static light scattering determines size of the particle and provides quick report on size distribution of NEs and also measures droplet size stability during storage (Araujo et al., 2011; Silva et al., 2012). They analyze the fluctuations in the intensity of scattering by droplets/particles due to Brownian movement which relates the size of the particles through Stockes–Einstein equation (Ruth et al., 1995). Light scattering was monitored at 25°C at a scattering angel of 90° using a neon laser wavelength of 632 nm.

1.6.1.2 *ZETA POTENTIAL*

ZetaPALS (phase analysis light scattering) instrumentation measures the charge on the surface of droplet in NE. It provides the z-average particle diameter of NE (Erol and Hans-Hubert, 2005). Zeta potential value in the magnitude of ±30 mV can be taken as the arbitrary value that separates low-changed surface from highly changed surface. Result of zeta potential values in the magnitude of ±30 mV reveals the stability of the nanodroplet formation. Zeta potential of NE is determined by factors such as surfactants, electrolyte concentration, size, pH, and particle morphology (Shakeel et al., 2013). Laser diffraction technique has been used to measure the particle size of the NE. In this case, the result is expressed in terms of the volume of equivalent spheres ($DN\%$) and mass of the mean of volume distribution (mass mean diameter). Particle polydispersity is given by uniformity and span. The span value is described by following expression:

$$\text{Span} = \frac{(D90\% - D10\%)}{D50\%}$$

where, $DN\%$ (N = 10%, 50%, and 90%), denotes the volume percentage of particles with diameters up to $DN\%$ equals to $N\%$. The smaller the span value the smaller the particle size distribution.

1.6.1.3   THERMAL ANALYSIS

Differential scanning calorimetry is used to analyze the crystallization temperature of a mixture of surfactants, phase transitions, melting of crystalline area, polymorphic behavior, and proportion of ice crystals in the NE. It is also used to evaluate the physical state of the unsaturated fatty acids present in NE (Uson et al., 2004). This thermoanalytical technique calculates the difference in the amount of heat required to increase the temperature of a sample and reference is measured as a function of temperature. The reference sample must have a well-defined heat capacity over different range of temperatures to be scanned (Venturini et al., 2011).

1.6.1.4   FOURIER-TRANSFORM INFRARED SPECTROSCOPY

Fourier-transform infrared (FTIR) spectroscopy is efficiently used to evaluate the crystalline structure and chemical groups present in the NE. FTIR is based on an infrared radiation that passes through a sample where it is mostly absorbed by the sample and some of it is transmitted. The resulting spectrum represents the molecular absorption and transmission, creating a molecular fingerprint of the sample. Each sample fingerprint presents its characteristic absorption peaks that correspond to the frequencies of vibrations between the bonds of the atoms of the material. Therefore, infrared spectroscopy results in a positive identification of different material additions to the size of the peaks in the spectrum, that is, a direct indication of the amount of material present in the sample. This technique allows infrared radiation to pass through the sample in the range of 4000–400 cm^{-1} at a resolution of 4 cm^{-1} and then provides fast and accurate spectrum. The spectra represent the molecular absorption and transmission of the NE. The absorption peak that corresponds to the frequencies of vibrations represents the functional groups to the size of the peaks in the spectrum, that is, a direct indication of the amount of functional groups present in the NE (Nicolet, 2001).

1.6.1.5   NUCLEAR MAGNETIC RESONANCE

Nuclear magnetic resonance (NMR) is the most sophisticated and efficient technique used to measure the self-diffusion coefficient of NE which is useful in the detection of types of NE (o/w or w/o) (Jenning et al., 2000). However, the application of NMR in characterization of NEs has been slightly exploited.

1.6.1.6   X-RAY DIFFRACTION

X-ray diffraction (XRD) provides the information regarding crystallographic structure, chemical composition, and physical properties of NE. The scattered intensity of the X-ray beam hitting the sample as a function of incident and scattered angles, polymerization, and wavelength. XRD is used to determine the crystal size from the analysis of peak broadening, determination of crystallite shape from the study of peak symmetry (Connolly, 2007).

1.6.1.7   SMALL-ANGLE-X-RAY SCATTERING

Small-angle-X-ray scattering (SAXS) technique is helpful in understanding the structural characteristic of colloidal size of particles in the range of 0.1–10°. The angular range gathers the information about shape and size of the macromolecules and pore size. This technique provides the structural information in the range between 5 and 25 nm. Only one-dimension scattering pattern can be obtained from SAXS which is the drawback of this technique (Luykx et al., 2008).

1.6.2 MORPHOLOGICAL ANALYSIS

Structural behavior and aggregation form of NEs can be determined by sophisticated instrumentation which includes transmission electron micros- copy (TEM), scanning electron microscopy (SEM), X-ray, that is, X-ray diffraction (XRD) or neutron scattering, atomic force microscopy (AFM), and cryoelectron microscopy (cryo-EM) (Mason et al., 2006). Most commonly used methods are discussed below.

1.6.2.1 TRANSMISSION ELECTRON MICROSCOPY

TEM analysis gives high-resolution micrographs of NE; it has the resolving power of 0.2 nm (Yuan, 2008; Luykx, 2008; Ribeiro, 2008). Qualita- tive measurement of sizes and size distribution of TEM micrographs can be examined with the help of digital image processing program. For TEM analysis, the sample should be negatively stained in 1% aqueous solution of phosphotungstic acid or by submerging in 2% uranyl acetate solution for 30 s and then the sample is placed on carbon-coated grid and is examined at appropriate voltage, usually 70 kV (Samah et al., 2010). However, this technique has some disadvantages that are listed below:

(i)   Extensive sample preparations are needed for some NE; it should be thin and electron transparent, which makes TEM analysis relatively time-consuming process.
(ii) Structure of NE may be changed during sample preparation.
(iii) Field of view is small and samples may get damaged during analysis due to electron beam.

1.6.2.3 SCANNING ELECTRON MICROSCOPY

SEM provides a three-dimensional micrograph of NE. An accelerating voltage usually at 15 kV can be used. Shape, size, and surface morphology of NE can be obtained by the use of automatic image analysis software. It is also capable of producing high-resolution micrographs at higher magnifications and allows ease in examination of the sample (Luykx, 2008; Barea, 2010).

1.6.2.4 ATOMIC FORCE MICROSCOPY

AFM is a very high-resolution (± 0.1 nm) type of scanning probe microscopy. It has been used to directly view the single atoms or molecules that have dimensions of a few nanometers. AFM is efficiently used to examine various kinds of compounds such as ceramics, polymers, composites, glass, and biological samples. Structural characterization for liposomes, proteins, and polysaccharides can be examined in AFM (Luykx, 2008). It has few drawbacks with analysis of sample surfaces. As the probe tip is in direct contact with sample surface, it is difficult to run on surfaces which are soft, sticky, or floating, loose particles.

1.7 EVALUATION PARAMETERS OF NANOEMULSION

1.7.1 POLYDISPERSITY

A particle size analyzer measures the polydispersity index of the NE, which measures the broadness of the size distribution derived from the cumulative analysis of dynamic light scattering. The result of polydispersity index indicates the homogeneity of the dispersion (Li et al., 2011). The higher the value of polydispersity, lower will be the uniformity of the dispersion.

1.7.2 VISCOSITY MEASUREMENT

Viscosity is an important factor for stability and efficient drug release. Brookfield-type rotary viscometer is used to determine the viscosity of NE; it can be used at different shear rates and also at different temperatures. The viscosity of NE is a function of NE components (surfactant, water, and oil) and their concentrations. High water content leads to the lower viscosity of the NE, whereas decreasing the concentration of surfactant and cosurfactant increases the interfacial tension between water and oil which results in high viscosity. Monitoring of viscosity change is a technique for evaluating the stability of liquid and semisolid preparations including NE formulation (Chiesa et al., 2008).

1.7.3 INTERFACIAL TENSION

Spinning drop apparatus has been generally used to measure the ultralow interfacial tension of NEs. NEs properties and formation can be calculated by interfacial tension. Ultralow values of interfacial tension are correlated with phase behavior, especially when the presence of surfactant or middle phase NEs is in equilibrium with oil and aqueous phase (Leong et al., 2009).

1.7.4 REFRACTIVE INDEX AND pH

Refractive index is determined using an Abbe-type refractometer at 25°C. pH of the NE can be measured by pH meter.

1.7.5 DYE SOLUBILIZATION

If a water-soluble dye is added in an o/w NE, the NE takes up the color uniformly. However, if the NE is w/o type and the dye is soluble in water, the emulsion takes up the color only in the dispersed phase and the emulsion is not uniformly colored. This can be revealed by microscopic examination.

1.7.6 DILUTION TEST

O/w NE can be evaluated by diluting with water, whereas w/o NE are not diluted with water and undergo phase inversion into o/w NE.

1.7.7 CONDUCTANCE MEASUREMENT

The conductance of NE is measured by a conductometer. External phase of o/w NE is water that is highly conducting, whereas in w/o, it is not because water is the internal or dispersal phase. Electrical conductivity measurements are efficient in determination of the characteristics of continuous phase and phase-inversion phenomena. In certain w/o NE, a sharp rise in conductivity was observed at low-volume fractions and such behavior is known as "percolative behavior" or exchange of ions between droplets before the formation of bicontinuous structures (Wulff-Perez et al., 2009; Burapapah et al., 2010).

1.7.8 RHEOLOGICAL MEASUREMENT

Rheological measurement is used in evaluation of flow behavior of the NE. The viscosity is assessed using a Bohlin rheometer analyzed at $25 \pm 0.1°C$ with a cone/plate apparatus 40 mm per 4° for each sample, the variation of shear rate γ used and the resulting shear stress σ will be determined (Edwards et al., 2006; Howe et al., 2008).

1.7.9 FILTER PAPER TEST

An o/w NE will spread out rapidly when dropped onto filter paper. In contrast, a w/o NE will migrate slowly. This method is not fit for highly viscous creams (Sharma and Jain, 1985).

1.7.10 FLUORESCENCE TEST

Many oils exhibit fluorescence when exposed to UV light. When a w/o NE is exposed to a fluorescent light under a microscope, the entire field fluoresces. An o/w NE exhibits spotty fluorescence.

1.7.11 PERCENTAGE TRANSMITTANCE

NE percentage transmittance is determined by an UV–visible spectrophotometer.

1.7.12 DRUG CONTENT

Preweighed NE is extracted by dissolving it in an appropriate solvent system and then the extracted sample is analyzed by spectrophotometer or high-performance liquid chromatography technique against standard solution of drug.

1.8   THERMODYNAMIC STABILITY STUDIES

During the thermodynamic stability of drug-loaded NE following stress tests are reported:

a) *Heating cooling cycle*: The NE formulations are subjected to six cycles at various refrigerator temperature ranges between 4°C and 45°C.

b) *Centrifugation*: The NE formulations are centrifuged at 3500 rpm and those which does not exhibit any phase are considered for the freeze-thaw stress test.

c) *Freeze-thaw cycle*: NE formulations are subjected to three cycles of freeze thaw between 21°C and ± 25°C. These thermodynamic stability studies are performed for a period of 3 months.

1.9   NANOEMULSIONS AS DRUG DELIVERY CARRIERS

NEs have greatest and promising future in drug delivery carriers by over-coming the obstacles in conventional methods. The use of NE in pharmaceutical field has emerged as a great extent from last decades. The tremendous advantages of NE is not only the drug absorption and drug targeting but also providing the opportunities for new drugs to be designed with bioavailability. Performance of the drugs has increased in NEs with the increasing retention time. NEs are efficient in drug delivery and drug targeting by various routes of administration such as oral, parenteral, ocular, pulmonary, transdermal, intravenous, and intranasal routes.

1.9.1   NANOEMULSIONS IN ORAL DRUG DELIVERY

NE formulation is a best tool for poorly permeable drugs and provides high oral bioavailability of hydrophobic drugs. To demonstrate the oral bioavailability of hydrophobic drug, paclitaxel (antineoplastic drug) was taken as a model. The o/w NEs were made with pine nut oil as internal phase, water as external phase, and egg lecithin as a primary emulsifier. Stearylamine and deoxycholic acid were used to supply positive and negative charges to the emulsions, respectively. Formulated NEs have a particle size range of 100–120 nm and zeta potential range from 34 to 245 mV. After oral administration of formulated NEs, when compared to aqueous solution, high amount of paclitaxel compound was noticed in the systemic circulation. This result confirms the oral bioavailability of hydrophobic drugs (Rutvji et al., 2011; Yashpal et al., 2013). Similarly, Khandavilli and Panchagnula (2007) prepared an NE containing paclitaxel drug and administered it orally to rats. They noticed that paclitaxel has been rapidly absorbed with 70.62% of bioavailability. The plasma drug concentration was 3 μg mL^{-1} between 0.5 and 18 h. This result strongly indicates that paclitaxel was efficiently absorbed throughout the gastrointestinal tract and also persisted up to 18 h in steady state. NEs have both hydrophilic and lipolytic units; and also NE of formulated drugs may even penetrate the cytoplasm and provide pharmacodynamic action inside the cells. Thus, it can be an excellent drug delivery system for steroids, hormones, antibiotics, diuretic, and proteins. When a drug, for example, primaquine incorporated into oral lipid NE showed efficient antimalarial activity against *Plasmodium berghei* infection in mice at 25% lower dose level as compared to conventional oral dose (Singh and Vingkar, 2008). Wang et al. (2008b) demonstrated study on mouse ear inflammation, when curcumin-loaded NE is orally administered to mice, they showed a higher anti-inflammatory effect than the surfactant solution containing curcumin. Lipid-based w/o NEs significantly enhance the oral bioavailability of hydrophilic and fragile compounds such as peptides and proteins. Hydrophilic drugs have been incorporated into aqueous phase of w/o NEs for protecting drugs from environmental factors such as acidic pH and enzymatic activity (Cilek et al., 2005).

1.9.2 NANOEMULSIONS IN OCULAR DRUG DELIVERY

An NE in ocular drug delivery is most promising and challenging in pharmaceutical industries. Eye drops are used as conventional drug delivery systems which results in poor bioavailability and less targeted due to lacrimal secretion and nasolacrimal drainage in eyes (Sieg and Robinson, 1977). The main drawback of ophthalmic conventional formulations is that they drain away from precorneal region in a few minutes after administration. Thus, constant instillation of drug is needed to enhance the required therapeutic dosage. However, due to tear drainage, the main compound of the administered drug is transported via nasolacrimal channel to gastric intestinal tract where it may be absorbed and results in side effect (Middleton et al., 1990). The o/w NEs can be used to overcome these problems and also highly enhance the corneal penetration. Ammar et al. (2009) conducted the study on formulated antiglaucoma drug dorzolamide hydrochloride as ocular NE with increased therapeutic efficiency and longer effect. Draize rabbit eye irritation test and histological examination revealed the efficiency along with enhanced drug bioavailability. Thus, the study confirms that the formulated dorzolamide hydrochloride drug-loaded NE is nonirritant compared to conventional eye drop. Calvo et al. (1996) examined ^{14}C-indomethacin-loaded NE for corneal penetration. The study showed 3.65 times higher penetration coefficient of the drug loaded in NE. The study revealed the long-term stability of formulated NE. Many researchers have demonstrated that the positively charged NEs are more appropriate for topical ocular drug delivery system (Hagigit et al., 2008). Researchers have conducted studies on positively-charged drug-loaded NEs for ocular applications and confirmed the high uptake of drug in ocular tissue (Klang et al., 2000; Tamilvanan and Benita, 2004). Lallemand et al. (2012) prepared a preservative-free cationic NE called "cationorm," designed for dry eye treatment. The toxicity and pharmacokinetic analysis revealed its therapeutic efficiency. Anionic o/w NE-containing ciclosporin (cyclosporine A; Restasis, Allergan) were developed to enhance the tear production for dry eye patient (Ding et al., 1995). Morsi et al. (2014) successfully formulated acetazolamide in NE for glaucoma treatment. The study revealed the acceptable physicochemical properties, thermodynamic stability, increasing penetration through cornea, high therapeutic efficacy, and prolonged effect of the drug.

1.9.3   NANOEMULSION IN PARENTERAL DRUG DELIVERY

NE in parenteral route of drug administration is simple and efficient for active drugs with low bioavailability and limited therapeutic index. NEs are potent tool in parenteral drug delivery systems; they have the ability to dissolve the large quantities of hydrophobic compounds and capability to protect them from hydrolysis and enzymatic degradation. Due to its sustained drug release characteristic, the continuous dosage of the drug is shortened; this suggests that NEs are ideal tool in parenteral drug administration. Furthermore, reduction in flocculation, sedimentation, and creaming, linked with large surface area and free energy, are the significant advantages of NE for parenteral drug delivery systems (Ravi and Padma, 2011). Khalil et al. (2015) developed benzimidazole-loaded NE for anticancer activity. The study sheds light on the solubility and concentration-dependent drug release. Chlorambucil is a lipophilic drug used as anticancer agent for breast cancer and ovarian cancer. The pharmacokinetic studies and anticancer activity results suggest the efficacy of chlorambucil-loaded NE as parenteral drug administration. Kelmann et al. (2007) prepared carbamazepine-loaded NE by spontaneous emulsification process as anticonvulsant drug for intravenous delivery administration. Wang et al. (2008a) prepared w/o NE loaded with morphine drug and administered subcutaneously in rats. The study revealed its sustained drug release in central nervous system and also *in vivo* analgesic period is prolonged from 1 to 3 h. Fast et al. (2008) prepared sevoflurane-loaded NE using perfluorooctyl bromide and a semifluorinated surfactant (F13M5) to increase the high solubility of the drug compared with intralipid. In addition, only 30% of the main compound (sevoflurane) is used to prepare stable NE with the six times higher solubility rate. Tan et al. (2015) demonstrated the study on encapsulating carbamazepine in NE for drug delivery carrier to brain, suggesting enhanced bioavailability characteristic through Higuchi's equation. Parenteral NE of carbamazepine obtained higher AUC0→∞, AUMC0→∞, and Cmax than carbamazepine solution in brain.

1.9.4 NANOEMULSION FOR TRANSDERMAL DRUG DELIVERY

NE in transdermal drug administration route is an exciting and challenging area due to its promising advantages such as high storage stability, low preparation cost, and thermodynamic stability. Lipid-based colloidal NEs are reported as one of the most potential formulations for improvement of transdermal permeation and bioavailability of poorly soluble drugs. Due to its extremely small droplet size, low interfacial tension, and low viscosity, it can enhance the penetration ability of active compounds and can deliver to deeper layer of the skin (Bouchemal et al., 2004). Patients can remove the drug input at any time just by removing the transdermal patch. Furthermore, the main advantage is the lack of adverse side effects like gastrointestinal side effects such as irritation and bowel ulcers which are linked with oral drug delivery. Transdermal drugs are prepared for various diseases and disorders such as Parkinson's disease, Alzheimer's disease, depression, anxiety, and cardiovascular conditions. However, the basic disadvantage which limits the use of transdermal route of administration is the barrier imposed by the skin for effective penetration of the bioactive. Hair follicles, sweat ducts, or directly across stratum corneum are the routes by which drugs can penetrate the skin but also these routes restrict drug absorption to a large extent and limit their bioavailability. In order to enhance the drug pharmacokinetics and targeting, main skin barriers need to be overcome. Also, locally applied drug redistribution through cutaneous blood and lymph vessels system needs to be controlled. NE is an appropriate tool in cosmetic products. It provides high hydration power than conventional emulsions and microemulsions (Guglielmini, 2008; Nohynek et al., 2007). When compared with other drug delivery systems, such as microemulsions, solid lipid nanoparticles, and liposomes, NE carrier is an elegant tool for topical delivery system which provides high permeation capacity, low skin irritation, low preparation cost, thermodynamic stability, good production feasibility, absence of organic solvents, and high drug loading capacity for topical drug delivery. In addition, they have also made plasma concentration profiles and bioavailability of drugs reproducible. Caffeine has been used to treat different types of cancer through oral drug delivery. Shakeel and Ramadan (2010) developed w/o NE formulated with caffeine for transdermal drug delivery and evaluated the *in vitro* skin permeation studies. The results showed the high permeability in NE formulation of caffeine than aqueous caffeine solution. Many researchers have found that NE formulations possess improved transdermal and dermal delivery properties *in vitro* and *in vivo* (Lee et al., 2003). Barakat et al. (2011) developed NE formulations of indomethacin by spontaneous emulsification technique for transdermal drug delivery. There was a significant increase in the permeability parameters such as permeability coefficient, steady-state flux, and enhancement ratio in the formulated NE than conventional indomethacin gel in rats with carrageenan-induced paw edema. A study was conducted to evaluate the antiinflammatory potential of curcumin in combination with emu oil from a nanoemulgel formulation for arthritic *in vivo* models. They have developed NE using emu oil, Cremophor RH40, and Labrafil M2125CS as oil phase, surfactant, and cosurfactant. Curcumin-loaded NE with emu oil was incorporated into carbopol gel for convenient application by topical route. The results revealed the efficacy of antiinflammatory activity of NE formulated with curcumin in combination with emu oil than pure curcumin carrageenan-induced paw edema and Freund's complete adjuvant-induced arthritic rat model (Jeengar et al., 2016). Pratap et al. (2012) prepared an o/w NE formulation of carvedilol by the spontaneous emulsification method for transdermal delivery to improve the water solubility as well as bioavailability of drug. Post application, plasma carvedilol significantly increased 6.41 times than the market dosage form. The study concluded that the NE has the ability to eliminate the first-pass metabolism and also improves the bioavailability of transdermally administered carvedilol. Increasing the concentration of soybean lecithin and glycerol resulted in smaller droplet size of NE with increased viscosity. When 10% of lecithin NE incorporated into o/w cream, it enhances 2.5 times of the skin hydration capacity of the cream. They have also demonstrated the penetrability of Nile red dye loaded with lecithin NE into abdominal skin of rat *in vivo*. They have noticed that arbitrary unit of fluorescence in the dermis layer that had received 9.9-fold higher NE loaded with Nile red dye than the cream with a Nile red-loaded general emulsion. Shakeel et al. (2008a) has done the experimental study on drug celecoxib which is lipophilic and poorly soluble with low oral bioavailability used to treat inflammation. NE loaded with celecoxib has showed 3.3-fold higher bioavailability of drug through transdermal administration than oral capsule formulation in rats. Aceclofenac is a nonsteroidal antiinflammatory drug. Ropinirole is a drug used to treat Parkinson's disease, which has low oral bioavailability and continuous dosing. Azeem et al. (2012) evaluated the antiparkinson activity by developing transdermal NE gel in rats induced with Parkinson lesioned brain by 6-OHDA. The histological investigation found nonirritant, nontoxic effects in rat models. The study has concluded the ropinirole NE gel afforded better activity when compared with conventional tablets in Parkinson-induced rats.

Modi et al. (2011) developed an NE formulated with aceclofenac for transdermal administration. Formulation was optimized using 2% w/w of aceclofenac, 10% w/w of Labrafac, 45% w/w surfactant mixture (Cremophor® EL: Ethanol), and 43% w/w of distilled water. The evaluation of *in vitro* skin permeation of aceclofenac-formulated NE with aceclofenac conventional gel and NE gel was compared. A highly significant percentage value of antiinflammatory effect was noted in aceclofenac formulated NE on carrageenan-induced paw edema in rats. Tamoxifen citrate-loaded NE was developed for breast cancer by Pathan and Setty (2011). The optimized formulation of tamoxifen citrate NE consisted of 5% w/w of drug, 4.12% w/w of oil phase, 37.15% w/w of surfactant, and 58.73% w/w of distilled water. Transdermal permeation of tamoxifen citrate through rat skin was determined by Keshary–Chien diffusion cell. The results of permeability parameter revealed that the developed tamoxifen citrate NE improves the transdermal efficacy.

1.9.5   NANOEMULSIONS FOR PULMONARY DRUG DELIVERY

NEs with drugs targeted toward lungs are gaining attention in recent years. Very few researches have been conducted in NE for pulmonary drug delivery system. Submicron emulsion system has not yet been fully exploited for pulmonary drug delivery systems. NEs are significant and potential vehicle in pulmonary drug delivery due to noninvasive administration through inhalation aerosols, avoidance of first-pass metabolism, site targeting delivery treatment for respiratory diseases, and availability of large surface area for local drug action and systemic absorption of drug. Nanocarrier systems in pulmonary drug delivery provide many advantages such as the ability to desire and uniform distribution of drug concentration among the alveoli, sustained drug release, enhanced solubility rate, improved patient compliance, limited side effects, and the promising drug internalization by cells (Heidi et al., 2009). Bivas-Benita et al. (2004) reported that cationic submicron emulsions are potential carrier for DNA vaccines to the lung; since, they are able to transfect pulmonary epithelial cells, which may induce cross priming of antigen-presenting cells and directly activate dendritic cells resulting in stimulation of antigen-specific T cells. Nevertheless, extensive research is needed to formulate inhalable submicron emulsions due to possible adverse effect of surfactants and oils on lung alveoli function. Nesamony et al. (2014) formulated various self-nanoemulsifying mixtures and prepared o/w NE and explored the potential in pulmonary drug delivery by nebulizing the NE in a commercial pediatric nebulizer. Ability of NE loaded with poor water-soluble substrates was evaluated using a drug model, ibuprofen. The study revealed the stability and acceptable physiochemical properties of ibuprofen, which suggests that the developed formulation could be used for inhalation for delivering compound possessing poor water solubility into lungs. A novel pressurized aerosol system has been devised for the pulmonary delivery of salbutamol using lecithin-stabilized microemulsions formulated in trichlorotrifluoroethane (Lawrence and Rees, 2000). Nasr et al. (2012) developed amphotericin B (AmB) lipid NE aerosols for peripheral respiratory airways through Pari Sprint jet nebulizer. The study concluded the potentiality of AmB lipid NE along with the nebulization performance. Carbamazepine-loaded NE mist was administered through nebulizer and also confirmed the employment of NE for active targeting of poorly water-soluble molecules into the lungs (Nesamony et al., 2013).

1.9.6 NANOEMULSIONS IN INTRANASAL DRUG DELIVERY

Intranasal drug delivery system has been considered as best route for the administration of the drugs next to parenteral and oral administration. Among other mucosal regions, nasal mucosa has been developed as a therapeutically viable channel for systemic drug delivery and also it has many advantages such as rapid drug absorption, moderately permeable epithelium, lack of drug degradation, better nasal bioavailability for smaller drugs, and quick target action (Pires et al., 2009). Nasal drug delivery administration is noninvasive, painless, and a good tolerated drug delivery system (Ugwoke et al., 2005). There are many limitations in targeting drugs to brain, particularly delivering hydrophilic drug and higher molecular compounds. Due to the impervious character of endothelium, this divides the systemic circulation and barrier between the blood and brain (Pardridge, 1999). The olfactory area of the nasal mucosa provides a direct connection between brain and the nose, and by the use of NEs loaded with drugs, for conditions such as Alzheimer's disease, depression, migraine, schizophrenia, Parkinson's diseases, meningitis, etc., which can be treated (Mistry et al., 2009). Csaba et al. (2009) prepared a NE loaded with risperidone and its administration for brain via nose has been reported. The nasal delivery indicates a suitable route to circumvent the obstacles for blood–brain barrier and provides direct drug delivery in the biophase of central nervous system (CNS)-active compound. It has also been recognized for administration of vaccines (Graff and Pollock, 2005). It is evident that this NE is more effective through the nasal administration rather than the intravenous administration. Immunity is achieved by the administration of mucosal antigen. Currently, the first intranasal vaccine has been marketed (Csaba et al., 2009). From last decades, the use of nano-based carriers is gaining better attention due to its protection to the biomolecules and also promotes nanocarrier interaction with mucosa and to direct antigen to the lymphoid tissues. Bhanushali et al. (2009) developed intranasal NE and gel formulations for rizatriptan benzoate for prolonged action. Various mucoadhesive agents were used in preparation of thermo-triggered mucoadhesive NEs. Mucoadhesive gel formulations of rizatriptan were prepared using various ratios of hydroxy propyl methylcellulose and Carbopol 980. Comparative evaluation of intranasal mucoadhesive gels and intranasal NEs indicated that greater brain targeting could be achieved with NEs. Insulin and testosterone are the drugs formulated for nasal administration (Tamilvanan, 2004).

1.9.7 NANOEMULSION AS GENE DELIVERY SYSTEM

Emulsion carriers have been introduced as alterative for gene transfer vectors to liposomes (Liu et al., 1996). Study on gene delivery in emulsion system showed that the binding of the emulsion/DNA complex was stronger when compared to liposomal carrier system. This stable NE enhances the efficiency of gene delivery than liposomes (Liu and Yu, 2010). Silvan et al. (2012) demonstrated the study on factors that influence DNA compaction in cationic lipid NE loaded with stearylamine. The study was conducted in various parameters such as time complexation, incubation of temperature, and the influence of the stearylamine incorporation phase. The study revealed that the excellent DNA compaction process occurs after 120 min of complexation, at low temperature ($4 \pm 1°C$), followed by the incorporation of the cationic lipid into the aqueous phase. The zeta potential of lipoplexes shows higher results compared to basic NEs; it confirms that the lipoplexes is appropriate tool for gene delivery.

1.9.8 NANOEMULSIONS IN VACCINE DELIVERY

NEs for vaccine delivery system have attracted great attention in pharmaceutical sector. This medication delivery tool uses nanotechnology to vaccinate human immunodeficiency virus (HIV). The recent study has proven that HIV can infect the mucosal immune system. Thus, developing mucosal immunity through NEs is an emerging scope in future against HIV (Bielinska et al., 2008). The oil-based emulsion is administered in the nose, as opposed to traditional vaccine route. Bivas-Benita et al. (2004) demonstrated the positively charged NE for pulmonary mucosal DNA vaccines to the lungs. The NE protected the DNA adsorbed to the cationic surface of the oil droplets from the degradation in the presence of fetal calf serum in an *in vitro* stability experiment. Consequently, DNA was able to enter the nucleus and resulted in protein expression in the Calu-3 cell line (human bronchial epithelial cell line). Therefore, the DNA vaccine is expected to be stable upon exposure to the physiological environment and can be used for pulmonary immunization. NEs are being used to transport inactivated microbes to a mucosal surface to produce an immune response. The first applications as vaccine, an influenza vaccine and an HIV vaccine, can proceed to clinical trials. The NE causes proteins applied to the mucosal surface to act as an adjuvant and thus facilitate the uptake by antigen-presenting cells. This results in the significant systemic and mucosal immune response because of the production of specific IgG and IgA antibody as well as cellular immunity. Research in influenza has indicated that animals can be prevented against influenza infection after a single mucosal exposure to the virus mixed with the NEs. Studies have also shown that animals exposed to recombinant gp120 in NEs on their nasal mucosa create a significant response to HIV, thus giving a basis to use of this material as an HIV vaccine. Additional research is carried out to complete the proof of concept in animal trails for other vaccines including anthrax and hepatitis B. The University of Michigan has licensed this technology to NanoBio (Rutvij et al., 2011).

1.9.9   NANOEMULSIONS IN COSMETICS

NEs are used as good vehicle in cosmetic field because the optimized dispersion of active compounds is easily absorbed to give effective results and reduce the water loss from the skin. In addition, it provides controlled delivery of cosmetics and for the optimized dispersion of active ingredients in particular skin layers. Due to their lipophilic interior, NEs are more appropriate for the transport of lipophilic drug than liposomes. Their smaller droplet size with high surface area permits effective delivery of the active ingredients to the skin. Moreover, NEs are gaining interest because of their bioactive nature. This may reduce the transepidermal water loss, indicating that the barrier function of the skin is strengthened. NEs are acceptable in cosmetics because there is no chance of creaming, sedimentation, flocculation, or coalescence, which is observed with microemulsions. The inclusion of the potentially irritating surfactants can be avoided by using high-energy equipments during manufacturing process. A polyethylene glycol-free NE for cosmetics has also been developed and formulations showed good stability (Sharma and Sarangdevot, 2012; Yashpal et al., 2013).

1.9.10   NANOEMULSION IN CANCER THERAPY

Towing to the submicron size of NE, they can be easily targeted to the tumor region. Even though NEs are predominately used for administering aqueous insoluble drugs, still they are receiving attention as colloidal carriers for targeted delivery of different anticancer drugs, photosensitizers, neutron capture therapy agents, or diagnostic agents. The progress of magnetic NEs is an innovative approach for anticancer therapy. Desai et al. (2008) investigated cytotoxicity and apoptosis in brain cancer cells (U-118 human glioblastoma cells) treated with paclitaxel and ceramide-C_6 in an o/w NE. Drug-loaded NEs are significantly more efficient than the aqueous solution. Paclitaxel and ceramide-C_6 were found to be delivering inside the U-118 cells and significantly decreased the cell viability up to 33%. Fang et al. (2009) prepared an NE for camptothecin, which is a potent anticancer active agent. Liquid perfluorocarbons and coconut oil, as an inner oily phase, are used to develop the NE. NE loaded with camptothecin was tested against melanoma cells and ovarian tumor cells. The drug-loaded NE showed cyto-toxicity against melanoma and ovary cancer cells *in vitro*. The NEs showed significantly inhibited growth and they also have entered tumor cells in a higher quantity. NE can deliver photosensitizers like Foscan® to deep tissue layers across the skin thereby inducing hyperthermia for subsequent free-radical generation. This technique can be used for cancer treatment in the form of photodynamic therapy (Primo et al., 2007). Camptothecin is a topoisomerase-I inhibitor which acts against a broad spectrum of cancers. However, its clinical application has been limited by its poor solubility, lack of stability, and toxicity behavior of the drug. Fang et al. (2009) developed an NE loaded with camptothecin using liquid perfluorocarbons and coconut oil as oil cores of the inner phase. Camptothecin-formulated NEs with a lower oil concentration exhibited anticancer activity against melanomas and ovarian cancer cells.

1.9.11   NANOEMULSION AS ANTIMICROBIAL AGENT

Antimicrobial NEs are o/w droplets size in the range from 200 to 600 nm. They are stabilized by using surfactants and alcohols. The NEs have a wide spectrum of activity against bacteria such as *E. coli*, *Salmonella*, *Staphylococcus aureus*, and spores such as anthrax; enveloped viruses such as HIV, herpes simplex; and fungi such as *Candida* and dermatophytes. The NEs are thermodynamically forced to fuse with lipid-containing microbes. This fusion is enhanced by the electrostatic attraction between the cationic charge of the emulsion and the anionic charge on the pathogenic microbes. When enough nanoparticles fuse with the pathogenic microbes, they release part of the energy trapped within the emulsion. Both the active compound and the energy released destabilize the pathogen lipid membrane, resulting in cell lysis and death. In case of spores, additional germination enhancers are added into the emulsion. Once germination beings, the germinating spores become susceptible to the antimicrobial action of the NE. An aspect of the NEs is their highly selective toxicity to microbes at concentration range that are nonirritating to skin or mucous membrane. The safety range of NE is because of the low amount of detergent in each droplet, yet when acting in concert, these droplets have enough energy concentration and surfactant to destabilize targeted microbes without affecting healthy cells. NEs with topical antimicrobial activity can only be previously achieved by systemic antibiotics (Subhashis et al., 2011).

1.9.12   PROPHYLAXIS IN BIOTERRORISM ATTACK

As they possess antimicrobial activity, much research has been done on the use of NEs as a prophylactic medicated dosage form, a human protective treatment, to prevent the people exposed to bioattack such as anthrax and ebola. The broad-spectrum NEs were checked on surfaces by the US Army (Rest Ops) in December, 1999 for decontamination of anthrax spore. It was checked again by Rest Ops in March, 2001 as a chemical decontamination agent. This technology has been tested on gangrene and *Clostridium botulism* spores, and can even be used on contaminated wounds to salvage limbs. The NEs can be formulated into a cream, foam, liquid, and spray to decontaminate a large number of materials, which is marketed as NANOSTAT™ (Nanobio Corp) (Yashpal et al., 2013).

1.9.13   NANOEMULSION AS NONTOXIC DISINFECTANT CLEANER

NEs have been employed as a disinfectant cleaner. A nontoxic cleaner for use in routine markets that include healthcare, travel, food processing, and military applications has been developed by Enviro Systems. They have been found to kill tuberculosis and large spectrum of viruses, bacteria, and fungi within 5–10 min without any of the hazards posed by other categories of disinfectants. The product requires no warning labels. It does not irritate eyes and can be absorbed through skin, inhaled, or swallowed with harmless effects. The disinfectant formulation is made up of nanospheres of oil droplets less than 100 μm which are suspended in water to produce an NE requiring only small amounts of the active ingredient, para-chloro-meta-xylenol. The nanospheres have surface charges that efficiently penetrate the surface charges on the membranes of microbes by breaking through electric barrier. Rather than drowning cells, the formulation allows para-chloro-meta-xylenol to target and penetrate cell walls. Therefore para-chloro-meta-xylenol is applicable at concentration ranges from one to two times lower than those of other disinfectants, so there are no toxic effects on human, animals, or the environment (Rutvij et al., 2011; Yashpal et al., 2013). Other microbial disinfectants need large doses of their respective active ingredients to surround pathogen's cell wall, which causes microbe to disintegrate, ideally "drowning" them in the disinfectant solution. The disinfectant is not flammable and so safe to store anywhere and to use in unstable conditions. It is nonoxidizing, nonacidic, and nonionic. It will not corrode plastic, metals, or acrylic, so it makes the product ideal for use on equipment and instruments. It is environmentally safe, so the economical cost and health risks associated with hazardous chemical disposal are removed. The preparation is a broad-spectrum disinfectant cleaner that can be applied to any hard surface, including equipment, walls, fixtures, counters, and floors. One product can now take the place of many other, decreasing product inventories and saving valuable space. Chemical disposal costs can be removed, and disinfection and cleaning can be reduced. Marketed as EcoTru™ (Enviro Systems) (Yashpal et al., 2013).

1.9.14 ADVANTAGE, DISADVANTAGE, AND MAJOR CHALLENGES OF NANOEMULSIONS AS DRUG DELIVERY SYSTEMS

1.9.14.1 ADVANTAGES OF NANOEMULSIONS AS DRUG DELIVERY SYSTEMS

- NEs are more convenient for efficient delivery of active components through the skin. It is nontoxic and nonirritant; hence, it can be employed on skin and mucous membranes.
- Transparent characteristics of NE, their fluidity, and also the absence of any thickeners give a pleasant aesthetic character and skin feel.
- Smaller droplet size allows NE to deposit uniformly on substrates. Low surface tension and low interfacial tension of the o/w droplets greatly enhance the wetting, spreading, and penetrating on the substrates.
- Extremely small droplet size of NE causes a large reduction in the gravity forces and the Brownian motion may be sufficient for overcoming gravity. This results in no creaming or sedimentation while storage.
- The small droplet size highly prevents flocculation of the droplets which allows the system to remain dispersed without separation. NEs are thermodynamically stable and their stability permits self emulsification of the system.
- The small-scale droplet size of NE prevents their coalescence because droplets are elastic and fluctuations of the system are prevented.
- NE can be formulated in numerous dosage forms such as cream, sprays, liquids, and foams.
- Enhances solubilization of lipophilic drugs and also helps in masking unpleasant taste of drugs.

- Greatly increases the absorption rate, bioavailability, and excludes the variability in absorption.
- Possibilities of controlled drug release and drug targeting, and the incorporation of a great variety of therapeutic actives.
- It enhances the stability of chemically unstable molecules by protecting them from oxidative degradation and photodegradation.
- Drug-loaded NE can be delivered by different routes such as oral, transdermal, parental, intranasal, ocular, and intravenous.
- NEs are more suitable for human and veterinary therapeutic purpose as they do not cause any harm to human and animal cells.
- Better uptakes of oil-soluble supplements in cell cultures. Improves growth and vitality of cultured cells. It allows toxicity studies of oil-soluble drugs in cell cultures.

1.9.14.2   MAJOR CHALLENGES

Although NEs offer several advantages as drug delivery systems, they have some limitation and major challenges that are discussed below:
- NE formulation is generally expensive technique as it requires achievement of the smaller droplet size thus, high financial backing.
- To enhance the stability of small droplet size, it requires large amount of surfactant and cosurfactant.
- NE formulations remain stable even for a few years; however, due to their extremely smaller droplet size, it has been reported that the Ostwald ripening could damage NEs which causes drawback in its application.
- It has less solubility capacity for high-melting compounds.
- Stability of NE is affected by environmental factors such as pH and temperature.
- Lack of understanding the mechanism of submicron production and interfacial chemistry of NE.

1.10   CONCLUSION

NE formulation offers several advantages such as delivery of drugs, cosmetics, and biological or diagnostic agents. As a drug delivery system,

NEs enhance the therapeutic efficiency of the targeted drug, reduce adverse effect, and also minimize the toxic reactions. Due to the choice of huge number of core materials as well as emulsifying surface components offered by NEs, they can be customized to provide different requirements. They have more recently received increasing attention as colloidal carriers for targeted delivery of various anticancer drugs, photosensitizers, and neutron capture therapy agents. The applications of NE are limited by the instability. Stability of formulation may be enhanced by controlling various factors such as type and concentration of surfactants, cosurfactants, types of oil phase, methods of preparation, process variables, and addition of additives used for the NE formulations. Ability in dissolving large quantities of low-soluble drugs, increased drug loading capacity, controlled drug release, and the capability to protect the targeted drugs from hydrolysis and enzymatic degradation suggests NEs as ideal drug delivery vectors. NEs promise significant efficiency than conventional drug delivery systems, and are a flexible tool for achieving desired drug delivery in different conditions. Overall, utilization of NEs as drug delivery system has been made as a possible strategy on new therapies and will have high impact in future research and development on therapeutic delivery systems.

KEYWORDS

- **nanoemulsions**
- **formulation**
- **characterization**
- **drug delivery carrier**
- **polydispersity**

REFERENCES

Amani, A.; York, P.; Chrystyn, H.; Clark, B. J.; Do, D. Q. Determination of Factors Controlling the Particle Size in Nanoemulsions Using Artificial Neural Networks. *Eur. J. Pharm. Sci.* **2008,** *35* (1), 42–51.

Ammar, H. O.; Salama, H. A.; Ghorab, M.; Mahmoud, A. A. Nanoemulsion as a Potential Ophthalmic Delivery System for Dorzolamide Hydrochloride. *AAPS Pharm. Sci. Tech.* **2009,** *10* (3), 808–19.

Anton, N.; Vandamme, T. F. The Universality of Low-energy Nano-emulsification. *Int. J. Pharm.* **2009,** *377* (1), 142–147.

Anton, N.; Benoit, J. P.; Saulnier, P. Design and Production of Nanoparticles Formulated from Nanoemulsion Templates: A Review. *J. Control. Release* **2008,** *128* (3), 185–99.

Araujo, F. A.; Kelmann, R. G.; Araujo, B. V.; Finatto, R. B.; Teixeira, H. F.; Koester, L. S. Development and Characterization of Parenteral Nanoemulsions Containing Thalidomide. *Eur. J. Pharm. Sci.* **2011,** *42* (3), 238–245.

Azeem, A.; Rizwan, M.; Ahmad, F. J.; Iqbal, Z.; Khar, R. K.; Aqil, M.; Talegaonkar, S. Nano-emulsion Components Screening and Selection: A Technical Note. *AAPS Pharm. Sci. Tech.* **2009,** *10* (1), 69–76.

Azeem, A.; Talegaonkar, S.; Negi, L. M.; Ahmad, F. J.; Khar, R. K.; Iqbal, Z. Oil Based Nano-carrier System for Transdermal Delivery of Ropinirole: A Mechanistic, Pharmacokinetic and Biochemical Investigation. *Int. J. Pharm.* **2012,** *422* (1), 436–444.

Barakat, N.; Fouad, E.; Elmedany, A. Formulation Design of Indomethacin-loaded Nano-emulsion for Transdermal Delivery. *Pharm. Anal. Acta* **2011,** *S2* (002), 2–8.

Barea, M. J.; Jenkins, M. J.; Gaber, M. H.; Bridson, R. H. Evaluation of Liposomes Coated with a pH Responsive Polymer. *Int. J. Pharm.* **2010,** *402* (1), 89–94.

Benita, S. Prevention of Topical and Ocular Oxidative Stress by Positively Charged Submicron Emulsion. *Biomed. Pharmacother.* **1999,** *53* (4), 193–206.

Bhanushali, R. S.; Gatne, M. M.; Gaikwad, R. V.; Bajaj, A. N.; Morde, M. A. Nanoemulsion Based Intranasal Delivery of Antimigraine Drugs for Nose to Brain Targeting. *Indian J. Pharm. Sci.* **2009,** *71* (6), 707.

Bhat, M. A.; Iqbal, M.; Al-Dhfyan, A.; Shakeel, F. Carvone Schiff Base of Isoniazid as a Novel Antitumor Agent: Nanoemulsion Development and Pharmacokinetic Evaluation. *J. Mol. Liq.* **2015,** *203,* 111–119.

Bielinska, A. U.; Janczak, K. W.; Landers, J. J.; Markovitz, D. M.; Montefiori, D. C.; Baker, J. R. Nasal Immunization with a Recombinant HIV gp120 and Nanoemulsion Adjuvant Produces Th1 Polarized Responses and Neutralizing Antibodies to Primary HIV Type 1 Isolates. *AIDS Res. Hum. Retroviruses* **2008,** *24* (2), 271–81.

Bilbao-Sainz, C.; Avena-Bustillos, R. J.; Wood, D. F.; Williams, T. G.; McHugh, T. H. Nano-emulsions Prepared by a Low-energy Emulsification Method Applied to Edible Films. *J. Agric. Food Chem.* **2010,** *58* (22), 11932–11938.

Bivas-Benita, M.; Oudshoorn, M.; Romeijn, S.; van Meijgaarden, K.; Koerten, H.; van der Meulen, H.; Lambert, G.; Ottenhoff, T.; Benita, S.; Junginger, H.; Borchard, G. Cationic Submicron Emulsions for Pulmonary DNA Immunization. *J. Control. Release* **2004,** *100* (1), 145–155.

Calvo, P.; Vila-Jato, J. L.; Alonso, M. J. Comparative *In vitro* Evaluation of Several Colloidal Systems, Nanoparticles, Nanocapsules, and Nanoemulsions, as Ocular Drug Carriers. *J. Pharm. Sci.* **1996,** *85* (5), 530–536.

Chiesa, M.; Garg, J.; Kang, Y. T.; Chen, G. Thermal Conductivity and Viscosity of Water-in-oil Nanoemulsions. *Colloids Surf. A* **2008,** *326* (1), 67–72.

Cilek, A.; Celebi, N.; Tırnaksız, F.; Tay, A. A Lecithin-based Microemulsion of Rh-insulin with Aprotinin for Oral Administration: Investigation of Hypoglycemic Effects in Non-diabetic and STZ-induced Diabetic Rats. *Int. J. Pharm.* **2005,** *298* (1), 176–85.

Connolly, J. R. Introduction to X-ray Powder Diffraction, 2007. http://epswww.unm.edu/xrd/xrdclass/01-XRD-Intro. (accessed October 6, 2017).

Constantinides, P. P.; Chaubal, M. V.; Shorr, R. Advances in Lipid Nanodispersions for Parenteral Drug Delivery and Targeting. *Adv. Drug Deliv. Rev.* **2008,** *60* (6), 757–767.

Csaba, N.; Garcia-Fuentes, M.; Alonso, M. J. Nanoparticles for Nasal Vaccination. *Adv. Drug Delivery Rev.* **2009,** *61* (2), 140–157.

Date, A. A.; Nagarsenker, M. S. Design and Evaluation of Self-nanoemulsifying Drug Delivery Systems (SNEDDS) for Cefpodoxime Proxetil. *Int. J. Pharm.* **2007,** *329* (1), 166–172.

Desai, A.; Vyas, T.; Amiji, M. Cytotoxicity and Apoptosis Enhancement in Brain Tumor Cells upon Coadministration of Paclitaxel and Ceramide in Nanoemulsion Formulations. *J. Pharm. Sci.* **2008,** *97* (7), 2745–2756.

Ding, S.; Tien, W. L.; Olejnik, O. Nonirritating Emulsions for Sensitive Tissue. United States Patent US 5,474,979, December 12, 1995.

Edwards, K. A.; Baeumner, A. J. Analysis of Liposomes. *Talanta* **2006,** *68* (5), 1432–1441.

El-Aasser, M. S.; Sudol, E. D. Miniemulsions: Overview of Research and Applications. *J. Coat.Technol. Res.* **2004,** *1* (1), 21–32.

Fang, J. Y.; Hung, C. F.; Hua, S. C; Hwang, T. L. Acoustically Active Perfluorocarbon Nanoemulsions as Drug Delivery Carriers for Camptothecin: Drug Release and Cytotoxicity Against Cancer Cells. *Ultrasonics* **2009,** *49* (1), 39–46.

Fasolo, D.; Schwingel, L.; Holzschuh, M.; Bassani, V.; Teixeira, H. Validation of an Isocratic LC Method for Determination of Quercetin and Methylquercetin in Topical Nanoemulsions. *J. Pharm. Biomed. Anal.* **2007,** *44* (5), 1174–1177.

Fast, J. P.; Perkins, M. G.; Pearce, R. A.; Mecozzi, S. Fluoropolymer-based Emulsions for the Intravenous Delivery of Sevoflurane. *Am. J. Anesthesiol.* **2008,** *109* (4), 651–656.

Fernandez, P.; Andre, V.; Rieger, J.; Kühnle, A. Nanoemulsion Formation by Emulsion Phase Inversion. *Colloid. Surf. A* **2004,** *251* (1), 53–58.

Giustini, M.; Palazzo, G.; Colafemmina, G.; Monica, M. D.; Giomini, M.; Ceglie, A. Microstructure and Dynamics of the Water-in-oil CTAB/N-pentanol/N-hexane/Water Microemulsion: A Spectroscopic and Conductivity Study. *J. Phys. Chem.* **1996,** *100* (8), 3190–3198.

Graff, C. L.; Pollack, G. M. Nasal Drug Administration: Potential for Targeted Central Nervous System Delivery. *J. Pharm. Sci.* **2005,** *94* (6), 1187–95.

Grigoriev, D. O.; Miller, R. Mono- and Multilayer Covered Drops as Carriers. *Curr. Opin. Colloid Interface Sci.* **2009,** *14* (1), 48–59.

Guglielmini, G. Nanostructured Novel Carrier for Topical Application. *Clin. Dermatol.* **2008,** *26* (4), 341–346.

Hagigit, T.; Nassar, T.; Behar-Cohen, F.; Lambert, G.; Benita, S. The Influence of Cationic Lipid Type on In-vitro Release Kinetic Profiles of Antisense Oligonucleotide from Cationic Nanoemulsions. *Eur. J. Pharm. Biopharm.* **2008,** *70* (1), 248–259.

Haritha, A.; Syed, P. B.; Koteswara, R. P.; Chakravarthi, V. A Brief Introduction to Methods of Preparation, Applications and Characterization of Nanoemulsion Drug Delivery Systems. *Indian J. Res. Pharm. Biotechnol.* **2013,** *1* (1), 25–28.

Hatziantoniou, S.; Deli, G.; Nikas, Y.; Demetzos, C.; Papaioannou, G. T. Scanning Electron Microscopy Study on Nanoemulsions and Solid Lipid Nanoparticles Containing High Amounts of Ceramides. *Micron* **2007,** *38* (8), 819–823.

Heidi, M. M.; Yun-Seok, R.; Xiao, W. Nanomedicine in Pulmonary Delivery. *Int. J. Nanomed.* **2009,** *4,* 299–319.

Holm, R.; Porter, C. J.; Müllertz, A.; Kristensen, H. G.; Charman, W. N. Structured Triglyc-eride Vehicles for Oral Delivery of Halofantrine: Examination of Intestinal Lymphatic Transport and Bioavailability in Conscious Rats. *Pharm. Res.* **2002,** *19* (9), 1354–1361.

Howe, A. M.; Pitt, A. R. Rheology and Stability of Oil-in-water Nanoemulsions Stabilised by Anionic Surfactant and Gelatin 2) Addition of Homologous Series of Sugar-based Co-surfactants. *Adv. Colloid Interface Sci.* **2008,** *144* (1), 30–37.

Huang, M.; Horwitz, T. S.; Zweiben, C.; Singh, S. K. Impact of Extractables/Leachables from Filters on Stability of Protein Formulations. *J. Pharm. Sci.* **2011,** *100* (11), 4617–4630.

Izquierdo, P.; Esquena, J.; Tadros, T. F.; Dederen, C.; Garcia, M. J.; Azemar, N.; Solans, C. Formation and Stability of Nano-emulsions Prepared Using the Phase Inversion Tempera-ture Method. *Langmuir* **2002,** *18* (1), 26–30.

Jaiswal, M.; Dudhe, R.; Sharma, P. K. Nanoemulsion: An Advanced Mode of Drug Delivery System. *3 Biotech.* **2015,** *5* (2), 123–127.

Jeengar, M. K.; Rompicharla, S. V.; Shrivastava, S.; Chella, N.; Shastri, N. R.; Naidu, V. G.; Sistla, R. Emu Oil Based Nano-emulgel for Topical Delivery of Curcumin. *Int. J. Pharm.* **2016,** *506* (1), 222–236.

Jenning, V.; Mader, K.; Gohla, S. H. Solid Lipid Nanoparticles (SLN™) Based on Binary Mixtures of Liquid and Solid Lipids: A ^{1}H-NMR Study. *Int. J. Pharm.* **2000,** *205* (1), 15–21.

Kawakami, K.; Yoshikawa, T.; Moroto, Y.; Kanaoka, E.; Takahashi, K.; Nishihara, Y.; Masuda, K. Microemulsion Formulation for Enhanced Absorption of Poorly Soluble Drugs. I. Prescription Design. *J. Control. Release* **2002,** *81* (1), 65–74.

Kelmann, R. G.; Kuminek, G.; Teixeira, H. F.; Koester, L. S. Carbamazepine Parenteral Nanoemulsions Prepared by Spontaneous Emulsification Process. *Int. J. Pharm.* **2007,** *342* (1), 231–239.

Khalil, R. M.; Basha, M.; Kamel, R. Nanoemulsions as Parenteral Drug Delivery Systems for a New Anticancer Benzimidazole Derivative: Formulation and In-vitro Evaluation. *Egypt Pharmaceut. J.* **2015,** *14* (3), 166.

Khandavilli, S.; Panchagnula, R. Nanoemulsions as Versatile Formulations for Paclitaxel Delivery: Peroral and Dermal Delivery Studies in Rats. *J. Invest. Dermatol.* **2007,** *27* (1), 154–62.

Klang, S.; Benita, S. For Intravenous Administration. Submicron Emulsions in Drug Targeting and Delivery. **1998,** *9,* 119.

Kotta, S.; Khan, A. W.; Pramod, K.; Ansari, S. H.; Sharma, R. K.; Ali, J. Exploring Oral Nanoemulsions for Bioavailability Enhancement of Poorly Water-soluble Drugs. *Expert Opin. Drug Deliv.* **2012,** *9* (5), 585–598.

Kotyla, T.; Kuo, F.; Moolchandani, V.; Wilson, T.; Nicolosi, R. Increased Bioavailability of a Transdermal Application of a Nano-sized Emulsion Preparation. *Int. J. Pharm.* **2008,** *347* (1), 144–148.

Kreilgaard, M.; Pedersen, E. J.; Jaroszewski, J. W. NMR Characterisation and Transdermal Drug Delivery Potential of Microemulsion Systems. *J. Control. Release* **2000,** *69* (3), 421–433.

Kumar, M.; Misra, A.; Babbar, A. K.; Mishra, A. K.; Mishra, P.; Pathak, K. Intranasal Nano-emulsion Based Brain Targeting Drug Delivery System of Risperidone. *Int. J. Pharm.* **2008,** *358* (1), 285–291.

Lallemand, F.; Daull, P.; Benita, S.; Buggage, R.; Garrigue, J. S. Successfully Improving Ocular Drug Delivery Using the Cationic Nanoemulsion, Novasorb. *J. Drug Deliv.* **2012,** *27,* 604204.

Lawrence, M. J.; Rees, G. D. Microemulsion-based Media as Novel Drug Delivery Systems. *Adv. Drug Deliv. Rev.* **2000,** *45* (1), 89–121.

Lee, P. J.; Langer, R.; Shastri, V. P. Novel Microemulsion Enhancer Formulation for Simultaneous Transdermal Delivery of Hydrophilic and Hydrophobic Drugs. *Pharm. Res.* **2003,** *20* (2), 264–269.

Lee, S. J.; McClements, D. J. Fabrication of Protein-stabilized Nanoemulsions Using a Combined Homogenization and Amphiphilic Solvent Dissolution/Evaporation Approach. *Food Hydrocoll.* **2010,** *24* (6), 560–569.

Leong, T. S.; Wooster, T. J.; Kentish, S. E.; Ashokkumar, M. Minimising Oil Droplet Size Using Ultrasonic Emulsification. *Ultrason. Sonochem.* **2009,** *16* (6), 721–727.

Li, X.; Anton, N.; Ta, T. M.; Zhao, M.; Messaddeq, N.; Vandamme, T. F. Microencapsulation of Nanoemulsions: Novel Trojan Particles for Bioactive Lipid Molecule Delivery. *Int. J. Nanomed.* **2011,** *6* (1), 1313–1325.

Liu, C. H.; Yu, S. Y. Cationic Nanoemulsions as Non-viral Vectors for Plasmid DNA Delivery. *Colloids Surf. B* **2010,** *79* (2), 509–515.

Liu, F.; Yang, J.; Huang, L.; Liu, D. Effect of Non-ionic Surfactants on the Formation of DNA/Emulsion Complexes and Emulsion-mediated Gene Transfer. *Pharm. Res.* **1996,** *13* (11), 1642–1646.

Liu, W.; Sun, D.; Li, C.; Liu, Q.; Xu, J. Formation and Stability of Paraffin Oil-in-water Nano-emulsions Prepared by the Emulsion Inversion Point Method. *J. Colloid Interface Sci.* **2006,** *303* (2), 557–563.

Luykx, D. M.; Peters, R. J.; van Ruth, S. M.; Bouwmeester, H. A Review of Analytical Methods for the Identification and Characterization of Nano Delivery Systems in Food. *J. Agric. Food Chem.* **2008,** *56* (18), 8231–8247.

Maali, A.; Mosavian, M. H. Preparation and Application of Nanoemulsions in the Last Decade (2000–2010). *J. Disper. Sci. Technol.* **2013,** *34* (1), 92–105.

Mason, T. G.; Wilking, J. N.; Meleson, K.; Chang, C. B.; Graves, S. M. Nanoemulsions: Formation, Structure, and Physical Properties. *J. Phys. Condens. Matter* **2006,** *18* (41), 635–643.

McClements, D. J.; Decker, E. A.; Weiss, J. Emulsion-based Delivery Systems for Lipophilic Bioactive Components. *J. Food Sci.* **2007,** *72* (8), 109–124.

Mehta, S. K. Isentropic Compressibility and Transport Properties of CTAB-alkanol-hydrocarbon-water Microemulsion Systems. *Colloids Surf. A* **1998,** *136* (1), 35–41.

Middleton, D. L.; Leung, S. S.; Robinson, J. R. Ocular Bioadhesive Delivery Systems. In *Bioadhesive Drug Delivery Systems*; CRC Press: Boca Raton, FL, 1990; pp 179–202.

Mistry, A.; Stolnik, S.; Illum, L. Nanoparticles for Direct Nose-to-brain Delivery of Drugs. *Int. J. Pharm.* **2009,** *379* (1), 146–157.

Modi, J. D.; Patel, J. K. Nanoemulsion-based Gel Formulation of Aceclofenac for Topical Delivery. *Int. J. Pharm. Pharm. Sci. Res.* **2011,** *1* (1), 6–12.

Morsi, N. M.; Mohamed, M. I.; Refai, H.; El Sorogy, H. M. Nanoemulsion as a Novel Ophthalmic Delivery System for Acetazolamide. *Int. J. Pharm. Pharm. Sci. Res.* **2014,** *6* (11), 227–236.

Mou, D.; Chen, H.; Du, D.; Mao, C.; Wan, J.; Xu, H.; Yang, X. Hydrogel-thickened Nanoemulsion System for Topical Delivery of Lipophilic Drugs. *Int. J. Pharm.* **2008,** *353* (1), 270–276.

Musa, S. H.; Basri, M.; Masoumi, H. R.; Karjiban, R. A.; Malek, E. A.; Basri, H.; Shamsuddin, A. F. Formulation Optimization of Palm Kernel Oil Esters Nanoemulsion-loaded with Chloramphenicol Suitable for Meningitis Treatment. *Colloids Surf. B* **2013,** *112,* 113–119.

Nesamony, J.; Kalra, A.; Majrad, M. S.; Boddu, S. H.; Jung, R.; Williams, F. E.; Schnapp, A. M.; Nauli, S. M.; Kalinoski, A. L. Development and Characterization of Nanostructured Mists with Potential for Actively Targeting Poorly Water-soluble Compounds into the Lungs. *Pharm. Res.* **2013,** *30* (10), 2625–2639.

Nesamony, J.; Shah, I. S.; Kalra, A.; Jung, R. Nebulized Oil-in-water Nanoemulsion Mists for Pulmonary Delivery: Development, Physico-chemical Characterization and *In vitro* Evaluation. *Drug Dev. Ind. Pharm.* **2014,** *40* (9), 1253–1263.

Nicolet, T. Introduction to Fourier Transform Infrared Spectrometry, 2001. http://mmrc.caltech.edu/FTIR/FTIRintro.pdf. [accessed December 2, 2017].

Nohynek, G. J.; Lademann, J.; Ribaud, C.; Roberts, M. S. Grey Goo on the Skin? Nanotechnology, Cosmetic and Sunscreen Safety. *Crit. Rev. Toxicol.* **2007,** *37* (3), 251–277.

Pardridge, W. M. Non-invasive Drug Delivery to the Human Brain Using Endogenous Blood–Brain Barrier Transport Systems. *Pharm. Sci. Technolo. Today* **1999,** *2* (2), 49–59.

Pathan, I. B.; Setty, C. M. Enhancement of Transdermal Delivery of Tamoxifen Citrate Using Nanoemulsion Vehicle. *Int. J. Pharm. Tech. Res.* **2011,** *3* (1), 287–297.

Pires, A.; Fortuna, A.; Alves, G.; Falcão, A. Intranasal Drug Delivery: How, Why and What For? *J. Pharm. Pharm. Sci.* **2009,** *12,* 288–311.

Porras, M.; Solans, C.; Gonzalez, C.; Gutierrez,; J.M. Properties of Water-in-oil (W/O) Nanoemulsions Prepared by a Low-energy Emulsification Method. *Colloids Surf. A* **2008,** *24* (1), 181–188.

Pratap, S. B.; Brajesh, K.; Jain, S. K.; Kausar, S. Development and Characterization of a Nanoemulsion Gel Formulation for Transdermal Delivery of Carvedilol. *Int. J. Drug Dev. Res.* **2012,** *4* (1), 151–161.

Primo, F. L.; Michieleto, L.; Rodrigues, M. A.; Macaroff, P. P.; Morais, P. C.; Lacava, Z. G.; Bentley, M. V.; Tedesco, A. C. Magnetic Nanoemulsions as Drug Delivery System for Foscan®: Skin Permeation and Retention *In vitro* Assays for Topical Application in Photodynamic Therapy (PDT) of Skin Cancer. *J. Magn. Magn. Mater.* **2007,** *311* (1), 354–357.

Rao, S. V.; Shao, J. Self-nanoemulsifying Drug Delivery Systems (SNEDDS) for Oral Delivery of Protein Drugs. I. Formulation Development. *Int. J. Pharm.* **2008a,** *362* (1), 2–9.

Rao, S. V.; Yajurvedi, K.; Shao, J. Self-nanoemulsifying Drug Delivery System (SNEDDS) for Oral Delivery of Protein Drugs. III. *In vivo* Oral Absorption Study. *Int. J. Pharm.* **2008b,** *362* (1), 16–9.

Ravi, T. P. U.; Padma, T. Nanoemulsions for Drug Delivery Through Different Routes. *Res. Biotechnol.* **2011,** *2* (3), 1–13.

Ribeiro, H. S.; Chu, B. S.; Ichikawa, S.; Nakajima, M. Preparation of Nanodispersions Containing β-carotene by Solvent Displacement Method. *Food Hydrocoll.* **2008,** *22* (1), 12–7.

Rutvij, J. P.; Gunjan, J. P.; Bharadia, P. D.; Pandya, V. M.; Modi, D. A. Nanoemulsion: An Advanced Concept of Dosage Form. *Int. J. Pharm. Cosmetol.* **2011,** *1* (5), 122–133.

Salager, J. L.; Marquez, L. Nanoemulsions: Where Are They Going To? *Colloidi Tpoint.* **2003,** *2,* 12–14.

Samah, N. A.; Williams, N.; Heard, C. M. Nanogel Particulates Located Within Diffusion Cell Receptor Phases Following Topical Application Demonstrates Uptake into and Migration Across Skin. *Int. J. Pharm.* **2010,** *401* (1), 72–78.

Shafiq, S.; Shakeel, F.; Talegaonkar, S.; Ahmad, F. J.; Khar, R. K.; Ali, M. Development and Bioavailability Assessment of Ramipril Nanoemulsion Formulation. *Eur. J. Pharm. Biopharm.* **2007,** *66* (2), 227–243.

Shah, P.; Bhalodia, D.; Shelat, P. Nanoemulsion: A Pharmaceutical Review. *Sys. Rev. Pharm.* **2010,** *1* (1), 24.

Shakeel, F; Haq, N; El-Badry, M; Alanazi, F. K; Alsarra, I. A. Ultra Fine Super Self-nano-emulsifying Drug Delivery System (SNEDDS) Enhanced Solubility and Dissolution of Indomethacin. *J. Mol. Liq.* **2013,** *180,* 89–94.

Shakeel, F.; Baboota, S.; Ahuja, A.; Ali, J.; Faisal, M. S.; Shafiq, S. Stability Evaluation of Celecoxib Nanoemulsion Containing Tween 80. *Thai. J. Pharm. Sci.* **2008a,** *32,* 4–9.

Shakeel, F.; Baboota, S.; Ahuja, A.; Ali, J.; Shafiq, S. Skin Permeation Mechanism of Aceclofenac Using Novel Nanoemulsion Formulation. *Die Pharmazie: An Int. J. Pharm. Sci.* **2008b,** *63* (8), 580–584.

Shakeel, F.; Baboota, S.; Ahuja, A.; Ali, J.; Shafiq, S. Skin Permeation Mechanism and Bioavailability Enhancement of Celecoxib from Transdermally Applied Nanoemulsion. *J. Nanobiotechnol.* **2008c,** *6* (1), 1.

Shakeel, F.; Ramadan, W. Transdermal Delivery of Anticancer Drug Caffeine from Water-in-oil Nanoemulsions. *Colloids Surf. B* **2010,** *75* (1), 356–62.

Shakeel, F.; Haq, N.; Alanazi, F. K.; Alsarra, I. A. Impact of Various Nonionic Surfactants on Self-nanoemulsification Efficiency of Two Grades of Capryol (Capryol-90 and Capryol-PGMC). *J. Mol. Liq.* **2013,** *182,* 57–63.

Sharma, S.; Sarangdevot, K. Nanoemulsions for Cosmetics. *Int. J. Adv. Res. Pharma. Bio. Sci.* **2012,** *2* (3), 408–415.

Sharma, S. N.; Jain, N. K. *A Textbook of Professional Pharmacy,* 1st ed.; Vallabh Prakashan: Delhi, India, 1985; p 201.

Sieg, J. W.; Robinson, J. R. Vehicle Effects on Ocular Drug Bioavailability II: Evaluation of Pilocarpine. *J. Pharm. Sci.* **1977,** *66* (9), 1222–1228.

Silva, A. L.; Alexandrino, F.; Verissimo, L. M.; Agnez-Lima, L. F.; Egito, L.; de Oliveira, A. G.; do Egito, E. S. Physical Factors Affecting Plasmid DNA Compaction in Stearyl-amine-containing Nanoemulsions Intended for Gene Delivery. *Pharmaceuticals* **2012,** *5* (6), 643–654.

Silvan, H. D.; Cerqueira, M. A.; Vicente, A. A. Nanoemulsions for Food Applications: Development and Characterization. *Food Bioprocess Technol.* **2012,** *5* (3), 854–867.

Singh, K. K.; Vingkar, S. K. Formulation, Antimalarial Activity and Biodistribution of Oral Lipid Nanoemulsion of Primaquine. *Int. J. Pharm.* **2008,** *347* (1), 136–143.

Solans, C.; Izquierdo, P.; Nolla, J.; Azemar, N.; Garcia-Celma, M. J. Nano-emulsions. *Curr. Opin. Colloid Interface Sci.* **2005,** *10* (3), 102–110.

Subhashis, D.; Satayanarayana, J.; Gampa, V. K. Nanoemulsion: A Method to Improve the Solubility of Lipophilic Drugs. *Int. J. Adv. Pharm. Sci.* **2011,** *2* (2–3), 72–83.

Subramanian, B.; Kuo, F.; Ada, E.; Kotyla, T.; Wilson, T.; Yoganathan, S.; Nicolosi, R. Enhancement of Anti-inflammatory Property of Aspirin in Mice by a Nano-emulsion Preparation. *Int. Immunopharmacol.* **2008,** *8* (11), 1533–1539.

Tadros, T.; Izquierdo, P.; Esquena, J.; Solans, C. Formation and Stability of Nanoemulsions. *Adv. Colloid Interface Sci.* **2004,** *108,* 303–318.

Tamilvanan, S. Submicron Emulsions as a Carrier for Topical (Ocular and Percutaneous) and Nasal Drug Delivery. *Indian J. Pharm. Educ.* **2004,** *38* (2), 73–80.

Tamilvanan, S.; Benita, S. The Potential of Lipid Emulsion for Ocular Delivery of Lipophilic Drugs. *Eur. J. Pharm. Biopharm.* **2004**, *58* (2), 357–368.

Tan, L. S.; Stanslas, J.; Basri, M.; Karjiban, R. A.; Kirby, P. B.; Sani, D.; Bin Basri, H. Nano-emulsion-based Parenteral Drug Delivery System of Carbamazepine: Preparation, Characterization, Stability Evaluation and Blood-Brain Pharmacokinetics. *Curr. Drug Deliv.* **2015**, *12* (6), 795–804.

Tarver, T. O. Food nanotechnology. Food Technology-Champaign Then Chicago. *Food Technol.* **2006**, *160* (11), 22–26.

Tenjarla, S. Microemulsions: An Overview and Pharmaceutical Applications. *Crit. Rev. Ther. Drug Carrier Syst.* **1999**, *16*, 461–521.

Tiwari, S. B.; Amiji, M. M. Nanoemulsion Formulations for Tumor Targeted Delivery. In: *Nanotechnology for Cancer Therapy.* Amiji, M. M. (Ed.), 1st ed, CRC Press: Boca Raton, 2006; pp 723–739.

Ugwoke, M. I.; Agu, R. U.; Verbeke, N.; Kinget, R. Nasal Mucoadhesive Drug Delivery: Background, Applications, Trends and Future Perspectives. *Adv. Drug Deliv. Rev.* **2005**, *57* (11), 1640–65.

Uson, N.; Garcia, M. J.; Solans, C. Formation of Water-in-oil (W/O) Nano-emulsions in a Water/Mixed Non-ionic Surfactant/Oil Systems Prepared by a Low-energy Emulsification Method. *Colloids Surf. A* **2004**, *250* (1), 415–421.

Venturini, C. G.; Jager, E.; Oliveira, C. P.; Bernardi, A.; Battastini, A. M.; Guterres, S. S.; Pohlmann, A. R. Formulation of Lipid Core Nanocapsules. *Colloids Surf. A* **2011**, *375* (1), 200–208.

Wang, L.; Li, X.; Zhang, G.; Dong, J.; Eastoe, J. Oil-in-water Nanoemulsions for Pesticide Formulations. *J. Colloid Interface Sci.* **2007**, *314* (1), 230–235.

Wang, J. J.; Hung, C. F.; Yeh, C. H.; Fang, J. Y. The Release and Analgesic Activities of Morphine and its Ester Prodrug, Morphine Propionate, Formulated by Water-in-oil Nano-emulsions. *J. Drug Target.* **2008a**, *16* (4), 294–301.

Wang, X.; Jiang, Y.; Wang, Y. W.; Huang, M. T.; Ho, C. T.; Huang, Q. Enhancing Anti-inflammation Activity of Curcumin Through O/W Nanoemulsions. *Food Chem.* **2008b**, *108* (2), 419–24.

Warisnoicharoen, W.; Lansley, A.; Lawrence, M. J. Nonionic Oil-in-water Microemulsions: The Effect of Oil Type on Phase Behavior. *Int. J. Pharm.* **2000**, *198* (1), 7–27.

Weiss, J.; Decker, E. A.; McClements, D. J.; Kristbergsson, K.; Helgason, T.; Awad, T. Solid Lipid Nanoparticles as Delivery Systems for Bioactive Food Components. *Food Biophys.* **2008**, *3* (2), 146–54.

Wulff-Perez, M.; Torcello-Gomez, A.; Galvez-Ruiz, M. J.; Martín-Rodríguez, A. Stability of Emulsions for Parenteral Feeding: Preparation and Characterization of O/W Nanoemulsions with Natural Oils and Pluronic f68 as Surfactant. *Food Hydrocoll.* **2009**, *23* (4), 1096–1102.

Yashpal, S.; Tanuj, H.; Harsh, K. Nanoemulsions: A Pharmaceutical Review. *Int. J. Pharma. Prof. Res.* **2013**, *4* (2), 928–935.

Yilmaz, E.; Borchert, H. H. Design of a Phytosphingosine-containing, Positively-charged Nanoemulsion as a Colloidal Carrier System for Dermal Application of Ceramides. *Eur. J. Pharm. Biopharm.* **2005**, *60* (1), 91–8.

Yilmaz, E.; Borchert, H. H. Effect of Lipid-containing, Positively Charged Nanoemulsions on Skin Hydration, Elasticity and Erythema—An *In vivo* Study. *Int. J. Pharm.* **2006**, *307* (2), 232–238.

Yuan, Y.; Gao, Y.; Mao, L.; Zhao, J. Optimisation of Conditions for the Preparation of β-carotene Nanoemulsions Using Response Surface Methodology. *Food Chem.* **2008a,** *107* (3), 1300–1306.

Yuan, Y.; Gao, Y.; Zhao, J.; Mao, L. Characterization and Stability Evaluation of β-carotene Nanoemulsions Prepared by High Pressure Homogenization Under Various Emulsifying Conditions. *Food Res. Int.* **2008b,** *41* (1), 61–8.

Zhao, D.; Gong, T.; Fu, Y.; Nie, Y.; He, L. L.; Liu, J.; Zhang, Z. R. Lyophilized Cheliensisin A Submicron Emulsion for Intravenous Injection: Characterization, *In vitro* and *In vivo* Antitumor Effect. *Int. J. Pharm.* **2008,** *357* (1), 139–47.

NANOEMULSIONS: FORMULATION INSIGHTS, APPLICATIONS, AND RECENT ADVANCES

MD. RIZWANULLAH[1], JAVED AHMAD[2*],
MOHAMMAD ZAKI AHMAD[2], and
MOHAMMED SAEED N. ALGAHTANI[2]

[1]*Department of Pharmaceutics, School of Pharmaceutical Education and Research, Jamia Hamdard, New Delhi 110062, India*

[2]*Department of Pharmaceutics, College of Pharmacy, Najran University, Kingdom of Saudi Arabia (KSA)*

Corresponding author. E-mail: jahmad18@gmail.com

ABSTRACT

Nanoemulsion effectively serves as a nanocarrier for the drug delivery as well as delivery of potential bioactive compounds in the management of various diseases. It is a colloidal dispersion with high kinetic stability, low viscosity, and optically transparent system with mean droplet size typically ranging from 10 to 200 nm. It has great potential for encapsulating the active therapeutics in its inner core and improves the stability profile of encapsulated moiety by providing protection against external milieu. Encapsulation of active therapeutics into inner core of nanoemulsions system can offer improvements in the chemical and/or enzymatic stability of active therapeutics as well. Nanoemulsions can be fabricated from generally recognized as safe ingredients as isotropic mixtures and exploiting simple and easy techniques such as vortexing and homogenization. Some of the potential advantages of nanoemulsions over conventional emulsions include significantly higher bioaccessibility and greater stability. This chapter focuses on the formulation insights, applications, and recent advancement in the field of nanoemulsion-based drug delivery systems. In addition, *in vivo* fate of nanoemulsion-based drug delivery systems as well as suitability of such carrier system for the delivery of active therapeutics through oral, parenteral, transdermal, ocular, and intranasal routes is also addressed.

2.1 INTRODUCTION

Nanoemulsions are stable, transparent, nanodispersions of oil and water phase, stabilized by an interfacial film of surfactant typically in combination with cosurfactant with droplet size in range of 10–200 nm (Solans et al., 2005; Anton and Vandamme, 2011) (Fig. 2.1). The reduction of surface tension near to zero resulted into the formation of stable nanoemulsion system. The specific surfactants, their combinations, or specific packing of the adsorbed layer with surfactant and cosurfactant help in achieving the same. Cosurfactants are generally short-chain alcohols (e.g., propanol or butanol) and considerably smaller in comparison to the principal surfactant. Interfacial conformational close-packing is supposed to be accountable for the minimum surface tensions and formation of nanoemulsions (Lawrence and Rees, 2000).

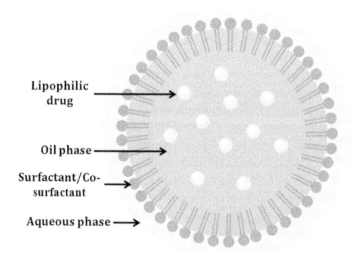

Lipophilic drug

Oil phase

Surfactant/Co-surfactant

Aqueous phase

FIGURE 2.1 (See color insert.) Lipophilic drug encapsulated inside nanoemulsion system.

They offer numerous advantages since: (1) they are highly favored when it is essential that the drug molecule to be encapsulated and protected from a harsh external environment, (2) they are simple to prepare in a consistent manner on an industrial scale, (3) they can be fabricated entirely from generally recognized as safe (GRAS) components using simple processing operations, (4) they may increase the bioavailability of encapsulated lipophilic bioactives due to their small droplet size, (5) they have relatively good stability to gravitational separation and aggregation (compared with emulsions) and may therefore be more stable in commercial products, and (6) they can be designed to have high optical clarity which is important for some commercial applications, for example, transparent foods, beverages, syrups, or lotions (Sarker, 2005; McClements, 2013).

Nanoemulsions have been widely exploited for their broad range of applications in drug delivery. Their typical characteristics such as improved drug loading capabilities, protection to the lipophilic drugs in the biological environment, improved oral bioavailability, and sustained drug release make them the suitable nanocarrier for drug delivery and related biomedical applications (Talegaonkar and Negi, 2015).

2.2 NANOEMULSION: PREPARATION AND EVALUATION

2.2.1 PRODUCTION AND CHARACTERIZATION OF NANOEMULSIONS

The primary objective for preparation of nanoemulsion is to achieve the droplet size in nanometer range (10–200 nm) and to impart stabilizing composition conditions in biological environment. Generally, two methods, high-energy emulsification or low-energy emulsification techniques, are used for the preparation of nanoemulsions (Jafari et al., 2007; Kentish et al., 2008; Shinoda et al., 1991). High-energy emulsification methods include high-pressure homogenization (HPH), microfluidization, and ultrasonication. HPH and microfluidization techniques can be carried out at the industrial scale, whereas ultrasonication is mainly used at bench level. Although high-energy emulsification methods are very effective in the reduction of droplet size, they are unsuitable for labile compounds and macromolecules including proteins and peptides. For these reasons, low-energy emulsification techniques are generally preferred for preparation of nanoemulsions. Low-energy emulsification methods include spontaneous emulsification (aqueous phase/oil phase titration method), the solvent diffusion method, and the phase-inversion temperature method (Shakeel et al., 2010; Shafiq and Shakeel, 2008; Anton et al., 2007). Table 2.1 lists the different methods and their descriptions for preparation of nanoemulsions.

TABLE 2.1   Different Preparation Techniques for Nanoemulsions.

Production techniques	Description
HPH	HPH is conducted by applying a high pressure over the system of primary emulsion by using a high-pressure homogenizer or a piston homogenizer to produce nanoemulsions of extremely low droplet size. Major disadvantage of these homogenizers are poor productivity, deterioration of components due to difficult mass production, and much heat generation. This technique can only be used for preparation of o/w liquid nanoemulsions containing less than 20% oil phase
Microfluidization	Microfluidization is a patented mixing technique, which makes use of apparatus called microfluidizer for the preparation of nanoemulsions. This apparatus uses a high-pressure positive displacement pump (500–20,000 psig), which forces the product through the interaction chamber. From the interaction chamber the product flows through the small channels called "microchannels" onto an impingement area resulting in very fine droplets
Ultrasonication	Ultrasonication provides another route for the preparation of nanoemulsions. In this method, the droplet size of conventional emulsion or even nanoemulsion is reduced with the help of sonication mechanism
Spontaneous emulsification	This is the simplest technique for the production of nanoemulsions. This technique simply involves blending the right proportions of oil, water, surfactant, and cosurfactant using mild agitation. During the nanoemulsification process, large transitory fluctuations in interfacial tension can occur. The formulation components are arrange accordingly so that resulting interfacial and bulk microstructures lead to an overall decrease in the free energy
PIT	In this technique, a fine dispersion is obtained by chemical energy as a result of phase transition. The phase transition is produced by varying the composition at constant temperature or vice versa

HPH, high-pressure homogenization; PIT, phase inversion temperature.

It is of utmost importance to have an in-depth knowledge about various characteristic properties that nanoemulsion must possess for their successful commercial exploitation. However, complete characterization of nanoemulsions is quite difficult due to their complexity, different structures and components involved in these systems, as well as the limitations associated with these techniques. Over the years, nanoemulsions have been characterized by different techniques but a complementary method is generally required to substantiate the results and overcome the limitations of the method used. At the macroscopic level, viscosity, conductivity, and dielectric methods provide useful information. Table 2.2 lists the different techniques and their descriptions for characterization of nanoemulsions (Shakeel et al., 2008; Shakeel et al., 2012; McClements, 2012).

TABLE 2.2   Different Characterization Techniques for Nanoemulsions.

Characterization techniques	Description
Viscosity measurements	These measurements are suitable to evaluate flow behavior of nanoemulsion
Conductivity measurements	It decides the type of nanoemulsion and identifies the phase inversion phenomenon
Dielectric measurements	These are important means of analyzing both structural and dynamic features of nanoemulsions
Phase behavior studies	Phase diagram are used to evaluate the phase behavior of components
NMR studies	NMR technique can be used to study the structure and dynamics of nanoemulsions
Electron microscopic studies: FFTEM and TEM	FFTEM and TEM are the most important techniques for the study of nanostructures because it directly produces images at high resolution and it can capture any coexistent structure and nanostructure transitions
Scattering techniques: SAXS, SANS, and PCS	Scattering techniques that have been utilized in the characterization of nanoemulsions include SAXS, SANS and static light scattering and dynamic light scattering or PCS. All these techniques are used to determine droplet size distribution
Interfacial tension measurement	Interfacial tension strongly affects the formulation and stability of nanoemulsions. Spinning-drop apparatus is used to determine interfacial tension
Thermal characterization	DSC is used for thermal characterization
Other methods	Dye solubility and dilution test, plane polarized light microscopy, refractive index, zeta potential measurements, etc.

DSC, differential scanning calorimetry; FFTEM, freeze fracture transmission electron microscopy; NMR, nuclear magnetic resonance; PCS, photon correlation spectroscopy; SANS, small angle neutron scattering; SAXS, small angle X-ray scattering; TEM, transmission electron microscopy.

2.2.2 THEORIES OF NANOEMULSION FORMATION

2.2.2.1 MIXED FILM THEORY

The theory deals with the spontaneous formation of nanoemulsion droplets because of the formation of interfacial film at the oil–water interface in presence of Smix (mixture of surfactant and cosurfactant). This results into the minimization of oil–water interfacial tension to minimum value. According to this theory, formation of mixed film was implicit to be liquid as well as duplex in nature having two-dimensional spreading pressure (π_i). This ultimately found to affect the interfacial tension (γ) as:

$$\gamma_i = \gamma_{o/w} - \pi_i$$

In the above equation, $\gamma_{o/w}$ indicates oil–water interfacial tension without the film. The absorption of huge amounts of Smix leads to the value of spreading pressure may become greater in comparison to $\gamma_{o/w}$ and give negative interfacial tension. Thus, provide the energy to amplify the interfacial area. This enhancement in interfacial area is responsible to minimize the droplet sizes. The formation of interfacial tension with negative value is a transient phenomenon (Eccleston, 1994).

2.2.2.2 SOLUBILIZATION THEORIES

According to this theory, nanoemulsion system is considered to be a thermodynamically stable monophasic dispersion of water-in-oil or oil-in-water micelles of spherical appearance. The solubilization of oil phase in normal micelles was found to be small. Moreover, the concentration of different components of nanoemulsion system was reported to be significant to solubilize the excess amount of hydrocarbon. It may swell directly into oil droplet devoid of forming low curvature various intermediate structures (Eccleston, 1994).

2.2.2.3 THERMODYNAMIC THEORY

As per the thermodynamic theory, oil droplets formation through oil phase bulk is resulted by an enhancement of interfacial area (ΔA), and therefore an interfacial energy ($\gamma \Delta A$). The entropy of oil droplets dispersion is found to be equivalent to $T\Delta S$. Therefore, free energy involved in the development of nanoemulsion system is expressed by:

$$\Delta G_f = \gamma \Delta A - T \Delta S$$

In the above equation, ΔG_f indicates the free energy in formation of nanoemulsion system, surface tension of the oil–water interface is denoted by γ, and temperature by T. The significant change in ΔA because of the production of huge tiny droplets is accompanied by the development of nanoemulsion system. The value of γ remains positive but very small in magnitude. Therefore, compensate easily by the entropic component. Mixing of one phase into the other phase in a form of huge small droplets due to the large entropy of dispersion system is responsible for dominancy of the entropic component. Therefore, it results in the formation of negative free energy, and nanoemulsification process is found to be spontaneous as well as thermodynamically stable (Bali et al., 2008).

2.2.3 PHYSICOCHEMICAL ASPECTS

The physicochemical characteristics required for nanoemulsion-based drug delivery system mainly depend on the specific needs of the final application. Nanoemulsions can be designed to have various optical, rheological, and stability characteristics by controlling their compositions and structures.

2.2.3.1 OPTICAL PROPERTIES

The optical property of nanoemulsions is basically dependent upon the droplet concentration. The degree of light scattering is increasing as the number of droplets increases in the system (Chantrapornchai et al., 1999). Furthermore, the degree of light scattering increases with increase in refractive index contrasts (i.e., difference in refractive index between oil phase and aqueous phase) (Chantrapornchai et al., 2001). The refractive index contrast therefore depends upon the compositions and physical states of the oil and aqueous phase system. The refractive index of the oil phase basically depends upon the presence of type and concentration of lipophilic components. It usually increases as the lipid phase crystallizes. Addition of water-soluble components such as sugars, salts, and polyols increases the refractive index of the aqueous phase (Chantrapornchai et al., 2001). A nanoemulsion system usually appears to be transparent when the refractive index of the oil and aqueous phases is found to be very close and resulting weak scattering of light.

2.2.3.2   RHEOLOGICAL PROPERTIES

Nanoemulsions may have a highly viscous or gel-like material for creams used for topical applications of bioactive components. The rheology of nanoemulsions may vary from low-viscosity fluids to gel-like materials depending on their composition and structure (Mason et al., 2006; McClements, 2011). To a first approximation, the viscosity of a nanoemulsion can be given by the following equation (Pal, 2011):

$$\eta = \eta_o \left(1 - \frac{\phi_e}{\phi_c} \right)^{-2}$$

Here, η is the viscosity of the nanoemulsion, η_o is the viscosity of the aqueous phase, f is the effective disperse-phase volume fraction, and ϕ_c (≈ 0.6) is a critical disperse-phase volume fraction, above which the droplets are so closely packed together that they cannot easily flow past each other (McClements, 2005). This equation demonstrates that the viscosity of a nanoemulsion is directly proportional to the viscosity of the aqueous phase, and it should increase as the concentration of the droplets increases. The viscosity usually increases moderately with increasing ϕ_e at low droplet concentrations, but then increases steeply when ϕ_e approaches ϕ_c (Pal, 2011). Therefore, rheological characteristics of nanoemulsions can be modulated in a number of ways such as (1) altering the viscosity of the aqueous phase (e.g., by adding cosolvents or polymers), (2) altering the total droplet concentration, (3) altering the droplet size, (4) altering the thickness of the interfacial layer surrounding the droplets, and (5) promoting droplet flocculation.

2.2.3.3 PHYSICAL STABILITY

The physical stability of nanoemulsions influences the shelf life; therefore, nanoemulsions should be designed to maintain their stability throughout the expected lifetime of a product. The stability of a nanoemulsion depends on its composition and structure, as well the various environmental conditions that it encounters within a product such as temperature, light, changes in pH, ionic strength, mechanical forces, and oxygen levels. Nanoemulsions may degrade through different physical mechanisms, including Ostwald ripening, gravitational separation, coalescence, and flocculation. The main factor determining the stability of nanoemulsions to these types of droplet aggregation is the balance of attractive and repulsive interactions operating between them. Nanoemulsions tend to be stable to aggregation when the repulsive forces dominate, but unstable when the attractive forces dominate (e.g., van der Waals', hydrophobic, depletion, or bridging) (McClements, 2005; Israelachvili, 2011). The tendency for flocculation and coalescence to occur usually decreases with decreasing droplet size, and so, nanoemulsions tend to be more stable to droplet aggregation than conventional emulsions (Lee et al., 2011). Ostwald ripening is one of the major forms of physical instability in many nanoemulsions, which is the process whereby small droplets shrink and large droplets grow due to diffusion of oil molecules through the intervening aqueous phase (Kabalnov, 2001). It can be retarded or arrested by adding substances known as "ripening inhibitors" to the oil phase prior to homogenization (Li et al., 2009; Lim et al., 2011; McClements et al., 2012).

2.2.3.4 CHEMICAL STABILITY

Many of the bioactive lipids that may be used for preparation of nanoemulsion are chemically unstable such as carotenoids, oil-soluble vitamins, conjugated linoleic acid, and ω-3-fatty acids (Qian et al., 2012; Belhaj et al., 2010). Light-catalyzed reactions may occur more rapidly in transparent nanoemulsions since more ultraviolet or visible light waves can penetrate into their interior than in opaque emulsions. Surface-catalyzed reactions (such as lipid oxidation or lipase digestion) may occur more rapidly in nanoemulsions because they have higher interfacial areas than conventional emulsions. It is therefore important to carefully design nanoemulsion-based delivery systems to retard any potential chemical degradation reactions that might occur. Therefore, a variety of strategies have been developed to retard the degradation of lipophilic drugs, including reducing oxygen and light levels, maintaining low storage temperatures, incorporating appropriate antioxidants, chelating transition metal catalysts, and controlling droplet interfacial properties (McClements and Decker, 2000; Boon et al., 2010).

2.2.4 IN VIVO FATE OF NANOEMULSIONS

2.2.4.1 INFLUENCE OF GASTROINTESTINAL MILIEU ON NANOEMULSIONS

The physicochemical properties of the nanoemulsions may be altered considerably as they pass through the various regions of the gastrointestinal

(GI) tract, for example, mouth, stomach, small intestine, and large intestine (Golding et al., 2011; Singh and Sarkar, 2011; van Aken, 2010). Nanoemulsions undergo very complex physicochemical and physiological situations in the GI tract, with alteration in temperature, fluid flows, different pH, ionic composition, enzyme activities, surface-active agents, biopolymers, and biological surfaces. These changes in GI environment may promote substantial change in the size, electrical characteristics, concentration, and composition of lipid nanoparticles, which will ultimately influence the oral bioavailability of any encapsulated lipophilic drug. For instance, the size of the droplets of nanoemulsion may either decrease or increase as they travel through the GI tract due to processes such as digestion and solubilization, flocculation and coalescence (McClements and Xiao, 2012).

2.2.4.2   BIOAVAILABILITY OF LIPOPHILIC DRUGS

Metabolism of lipophilic drugs may occur in different locations in the human body such as within the lumen of the GI tract, within the epithelium cells, or within the liver (first-pass metabolism). The initial characteristics of a nanoemulsion-based delivery system may alter the overall bioavailability of an encapsulated lipophilic component by affecting each of these physicochemical mechanisms (Acosta, 2009).

2.2.4.3   BIOACCESSIBILITY

For conventional emulsions, the bioaccessibility is generally assumed to be the fraction of the drugs released from the lipid nanoparticles and incorporated into the mixed micelles present the small intestine, because the oil droplets in the emulsions are considered to be too large to penetrate through the mucous layer. The incorporation of lipophilic drugs into mixed micelles usually requires the digestion of the lipid phase (typically triacylglycerols) surrounding the lipophilic drugs in the oil droplets (Porter et al., 2007; Nik et al., 2011), because this facilitates the drug release and helps in the development of mixed micelles with a high solubilization capacity. In addition, a limited amount of the lipophilic drugs may be directly transferred from the oil droplets into the micelles through diffusion (even without the action of lipase) (Yu and Huang, 2012; Wang et al., 2012). The same mechanisms will also be important for nanoemulsions, but it may also be possible for the small oil droplets within nanoemulsions to be directly absorbed by the epithelium cells. Nanoemulsions can travel through the mucous layer since their droplet size is smaller than its pore size (diameter <300 nm), and then reach the epithelium cells where they could be absorbed by active or passive transport mechanisms (Cone, 2009; Ensign et al., 2012). Nevertheless, this opportunity depends on the time for the lipid nanoparticles to be digested, relative to the time to penetrate through the mucous layer.

2.2.4.4 ABSORPTION

Lipophilic drugs present within the lumen of the GI tract are generally transported through the digestive fluids, and then pass through the mucous layer, where they are then absorbed by the GI epithelia (Pouton and Porter, 2008; Pouton, 2000). For nanoemulsions fabricated using biodegradable lipids, the lipophilic drugs are usually solubilized and transported by the mixed micelles produced by lipid digestion. Indeed, a recent study revealed that the absorption of curcumin by Caco-2 cells was significantly higher from a digested nanoemulsion (with lipase) than from an undigested one (no lipase), suggesting that the major absorption mechanism was via mixed micelles rather than via nanoemulsion droplets (Yu and Huang, 2012). In principle, lipophilic drugs may be transported to the epithelium cells within lipid droplets for nanoemulsions prepared using nonbiodegradable lipids or for nanoemulsions where incomplete lipid digestion occurs.

The physicochemical characteristics of the lipid nanoparticles in the small intestine may affect the absorption process. The transport rate of lipid nanoparticles through the mucous layer depends upon the particle size relative to the pore size of mucous layer (Ensign et al., 2012; Florence, 2012). Relatively small nanoparticles (diameter <300 nm) may be able to penetrate into and through the mucous layer, whereas larger ones cannot (Ensign et al., 2012). Whether whole lipid nanoparticles can travel across the mucous layer will depend on the mass transport rate as compared to digestion rate. Biodegradable lipid nanoparticles may be totally digested before they reach the epithelium cells. The ability of lipid nanoparticles to become confined within the mucous layer will increase their retention time within the GI tract (Ensign et al., 2012), which should lead to higher lipid digestion and greater oral bioavailability. The surface characteristics of the nanoemulsions may also affect their ability to travel into and through the mucous layer (Crater and Carrier, 2010). For example, cationic lipid nanoparticles may bind to anionic biopolymer molecules within the mucous layer, thereby increasing their retention time but inhibiting their transport to the epithelium cells (Crater and Carrier, 2010).

Once the lipophilic drugs have reached to the surface of the GI epithelial cells, they may be absorbed by a variety of passive and active transport mechanisms depending on their molecular structure (Cai et al., 2010; Florence and Hussain, 2001). At this stage, the absorption of lipophilic drugs may be limited by its permeation through the cell membrane (Metcalfe and Thomas, 2010; Sugano, 2009a, 2009b). If intact lipid nanoparticles are able to penetrate through the mucous layer and reach the layer of GIepithelial cells, they may be absorbed through a number of different mechanisms including (Powell et al., 2010; Frohlich and Roblegg, 2012): paracellular—transported through the narrow gaps ("tight junctions") between the epithelium cells; transcellular—transported directly across the epithelium cell walls by various endocytosis processes; and persorption—transported through holes formed in the epithelium layer due to cell turnover. The route taken, and the rate and extent of lipid droplet uptake, will depend on particle properties such as size, charge, and interfacial chemistry (Hu et al., 2009; des Rieux et al., 2006).

2.2.4.5 METABOLISM

Ingested lipophilic drugs may undergo changes in their chemical structure due to metabolism by various chemical and enzymatic processes occurring within the GI tract, the epithelium cells, and the systemic circulation (Payne et al., 2012; Xie et al., 2012; Scalbert and Williamson, 2000). The extent and nature of these changes can be controlled somewhat by careful design of delivery systems. For instance, it is possible to develop novel drug delivery systems that facilitate the uptake of lipophilic drugs either through the lymphatic route or through the portal vein route by changing the lipid-phase composition (Kotta et al., 2012; Iqbal and Hussain, 2009). Lipophilic components that travel through the lymphatic route do not undergo first-pass metabolism within the liver, which may alter their subsequent bioactivity (Trevaskis et al., 2008; Yanez et al., 2011). Further research is needed to establish the role of lipid droplet properties on the metabolism of lipophilic components encapsulated within nanoemulsion-based delivery systems.

2.3   APPLICATIONS AND FUTURE DIRECTIONS

2.3.1   *DRUG DELIVERY APPLICATIONS: RECENT PROGRESS*

The most popular application of nanoemulsions is that of a drug nanocarrier. Nanoemulsions have been exploited to deliver the drugs through different route of administration, including oral, transdermal, parenteral, inhalational, ocular, and intranasal (Table 2.3). Among them, parenteral, oral, and trans-dermal routes of drug delivery are the most investigated pathways in case of nanoemulsion applications.

2.3.1.1   *ORAL DRUG DELIVERY*

Oral drug delivery is the most common route of administration among all other routes that have been extensively explored for systemic delivery of therapeutics. Oral route is considered natural, most convenient, and safe due to its ease of administration, patient compliance, and cost-effective manu-facturing process (Rizwanullah et al., 2017; Ahmad et al., 2013; Ahmad et al., 2015). Nanoemulsion represents extremely low interfacial tensions and great o/w interfacial areas. Nanoemulsions have significantly higher solu-bilization capacity and their high thermodynamic stability provides advan-tages over unstable dispersions, such as emulsions and suspensions, and has long shelf life. Nanoemulsions showed great enhancement in oral bioavail-ability and therapeutic efficacy of lipophilic compounds. In this context, Akhtar et al. (2016) investigated the efficacy of repaglinide-loaded nano-emulsion against diabetes. *In vitro* dissolution studies of developed nano-emulsion exhibited significantly higher drug release (98.22%), finest droplet size (76.23 nm), slightest polydispersity value (0.183), and least viscosity (21.45 cps) compared to oral tablet formulation. Furthermore, the optimized repaglinide-loaded nanoemulsion exhibited excellent antidiabetic activity in experimental diabetic rats for 12 h. Wan et al. (2016) developed curcumin-loaded nanoemulsion for improved oral bioavailability and enhanced anti-cancer activity. Their results showed that the developed nanoemulsion exhibited 2.29–4.04- and 4.06–8.27-folds higher absorption constants and effective permeabilities, respectively, compared to free curcumin. Curcumin-loaded nanoemulsion showed 733.59% relative bioavailability compared to free curcumin. Furthermore, curcumin-loaded nanoemulsion exhibited significantly enhanced inhibition against A549 cell line compared to free curcumin. In another study, Pangeni et al. (2016) developed multiple nanoemulsions for combinational delivery of oxaliplatin (OXA) and 5-fluorouracil (5-FU) for treatment of colorectal cancer after oral administration. Their results showed that OXA/5-FU-loaded nanoemulsion exhibited 4.80- and 4.30-fold higher *in vitro* permeabilities through a Caco-2 cell monolayer compared to free OXA and 5-FU, respectively. The developed nanoemulsion exhibited significantly enhanced oral absorption and exhibited 9.19- and 1.39-fold higher oral bioavailability compared to free OXA and 5-FU, respectively. Furthermore, developed nanoemulsion exhibited significantly higher tumor growth inhibition in CT26 tumor-bearing mice after oral administration compared to free OXA and 5-FU, respectively.

2.3.1.2 PARENTERAL DRUG DELIVERY

The concept of microemulsion for parenteral delivery was developed from submicron sized fat o/w emulsion used for parenteral administration of nutrition, namely, Intralipid® in 1960s (Wretlind, 1981). Intravenous drug delivery shows the highest bioavailability (almost 100%) among all administration routes with superior advantages in immediate effect, targeting effect, and overcoming the first-pass effect (Gao et al., 2008). In this context, Li et al. (2011) developed docetaxel-loaded nanoemulsions and investigated anticancer efficacy against lung cancer after intravenous (i.v.) administration. The developed nanoemulsion showed droplet size and zeta potential of 169 nm and −33.9mV, respectively. Acute toxicity studies after i.v. administration showed no toxicity for the docetaxel-loaded nanoemulsion. The docetaxel-loaded nanoemulsion exhibited significantly higher apparent volume of distribution and area under curve compared to docetaxel after i.v. administration to rats. Moreover, developed nanoemulsion exhibited significantly higher anticancer efficiency compared to docetaxel after i.v. administration into Lewis lung carcinoma-bearing male Sprague Daley rats. Zhang et al. (2011) developed doxorubicin–oleic acid ionic complex-loaded nanoemulsion to investigate their pharmacokinetic behavior and efficacy against HepG2 hepatoma cells after i.v. administration. Pharmacokinetics and *in vivo* distribution studies in mice showed that the developed nanoemulsion exhibited significantly higher doxorubicin concentration in blood and longer mean resident time (MRT) compared to free doxorubicin after i.v. administration. Moreover, the developed nanoemulsion showed significantly decreased concentration of doxorubicin in kidney, heart, and lung. In addition, the developed nanoemulsion exhibited significantly higher cytotoxicity against HepG2 hepatoma cells compared to free doxorubicin. Similarly, Zhao et al. (2013) developed lycobetaine–oleic acid ionic complex-loaded nanoemulsion to investigate their pharmacokinetic behavior and improved efficacy against lung cells after i.v. administration. Pharmacokinetic study in male albino Wistar rats showed that developed nanoemulsion displayed ~30-fold higher area under the concentration (AUC) after i.v. administration. Moreover, the developed nanoemulsion showed significantly lower lycobetaine concentration in the kidney, heart, and liver. In addition, lycobetaine–oleic acid ionic complex-loaded nanoemulsion exhibited significantly higher anticancer efficacy and longer survival time compared to free lycobetaine in both heterotopic and lung metastatic tumor models.

2.3.1.3 TRANSDERMAL DRUG DELIVERY

Nanoemulsion-based transdermal drug delivery approach has been reported to offer various advantages such as high permeation of drug across the skin, low skin irritation, and high drug loading efficiency in comparison to other nanocarrier systems for percutaneous absorption of drug. In addition, the transdermal administration of drug provides sustained drug release for longer period of time, no first-pass metabolism, as well as high patient compliance in comparison to oral or parenteral drug delivery (Sonneville-Aubrun et al., 2004). In this context, El-Leithy et al. (2015) developed and evaluated indomethacin-loaded nanoemulsion as a transdermal delivery system. The developed nanoemulsions showed 610-fold higher drug solubility compared to water. The developed nanoemulsions exhibited controlled release of indomethacin through semipermeable membrane and significantly enhanced indomethacin permeation through rat skin. Moreover, indomethacin-loaded nanoemulsion exhibited significantly higher concentration of indomethacin up to 32 h in rats after transdermal application. Husain et al. (2016) developed amphotericin B-loaded nanoemulsion gel and evaluated antifungal activity after transdermal administration. Their results showed that the developed nanoemulsion gel exhibited ~4.5-fold higher skin percutaneous permeation flux rate compared to plain drug solution. Hemolytic and histopathological examination suggested that the developed nanoemulsion gel for cutaneous infection is safe and efficacious than oral administration. Furthermore, Negi et al. (2015) developed lidocaine- and prolocaine-loaded nanoemulsion for improved percutaneous absorption after transdermal administration. Ex vivo skin permeation studies showed that the developed nanoemulsions exhibited significantly higher skin permeation rate compared to conventional nanoemulsion. Dermatokinetic studies showed the presence of significantly higher concentration of anesthetic drugs in the dermis layer of skin compared to conventional formulation. The *in vivo* efficacy studies in rabbits showed that the developed nanoemulsions exhibited significantly improved onset and duration of local anesthetic action compared to the marketed cream.

2.3.1.4 OCULAR DRUG DELIVERY

For the treatment of various ophthalmic disorders, most commonly used vehicle in clinical practice are aqueous solutions. Indeed, this vehicle for ocular drug delivery leads to a low ocular bioavailability due to the rapidly diluted and eliminated because of lacrimation, blinking action, as well as nasolacrimal duct elimination before the drug permeated into the cornea and sclera (Cotlier et al., 1975). Therefore, various oil-in-water nanoemulsion-based drug delivery vehicle has been explored for release of drug at ophthalmic level. Patel et al. (2016) developed loteprednol etabonate-encapsulated cationic nanoemulsion gel for improved ocular bioavailability. Their results showed that the developed in situ nanoemulsion gel was clear, had shear thinning in nature, and exhibited zero-order release kinetics. The developed cationic nanoemulsified in situ gel exhibited significantly lower ocular irritation potential and significantly higher maximum plasma concentration and area under curve, MRT, 2.54-fold improved ocular bioavailability compared to marketed formulation. Similarly, Katzer et al. (2014) developed nanoemulsion of castor and mineral oil and investigated its compatibility with soft contact lens. The developed nanoemulsion showed droplet size, polydispersity index, and zeta potential of 234 nm, 0.16, −8.56 mV, respectively. The developed nanoemulsion was nonirritant. After ocular administration, SEM analysis confirmed that the nanoemulsion showed almost complete transparency. Furthermore, no significant differences were found between lens ion permeability coefficients before and after administration.

2.3.1.5 INTRANASAL DRUG DELIVERY

The intranasal route of drug administration has great impact due to the noninvasive nature, potential to nose to brain targeting, avoid first-pass metabolism as well as high vascularity and relatively large surface area for drug absorption. Nanoemulsions system for intranasal drug delivery reported to be encapsulating the drugs with high loading efficiency, penetrate the drug effectively through anatomical barriers as well as efficient drug release at the site of action with good stability and biocompatibility (Marianecci et al., 2015). In this context, Mahajan et al. (2014) developed saquinavir mesylate-loaded nanoemulsion for brain targeting after intranasal administration. The developed nanoemulsion showed significantly enhanced drug permeation rate compared to plain drug suspension. No significant adverse effect was observed in cilia toxicity study using sheep nasal mucosa. *In vivo* biodistribution studies showed that the developed nanoemulsion exhibited significantly higher concentration in the brain compared to plain drug suspension. Moreover, optimized nanoemulsion showed significantly higher targeting efficiency and nose-to-brain direct drug transport after intranasal administration. Furthermore, Pangeni et al. (2014) developed vitamin E-encapsulated resveratrol nanoemulsion for the treatment of Parkinson's disease after intranasal administration. The developed nanoemulsion showed droplet size, polydispersity index, and zeta potential of 102 ± 1.46 nm, 0.158 ± 0.02, and -35 ± 0.02 mV, respectively. Pharmacokinetic studies showed that the developed nanoemulsion exhibited significantly higher concentration of the drug in the brain compared to free resveratrol after intranasal administration. Histopathological studies showed that the developed nanoemulsion exhibited significantly decreased degenerative changes. Furthermore, nanoemulsion-treated group showed significantly higher levels of glutathione and superoxide dismutase and significantly lower level of malondialdehyde.

TABLE 2.3 Nanoemulsions Exploited to Deliver the Drugs Through Different Routes of Administration.

Route	Encapsulated drug	Excipients	Outcome	References
Oral	Repglinide	Sefsol 218, Tween 80, transcutol	Significantly improved antidiabetic activity in experimental diabetic rats	Akhtar et al., 2016
	Curcumin	Ethyl oleate, cremorphor EL 35, polyethylene glycol 400	2.29–4.04- and 4.06–8.27-folds higher absorption constants and effective permeabilities, 733.59% relative bioavailability, and significantly enhanced cytotoxicity against A549 cell line	Wan et al., 2016

TABLE 2.3   *(Continued)*

Route	Encapsulated drug	Excipients	Outcome	References
	Oxali-platin and 5-fluorouracil	Labrasol, transcutol HP, cremophor EL	9.19- and 1.39-fold higher oral bioavail-ability compared to free OXA and 5-FU respectively and exhibited signifi-cantly higher tumor growth inhibition in CT26 tumor-bearing mice	Pangeni et al., 2016
Parenteral	Docetaxel	Oleic acid, glycerin, polox-amer 188	Significantly improved pharmaco-kinetic behavior and enhanced anticancer activity	Li et al., 2011
	Doxorubicin hydrochloride	Oleic acid, leci-thin, vitamin E	Significantly enhanced AUC and MRT and significantly higher cytotoxicity against HepG2 hepatoma cells	Zhang et al., 2011
	Lycobetaine	Oleic acid, lecithin, mPEG2000-DSPE	~30-fold higher AUC and significantly higher anticancer efficacy and longer survival time	Zhao et al., 2013
Trans-dermal	Indomethacin	Capryol 90, labrafil, Tween 80, pluronic F127, transcutol HP	610-fold higher drug solubility compared to water and signifi-cantly enhanced indomethacin permeation through rat skin	El-Leithy et al., 2015
	Amphotericin B	Sefsol 218, Tween 80, Transcutol P	~4.5-fold higher skin percutaneous permeation flux rate and for cutaneous infection it is safe and efficacious than oral administration	Hussain et al., 2016

TABLE 2.3 *(Continued)*

Route	Encapsulated drug	Excipients	Outcome	References
	Lidocaine and prolocaine	Soybean oil, Phospholipon 90G, Tween 80, poloxamer 407	Significantly higher skin permeation rate, significantly higher concentration of anesthetic drugs in the dermis layer of skin and significantly improved onset and duration of local anesthetic action	Negi et al., 2015
Ocular	Loteprednol etabonate	Capryol 90, Acrysol EL135, Transcutol P	Exhibited zero order release kinetics, significantly lower ocular irritation potential, and 2.54-fold improved ocular bioavailability	Patel et al., 2016
	Castor oil and mineral oil	Sorbitan mono-stearate, polysorbate 80,	nonirritant, 100% transparency, and improved ion permeability coefficients	Katzer et al., 2014
Intranasal	Saquinavir mesylate	Capmul MCM, Tween 80, PEG 400	Significantly higher concentration in the brain, significantly higher targeting efficiency and nose-to-brain drug direct transport	Mahajan et al, 2014
	Resveratrol	Vitamin E, sefsol 218, Tween 80	Significantly higher concentration of the drug in the brain, significantly decreased degenerative changes, significantly higher levels of glutathione and superoxide dismutase, and significantly lower level of malondialdehyde	Pangeni et al., 2014

5-FU, 5-fluorouracil; AUC, area under the concentration–time curve; MRT, mean resident time; OXA, oxaliplatin.

2.3.2 FUTURE DIRECTIONS

Nanoemulsions have significant potential for exploitation as delivery systems for lipophilic drugs. These delivery systems can be prepared using high-energy or low-energy methods from GRAS components. High-energy methods require relatively expensive mechanical devices (homogenizers), but can be used to prepare nanoemulsions at low surfactant-to-oil ratios (SORs), and can be prepared using natural emulsifiers (such as proteins and polysaccharides). Low-energy methods are simple and easy to implement and do not require any expensive equipment, but at present they can only produce nanoemulsions at relatively high SORs and require the use of synthetic surfactants. Nanoemulsions have some advantages over conventional emulsions for encapsulating bioactive lipophilic components due to their good physical stability, high optical clarity, and ability to increase the bioavailability of encapsulated components.

In particular, there are still many areas where further research is required to ensure their safe and effective utilization. More efficient studies are needed to understand the biological fate of lipophilic drug-loaded nanoemulsion. An increasing number of studies should be conducted using simulated GI models, but more research is needed to be carried out in suitable preclinical and clinical model. More comprehensive studies related to the effect of droplet size, charge, and composition on the bioaccessibility, absorption, and metabolisms of incorporated lipophilic drugs are required for future directions.

2.4 CONCLUDING REMARKS

The present chapter addresses the current state of nanoemulsions system along with its recent progress in drug delivery applications. It encompasses the production and characterization techniques involved in the development of nanoemulsion system, their physicochemical behavior, *in vivo* fate, as well as various theories dealing the process of nanoemulsification. Furthermore, various preclinical and *in vitro* studies clearly indicate the significant outcome exploiting nanoemulsion-based drug delivery through different routes (such as oral, parenteral, transdermal, ocular, and intranasal) to improve the efficacy of encapsulated active therapeutics. Indeed, nano-emulsion possesses great potential to improve drug delivery aspects; hence, more emphasis should be given to exploit the present understanding for the development of future nanomedicine for various chronic deadly diseases such as cancer, HIV, etc.

KEYWORDS

- **nanoemulsion**
- **colloidal dispersion**
- **GRAS ingredients**
- **isotropic mixture**
- *in vivo* **fate**
- **route of administration**

REFERENCES

Acosta, E. Bioavailability of Nanoparticles in Nutrient and Nutraceutical Delivery. *Curr. Opin. Colloid Interface Sci.* **2009**, *14*, 3–15.

Ahmad, J.; Kohli, K.; Mir, S. R.; Amin, S. Lipid Based Nanocarriers for Oral Delivery of Cancer Chemotherapeutics: An Insight in the Intestinal Lymphatic Transport. *Drug Deliv. Lett.* **2013**, *3*, 38–46.

Ahmad, J.; Amin, S.; Rahman, M.; Rub, R. A.; Singhal, M.; Ahmad, M. Z.; Rahman, Z.; Addo, R. T.; Ahmad, F. J.; Mushtaq, G.; Kamal, M. A.; Akhter, S. Solid Matrix Based Lipidic Nanoparticles in Oral cancer Chemotherapy: Applications and Pharmacokinetics. *Curr. Drug Metab.* **2015**, *16*, 633–644.

Akhtar, J.; Siddiqui, H. H.; Fareed, S.; Badruddeen.; Khalid, M.; Aqil, M. Nanoemulsion: For Improved Oral Delivery of Repaglinide. *Drug Deliv.* **2016**, *23*, 2026–2034.

Anton, N.; Vandamme, T. F. Nanoemulsions and Micro-emulsions: Clarifi Cations of the Critical Differences. *Pharm. Res.* **2011**, *28*, 978–985.

Anton, N.; Gayet, P.; Benoit, J. P.; Saulnier, P. Nanoemulsions and Nanocapsules by PIT Methods: An Investigation on the Role of Temperature Cycling on the Emulsion Phase Inversion. *Int. J. Pharm.* **2007**, *344*, 44–52.

Bali, V.; Bhavna.; Ali, M.; Baboota, S.; Ali, J. Potential of Microemulsions in Drug Delivery and Therapeutics: A Patent Review. *Rec. Pat. Drug Deliv. Form.* **2008**, *2*, 136–144.

Belhaj, N.; Arab-Tehrany, E.; Linder, M. Oxidative Kinetics of Salmon Oil in Bulk and in Nanoemulsion Stabilized by Marine Lecithin. *Process Biochem.* **2010**, *45*, 187–195.

Boon, C. S.; McClements, D. J.; Weiss, J.; Decker, E. A. Factors Influencing the Chemical Stability of Carotenoids in Foods. *Crit. Rev. Food Sci. Nutr.* **2010**, *50*, 515–532.

Cai, Z.; Wang, Y.; Zhu, L. J.; Liu, Z. Q. Nanocarriers: A General Strategy for Enhancement of Oral Bioavailability of Poorly Absorbed or Pre-systemically Metabolized Drugs. *Curr. Drug Metab.* **2010**, *11*, 197–207.

Chantrapornchai, W.; Clydesdale, F.; McClements, D. J. Influence of Droplet Characteristics on the Optical Properties of Colored Oil-in-water Emulsions. *Colloids Surf. A Physicochem. Eng. Asp.* **1999,** *155*, 373–382.

Chantrapornchai, W.; Clydesdale, F. M.; McClements, D. J. Influence of Relative Refractive Index on Optical Properties of Emulsions. *Food Res. Int.* **2001,** *34*, 827–835.

Cone, R. A. Barrier Properties of Mucus. *Adv. Drug Delivery Rev.* **2009,** *61*, 75–85.

Cotlier, E.; Baskin, M.; Kresca, L. Effects of Lysophosphatidyl Choline and Phospholipase A on the Lens. *Invest. Ophthalmol.* **1975,** *14*, 697–701.

Crater, J. S.; Carrier, R. L. Barrier Properties of Gastrointestinal Mucus to Nanoparticle Transport. *Macromol. Biosci.* **2010,** *10*, 1473–1483.

des Rieux, A.; Fievez, V.; Garinot, M.; Schneider, Y. J.; Preat, V. Nanoparticles as Potential Oral Delivery Systems of Proteins and Vaccines: A Mechanistic Approach. *J. Control. Release* **2006,** *116*, 1–27.

Eccleston, J. Microemulsions. In *Encyclopedia of Pharmaceutical Technology*; Swarbrick, J., Boylan, J. C., Eds.; Marcel Dekker: New York, 1994; Vol. 9, pp 375–421.

El-Leithy, E. S.; Ibrahim, H. K.; Sorour, R. M. *In vitro* and *In vivo* Evaluation of Indomethacin Nanoemulsion as a Transdermal Delivery System. *Drug Deliv.* **2015,** *22*, 1010–1017.

Ensign, L. M.; Cone, R.; Hanes, J. Oral Drug Delivery with Polymeric Nanoparticles: The Gastrointestinal Mucus Barriers. *Adv. Drug Deliv. Rev.* **2012,** *64*, 557–570.

Florence, A. T.; Hussain, N. Transcytosis of Nanoparticle and Dendrimer Delivery Systems: Evolving Vistas. *Adv. Drug Deliv. Rev.* **2001,** *50*, 69–89.

Florence, A. T. "Targeting" Nanoparticles: The Constraints of Physical Laws and Physical Barriers. *J. Control. Release* **2012,** *164*, 115–124.

Frohlich, E.; Roblegg, E. Models for Oral Uptake of Nanoparticles in Consumer Products. *Toxicology* **2012,** *291*, 10–17.

Gao, L.; Zhang, D.; Chen, M. Drug Nanocrystals for the Formulation of Poorly Soluble Drugs and Its Application as a Potential Drug Delivery System. *J. Nanopart. Res.* **2008,** *10*, 845–862.

Golding, M.; Wooster, T. J.; Day, L.; Xu, M.; Lundin, L.; Keogh, J.; Clifton, P. Impact of Gastric Structuring on the Lipolysis of Emulsified Lipids. *Soft Matter* **2011,** *7*, 3513–3523.

Hu, L.; Mao, Z. W.; Gao, C. Y. Colloidal Particles for Cellular Uptake and Delivery. *J. Mater. Chem.* **2009,** *19*, 3108–3115.

Hussain, A.; Samad, A.; Singh, S. K.; Ahsan, M. N.; Haque, M. W.; Faruk, A.; Ahmed, F. J. Nanoemulsion Gel-based Topical Delivery of an Antifungal Drug: *In vitro* Activity and *In vivo* Evaluation. *Drug Deliv.* **2016,** *23*, 642–647.

Iqbal, J.; Hussain, M. M. Intestinal Lipid Absorption. *Am. J. Physiol. Endocrinol. Metab.* **2009,** *296*, 1183–1194.

Israelachvili, J. *Intermolecular and Surface Forces*, 3rd ed.; Academic Press: London, UK, 2011.

Jafari, S. M.; He, Y.; Bhandari, B. Optimization of Nanoemulsion Production by Microfluidization. *Eur. Food Res. Technol.* **2007,** *225*, 733–741.

Kabalnov, A. Ostwald Ripening and Related Phenomena. *J. Dispers. Sci. Technol.* **2001,** *22*, 1–12.

Katzer, T.; Chaves, P.; Bernardi, A.; Pohlmann, A. R.; Guterres, S. S.; Beck, R. C. Castor Oil and Mineral Oil Nanoemulsion: Development and Compatibility with a Soft Contact Lens. *Pharm. Dev. Technol.* **2014,** *19*, 232–237.

Kentish, S.; Wooster, T. J.; Ashokkumar, M.; Balachandran, S.; Mawson, R.; Simons, L. The Use of Ultrasonics for Nanoemulsion Preparation. *Innvot. Food Sci. Emerg. Technol.* **2008,** *9,* 170–175.

Kotta, S.; Khan, A. W.; Pramod, K.; Ansari, S. H.; Sharma, R. K.; Ali, J. Exploring Oral Nanoemulsions for Bioavailability Enhancement of Poorly Water-soluble Drugs. *Expert Opin. Drug Deliv.* **2012,** *9,* 585–598.

Lawrence, M. J.; Rees, G. D. Microemulsion-based Media as Novel Drug Delivery Systems. *Adv. Drug Deliv. Rev.* **2000,** *45,* 89–121.

Lee, S. J.; Choi, S. J.; Li, Y.; Decker, E. A.; McClements, D. J. Protein-stabilized Nano-emulsions and Emulsions: Comparison of Physicochemical Stability, Lipid Oxidation, and Lipase Digestibility. *J. Agric. Food Chem.* **2011,** *59,* 415–427.

Li, Y.; Le Maux, S.; Xiao, H.; McClements, D. J. Emulsion-based Delivery Systems for Tributyrin, a Potential Colon Cancer Preventative Agent. *J. Agric. Food Chem.* **2009,** *57,* 9243–9249.

Li, X.; Du, L.; Wang, C.; Liu, Y.; Mei, X.; Jin, Y. Highly Efficient and Lowly Toxic Docetaxel Nanoemulsions for Intravenous Injection to Animals. *Pharmazie* **2011,** *66,* 479–483.

Lim, S. S.; Baik, M. Y.; Decker, E. A.; Henson, L.; Popplewell L. M.; McClements, D. J.; Choi S. J. Stabilization of Orange Oil-in-water Emulsions: A New Role for Ester Gum as an Ostwald Ripening Inhibitor. *Food Chem.* **2011,** *128,* 1023–1028.

Mahajan, H. S.; Mahajan, M. S.; Nerkar, P. P.; Agrawal, A. Nanoemulsion-based Intranasal Drug Delivery System of Saquinavir Mesylate for Brain Targeting. *Drug Deliv.* **2014,** *21,* 148–154.

Marianecci, C.; Rinaldi, F.; Hanieh, P. N.; Paolino, D.; Marzio, L. D.; Carafa, M. Nose to Brain Delivery: New Trends in Amphiphile-based "Soft" Nanocarriers. *Curr. Pharm. Des.* **2015,** *21,* 5225–5232.

Mason, T. G.; Wilking, J. N.; Meleson, K.; Chang, C. B.; Graves, S. M. Nanoemulsions: Formation, Structure, and Physical Properties. *J. Phys. Condens. Matter* **2006,** *18,* 635–666.

McClements, D. J. *Food Emulsions: Principles, Practice, and Techniques,* 2nd ed.; CRC Press: FL, USA, 2005.

McClements, D. J. Edible Nanoemulsions: Fabrication, Properties, and Functional Perfor-mance. *Soft Matter* **2011,** *7,* 2297–2316.

McClements, D. J. Nanoemulsions Versus Microemulsions: Terminology, Differences and Similarities. *Soft Matter* **2012,** *8,* 1719–1729.

McClements, D. J. Nanoemulsion-based Oral Delivery Systems for Lipophilic Bioactive Components: Nutraceuticals and Pharmaceuticals. *Ther. Deliv.* **2013,** *4,* 841–857.

McClements, D. J.; Decker, E. A. Lipid Oxidation in Oil-in-water Emulsions: Impact of Molecular Environment on Chemical Reactions in Heterogeneous Food Systems. *J. Food Sci.* **2000,** *65,* 1270–1282.

McClements, D. J.; Xiao, H. Potential Biological Fate of Ingested Nanoemulsions: Influence of Particle Characteristics. *Food Func.* **2012,** *3,* 202–220.

McClements, D. J.; Henson, L.; Popplewell, L. M.; Decker, E. A.; Choi, S. J. Inhibition of Ostwald Ripening in Model Beverage Emulsions by Addition of Poorly Water Soluble Triglyceride Oils. *J. Food Sci.* **2012,** *77,* 33–38.

Metcalfe, P. D.; Thomas, S. Challenges in the Prediction and Modeling of Oral Absorption and Bioavailability. *Curr. Opin. Drug Discov. Develop.* **2010,** *13,* 104–110.

Negi, P.; Singh, B.; Sharma, G.; Beg, S.; Katare, O. P. Biocompatible Lidocaine and Prilo-caine Loaded-nanoemulsion System for Enhanced Percutaneous Absorption: QbD-based

Optimisation, Dermatokinetics and *In vivo* Evaluation. *J. Microencapsul.* **2015**, *32*, 419–431.

Nik, A. M.; Corredig, M.; Wright, A. J. Release of Lipophilic Molecules During *In vitro* Digestion of Soy Protein-stabilized Emulsions. *Mol. Nutr. Food Res.* **2011**, *55*, 278–289.

Pal, R. Rheology of Simple and Multiple Emulsions. *Curr. Opin. Colloid Interface Sci.* **2011**, *16*, 41–60.

Pangeni, R.; Sharma, S.; Mustafa, G.; Ali, J.; Baboota, S. Vitamin E Loaded Resveratrol Nanoemulsion for Brain Targeting for the Treatment of Parkinson's Disease by Reducing Oxidative Stress. *Nanotechnology* **2014**, *25*, 485102.

Pangeni, R.; Choi, S. W.; Jeon, O. C.; Byun, Y.; Park, J. W. Multiple Nanoemulsion System for an Oral Combinational Delivery of Oxaliplatin and 5-fluorouracil: Preparation and *In vivo* Evaluation. *Int. J. Nanomedicine* **2016**, *11*, 6379–6399.

Patel, N.; Nakrani, H.; Raval, M.; Sheth, N. Development of Loteprednol Etabonate-loaded Cationic Nanoemulsified In-situ Ophthalmic Gel for Sustained Delivery and Enhanced Ocular Bioavailability. *Drug Deliv.* **2016**, *23*, 3712–3723.

Payne, A. N.; Zihler, A.; Chassard, C.; Lacroix, C. Advances and Perspectives in *In vitro* Human Gut Fermentation Modeling. *Trends Biotechnol.* **2012**, *30*, 17–25.

Porter, C. J. H.; Trevaskis, N. L.; Charman, W. N. Lipids and Lipid-based Formulations: Optimizing the Oral Delivery of Lipophilic Drugs. *Nat. Rev. Drug Discov.* **2007**, *6*, 231–248.

Pouton, C. W. Lipid Formulations for Oral Administration of Drugs: Non-emulsifying, Self-emulsifying and 'Self-microemulsifying' Drug Delivery Systems. *Eur. J. Pharm. Sci.* **2000**, *11*, 93–98.

Pouton, C. W.; Porter, C. J. H. Formulation of Lipid-based Delivery Systems for Oral Administration: Materials, Methods and Strategies. *Adv. Drug Deliv. Rev.* **2008**, *60*, 625–637.

Powell, J. J.; Faria, N.; Thomas-McKay. E.; Pele, L. C. Origin and Fate of Dietary Nanoparticles and Microparticles in the Gastrointestinal Tract. *J. Autoimmun.* **2010**, *34*, 226–233.

Qian, C.; Decker, E. A.; Xiao, H.; McClements, D. J. Physical and Chemical Stability of Betacarotene-enriched Nanoemulsions: Influence of pH, Ionic Strength, Temperature, and Emulsifier Type. *Food Chem.* **2012**, *132*, 1221–1229.

Rizwanullah, M.; Amin, S.; Ahmad, J. Improved Pharmacokinetics and Antihyperlipidemic Efficacy of Rosuvastatin-loaded Nanostructured Lipid Carriers. *J. Drug Target.* **2017**, *25*, 58–74.

Sarker, D. K. Engineering of Nanoemulsions for Drug Delivery. *Curr. Drug Deliv.* **2005**, *2*, 297–310.

Scalbert, A.; Williamson, G. Dietary Intake and Bioavailability of Polyphenols. *J. Nutr.* **2000**, *130*, 2073–2085.

Shafiq, S.; Shakeel, F. Enhanced Stability of Ramipril in Nanoemulsion Containing Cremophor-EL: A Technical Note. *AAPS Pharm. Sci. Tech.* **2008**, *9*, 1097–1101.

Shakeel, F.; Baboota, S.; Ahuja, A.; Ali, J.; Shafiq, S. Celecoxib Nanoemulsion: Skin Permeation Mechanism and Bioavailability Assessment. *J. Drug Target.* **2008**, *16*, 733–740.

Shakeel, F.; Ramadan, W.; Faisal, M. S.; Rizwan, M.; Faiyazuddin, M.; Mustafa, G.; Shafiq, S. Transdermal and Topical Delivery of Anti-inflammatory Agents Using Nanoemulsion/Microemulsion: An Updated Review. *Curr. Nanosci.* **2010**, *6*, 184–198.

Shakeel, F.; Shafiq, S.; Haq, N.; Alanazi, F. K.; Alsarra, I. A. Nanoemulsions as Potential Vehicles for Transdermal and Dermal Delivery of Hydrophobic Compounds: An Overview. *Expert Opin. Drug Deliv.* **2012**, *9*, 953–974

Shinoda, K.; Araki, M.; Sadaghiani, A.; Khan, A.; Lindman, B. Lecithin-based Microemulsions: Phase Behaviour and Microstructure. *J. Phys. Chem.* **1991,** *95,* 989–993.

Singh, H.; Sarkar, A. Behaviour of Protein Stabilized Emulsions Under Various Physiological Conditions. *Adv. Colloid Interface Sci.* **2011,** *165,* 47–57.

Solans, C; Izquierdo, P.; Nolla, J.; Azemar, N.; Garcia-Celma, M. J. Nanoemulsions. *Curr. Opin. Colloid. Interface Sci.* **2005,** *10,* 102–110.

Sonneville-Aubrun, O.; Simonnet, J. T.; L'Alloret, F. Nanoemulsions: A New Vehicle for Skincare Products. *Adv. Colloid Interface Sci.* **2004,** *108–109,* 145–149.

Sugano, K. Estimation of Effective Intestinal Membrane Permeability Considering Bile Micelle Solubilisation. *Int. J. Pharm.* **2009a,** *368,* 116–122.

Sugano, K. Computational Oral Absorption Simulation for Low-solubility Compounds. *Chem. Biodivers.* **2009b,** *6,* 2014–2029.

Talegaonkar, S.; Negi, L. M. Nanoemulsion in Drug Targeting. In *Targeted Drug Delivery: Concepts and Design*; Devarajan, P. V., Jain, S., Eds.; Springer: Cham, Heidelberg, New York, 2015; pp 433–460.

Trevaskis, N. L.; Charman, W. N.; Porter, C. J. H. Lipid-based Delivery Systems and Intestinal Lymphatic Drug Transport: A Mechanistic Update. *Adv. Drug Deliv. Rev.* **2008,** *60,* 702–716.

van Aken, G. A. Relating Food Emulsion Structure and Composition to the Way It Is Processed in the Gastrointestinal Tract and Physiological Responses: What are the Opportunities? *Food Biophys.* **2010,** *5,* 258–283.

Wan, K.; Sun, L.; Hu, X.; Yan, Z.; Zhang, Y.; Zhang, X.; Zhang, J. Novel Nanoemulsion Based Lipid Nanosystems for Favorable *In vitro* and *In vivo* Characteristics of Curcumin. *Int. J. Pharm.* **2016,** *504,* 80–88.

Wang, P.; Liu, H. J.; Mei, X. Y.; Nakajima, M.; Yin, L. J. Preliminary Study into the Factors Modulating Beta-carotene Micelle Formation in Dispersions Using an *In vitro* Digestion Model. *Food Hydrocoll.* **2012,** *26,* 427–433.

Wretlind, A. Development of Fat Emulsions. *J. Parenter. Enteral. Nutr.* **1981,** *5,* 230–235.

Xie, G.; Zhao, A.; Zhao, L.; Chen, T.; Chen, H.; Qi, X.; Zheng, X.; Ni, Y.; Cheng, Y.; Lan, K.; Yao, C.; Qiu, M.; Jia W. Metabolic Fate of Tea Polyphenols in Humans. *J. Proteome Res.* **2012,** *11,* 3449–3457.

Yanez, J. A.; Wang, S. W. J.; Knemeyer, I. W.; Wirth, M. A.; Alton, K. B. Intestinal Lymphatic Transport for Drug Delivery. *Adv. Drug Deliv. Rev.* **2011,** *63,* 923–942.

Yu, H. L.; Huang, Q. R. Improving the Oral Bioavailability of Curcumin Using Novel Organogel-based Nanoemulsions. *J. Agric. Food Chem.* **2012,** *60,* 5373–5379.

Zhang, X.; Sun, X.; Li, J.; Zhang, X.; Gong, T.; Zhang Z. Lipid Nanoemulsions Loaded with Doxorubicin–Oleic Acid Ionic Complex: Characterization, *In vitro* and *In vivo* Studies. *Pharmazie* **2011,** *66,* 496–505.

Zhao, H.; Lu, H.; Gong, T.; Zhang, Z. Nanoemulsion Loaded with Lycobetaine–Oleic Acid Ionic Complex: Physicochemical Characteristics, *In vitro, In vivo* Evaluation, and Antitumor Activity. *Int. J. Nanomed.* **2013,** *8,* 1959–1973.

CHAPTER 3

NANOEMULSIONS: ROUTES OF ADMINISTRATION AND APPLICATIONS

KIFAYAT ULLAH SHAH[1], BEY HING GOH[2], LEARN-HAN LEE[2], and TAHIR MEHMOOD KHAN[2,3*]

[1]*Department of Pharmacy, Quaid-i-Azam University, Islamabad, Pakistan*

[2]*School of Pharmacy, Monash University, Bandar Sunway, Selangor 45700, Malaysia*

[3]*The Institute of Pharmaceutical Sciences, University of Veterinary and Animal Science, Lahore, Pakistan*

Corresponding author. E-mail: tahir.mehmood@monash.edu; tahir.khan@uvas.edu.pk

ABSTRACT

With the discovery of nanoemulsions, the delivery of nonmiscible substance can be done more efficiently to get the desired therapeutic benefit. Nanoemulsions are the conventional emulsions having a size range of firmly below 200 nm and below 300 nm, and with this size they have vast benefits and advantages in modern health science for the treatment of various disorders. Their importance is depicted from their promising compatibility and capability to solubilize, carry, and target the encapsulated bioactive compounds to their active site of action. These colloidal systems can be applied to almost all the drug delivery routes and consequently holding the most potential for various fields such as cosmetology, therapeutics, and biotechnology.

3.1 INTRODUCTION

Poor oral bioavailability of various bioactive species in the maintenance and promotion of human health confines their effectiveness. These agents may be of various origins such as vitamins, minerals, and nutraceuticals in supplements or foods, or may be drugs within pharmaceutical products. The physiological or physicochemical factors that affect the bioavailability profiles of these bioactive agents include chemical/biochemical transformation before or after ingestion, low solubility in gastrointestinal tract (GIT) fluids, limited absorption through epithelium cell wall and first pass metabolism, etc. Among these the low water solubility is the utmost difficult task in the growth of new enhanced medicinal products of nearly half of the new biological compounds discovered so far. For an improvement in the oral bioavailability profiles of these agents, effective knowledge-based strategies are required to overcome such limitations (Aboalnaja et al., 2016; McClements et al., 2015; Williams et al., 2013).

Recently, lipid-based drug delivery systems (LBDDS) have appeared as the most favorable approach in the improvement of solubility, absorption, and subsequently oral bioavailability of bioactive agents with a lower aqueous solubility (Jain et al., 2015). There is a lot of variety in lipid-based oral formulations extending from common oil solutions to very complex surfactant, cosurfactant, and oil mixtures. LBDDS can be modified according to needs either as hydrophilic or hydrophobic by changing their components or concentration (Rehman et al., 2016). Their bioavailability enhancement mechanisms include prolonged retardation time in the stomach, alteration in the physical barriers (Brüsewitz et al., 2007), alterations in biochemical barrier (Porter et al., 2007), enhanced solubility (Mohsin et al., 2012), and decreased drug metabolism (Ren et al., 2008). These systems, with time, have been geared up to micro- and nanoscales consequently leading to upgraded therapeutic potential for low aqueous-soluble bioactive compounds primarily belonging to Biopharmaceutics Classification System class II and III (Pathak and Raghuvanshi, 2015). Lipid-based nanocarriers among the many multifunctional nanocarriers developed recently are the least toxic *in vivo* (Puri et al., 2009). These lipid-based nanocarriers comprise liposomes, solid-lipid nanoparticles, nano-structured lipid carrier, microemulsions, nanoemulsions, self-microemulsifying drug delivery systems and self-nano-emulsifying drug delivery systems (SNEDDS) (Hadinoto et al., 2013), etc.

3.2 NANOEMULSIONS

An emulsion may be defined as "thermodynamically unstable colloidal dispersion having two or more immiscible liquids, with the one liquid being dispersed as small spherical droplets in the other and may be oil-in-water or water-in-oil type" (McClements, 2012). Surfactants that are approved by Food and Drug Administration for human consumption are used to develop nanoemulsions (Lovelyn and Attama, 2011).

In the literature, the emulsions with particle size ranging from 20 to 200 nm are typically termed as miniemulsions, nanoemulsions, ultrafine emulsions, and submicron emulsions as shown in Figure 3.1. Due to the fact that these particles are in nanoscale and to avoid any misunderstanding with microemulsions (thermodynamically stable), the term nanoemulsions is preferred. Nanoemulsions, because of their characteristic size are transparent and have an enhanced stability against creaming or sedimentation (Solans et al., 2005). According to phase behavior studies, the droplet size is mainly controlled by surfactant-phase-structure (lamellar or bicontinuous microemulsion) at inversion point and triggered either by composition or temperature. Consequences of the reduction in the droplet size to nanorange results in some very interesting modifications in the physical properties such as visual clearness and rare elastic behavior. Moreover, nanoemulsions in comparison to conventional emulsions offer obvious advantages due to lack of flocculation, creaming, and sedimentation. Their large interfacial area has a positive impact on drug transport and its delivery as well as its targeting to a specific site (Lovelyn and Attama, 2011). These nanoemulsions not only have amazing wetting, distribution, and penetrating capabilities but could also be scaled up due to the comfort in production (Date et al., 2010). Recently some solid nanoemulsions formulations have been reported that further enhance its stability and advantageous properties (Singh et al., 2012). Evidences prove that as the droplet size reaches 100, 90, and 30 nm, the bioavailability of the emulsions improves substantially, stability strangely rises, and transparency increases, respectively (McClements, 2012).

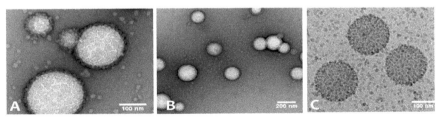

FIGURE 3.1 Silicate particle stabilized oil/water nanoemulsions visualized via (A), (B) negative-stain TEM, and (C) cryo TEM. (Reprinted with permission. © 2014 Chime SA, Kenechukwu FC, Attama AA. Published in [short citation] under CC BY 3.0 license. Available from: http://dx.doi.org/10.5772/58673)

3.3   DIFFERENCES BETWEEN MICROEMULSIONS AND NANOEMULSIONS

Nanoemulsions are the conventional emulsions having a size range of firmly below 200 nm and below 300 nm in general; however, microemulsions are basically swollen micelles system where the discrete phase is incorporated into surfactants micelle central under particular environmental conditions and compositions (Rehman et al., 2016). Both systems are different physicochemically but in case of their molecular or structural aspects they have much resemblance. Under special conditions and compositions, the microemulsions have a great resemblance with nanoemulsions, approaching almost the similar nanoranged spherical droplets (Anton and Vandamme, 2011).

Almost same materials are being used in the preparation of both microemulsions and nanoemulsions aqueous phase, oil phase, surfactants, and cosurfactants. Both the particles have identical basic configuration comprising hydrophilic and hydrophobic fragments with particles ranging from <300 nm for nanoemulsion and <100 nm for microemulsion. In contrast to microemulsions which are made naturally, the nanoemulsions fabrication need high-energy techniques; however, both the systems require some external energy source in order to support the mass transport by overwhelming the kinetic energy barriers. Nanoemulsions in comparison to microemulsions need low activation energy and thus low external energy and mainly are produced using spontaneous emulsification techniques that creates uncertainty and perpetuations of the matter. The main bases of misunderstanding amid the two substantially dissimilar colloidal systems include similar molecular characteristics the formulation procedures with additional incorrect understanding of the nanoemulsion systems as ultrafine, mini- or submicron emulsions (Anton and Vandamme, 2011; Kong et al., 2011). Nanoemulsions in comparison to microemulsions which are at an equilibrium phase (surfactant molecules distributed in the form of micelles at critical micelle concentration, are at nonequilibrium and are kinetically stable.

It is most important to distinguish between the two systems in order to eliminate the basic concepts that may lead to unacceptable characterization of the colloidal system manipulating their preparation methods, physicochemical properties, and various stability factors. Following are some key points to stipulate the accurate nature of the colloidal system (Rehman et al., 2017):

(1)   Keeping in mind the particle size distribution of the two systems, microemulsions have one thin peak whereas nanoemulsions have thin and wide peaks (may be one or many) Figure 3.3.

(2) In case of microemulsions the particle shape is somewhat irregular, it may be spherical or nonspherical in contrast to nanoemulsions with a spherical particle shape Figure 3.2.

(3) At specific conditions of temperature, pressure, and composition, nanoemulsion change their properties during prolonged shelf life but the same properties do not change in the case of microemulsions.

(4) Droplet size of nanoemulsion remains the same when diluted with continuous phase but in case of microemulsions, surfactant concentration is disturbed, droplet size changes, and micelle can even be destroyed occasionally.

(5) Whatever method is used for the development of microemulsions, the droplet structure and size remain the same but changes in case of nanoemulsions because these are thermodynamically unstable systems.

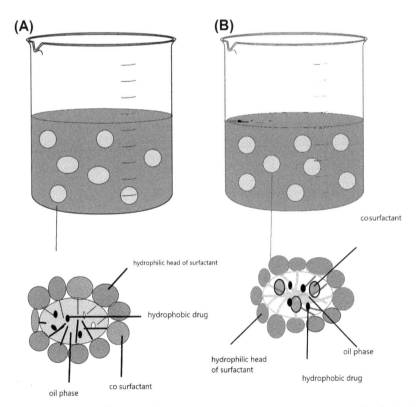

FIGURE 3.2 (See color insert.) Depicting (A) nanoemulsion droplet; round and showing polydispersity in size and (B) microemulsion droplet; shape which may be or may be not sphere-shaped and entire droplets of equal diameter and shape.

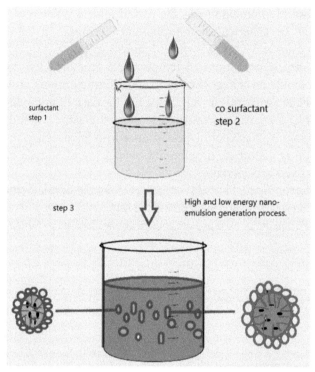

FIGURE 3.3 (See color insert.) Nanoemulsion preparation requires a specific mixing order where oil phase is first mixed with surfactant; however, microemulsions do not require any particular mixing order.

3.4 FABRICATION OF NANOEMULSIONS

Nanoemulsions in general can be fabricated encompassing two main approaches (1) high energy technique and (2) low-energy method as shown in Figure 3.4 (Koroleva and Yurtov, 2012). The most common method adopted nowadays for the fabrication of nanoemulsions is the high-energy method but low-energy method is also sometimes used for certain applications such as formation of SNEDDS (Čerpnjak et al., 2013). Intense localized energy fields are produced by mechanical devices in high-energy approach that in the presence of an emulsifying agent breaks and intermixes the water and oil phases such as sonicators, microfluidizers, high-pressure homogenizers, etc. The energy required for the oil droplets to disrupt depends on its interfacial tension that in turns depends on nature of the emulsifier used. Contrarily, the low-energy technique depends on spontaneous creation of minute oil droplets when there is a specific modification in either temperature or composition of the system (Aboalnaja et al., 2016; Perazzo et al., 2015). Each method for nanoemulsions production has both advantages and disadvantages as well, also range of oil and surfactant type is crucial for proper homogenization. High-energy methods usually are more preferable for development of nano-emulsions from a wider range of emulsifiers and oils, etc. in comparison to low-energy methods. These approaches require little surfactant/oil ratios but involve costly mechanical homogenizers. On the other hand, the low-energy approaches are simple and do not involve highly expensive sophisticated equipment but the surfactant to oil ratios are comparatively higher than high-energy techniques and are usually appropriate for the synthesis of small molecule surfactants. For the fabrication of nanoemulsions, the choice of proper method is crucial and needs to be selected on case-by-case basis depending on bioactive material's nature and the final product (Aboalnaja et al., 2016). Some nanoemulsification techniques reported in the literature are illustrated in Table 3.1.

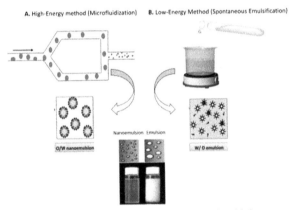

FIGURE 3.4 (A) Shows fabrication of nanoemulsions using high-energy technique while (B) shows low-energy technique.

TABLE 3.1 Nanoemulsification Techniques Reported in Literature.

Sr. no.	Nanoemulsifica-tion techniques	Emulsion system/ ingredients	Size distribution	References
1	Rotor and stator (ultraturrax)	n-hexane, Tween 80, and β-carotene	150 nm	Silva et al., 2011
2	High-pressure homogenization	MCT, Tween 80, and curcumin	82 nm	Wang et al., 2008
3	High-pressure homogenization	n-hexane, Tween 80, and β-carotene		Tan and Nakajima, 2005

TABLE 3.1   *(Continued)*

Sr. no.	Nanoemulsifica-tion techniques	Emulsion system/ ingredients	Size distribution	References
4	Ultrasound	Tween 40 and flax seed oil	135 nm	Kentish et al., 2008
5	Ultrasound	Ethanolic solution comprising 10% (W/W) of MCT, Poloxamer or Tween 80, and 2-(butylamino)-1-phenyl-1-ethanethiosulfuric acid (BphEA)	200–251 nm	de Araújo et al., 2007
6	Phase-inversion temperature	Tetradecane and Brij30	80–120 nm	Rao and McClements, 2010
7	Spontaneous emulsification	Ethanol, egg-lecithin, and Quercetin or Methyl-quercetin	300 nm	Fasolo et al., 2007

MCT, medium-chain triacylglycerols.

3.5   NANOEMULSIONS AS DRUG DELIVERY SYSTEMS; ADVANTAGES

The following applications of nanoemulsions in drug delivery systems make them of imminence attraction (Aboofazeli, 2010; Lovelyn and Attama, 2011; Tadros et al., 2004).

1. No creaming and sedimentation occurs because of the fact that the small particle size causes a great decrease in gravity and Brownian movement could be enough to overwhelm gravity.
2. The system remains dispersed because of weak flocculation induced by small droplet size of the nanoemulsion.
3. Small droplets size results in preventing coalescence, as these droplets are elastic in nature thus surface flocculation is prevented.
4. Due to large surface area of the nanoemulsions, these systems trigger rapid penetration and subsequently increased delivery of bioactive compounds via skin.
5. Due to the systems' transparent nature, reasonable fluidity and absence of thickeners give them aesthetic properties and feeling on the site of application.

6. Nanoemulsions encompass a reasonable low amount of surfactants as compared to microemulsions comprising 20% or higher surfactant concentration.
7. Because of very small droplet size, allow nanoemulsions to evenly deposit on different substrates and also distribution, wetting, and permeation are enhanced due to low surface tension.
8. Nanoemulsions are a good alternative to liposomes and other vesicles due to their low stability.

3.6 NANOEMULSION DRUG DELIVERY SYSTEMS; DISADVANTAGES

In spite of many advantages, nanoemulsions achieved attention recently for some reasons which are given as follows (Lovelyn and Attama, 2011):

1. Nanoemulsion preparation needs special attention and techniques such as the use of high-pressure homogenizers, for example, microfluidizers which have emerged recently.
2. Due to the fact that the nanoemulsions encompass the use of expensive equipment as well as high emulsifier concentration therefore, there is a common perception of these being expensive in the personal care.
3. Lack of benefit demonstrations acquired by the use of nanoemulsions in comparison to classical macroemulsions.
4. Absence of interfacial chemistry understandings involved in nanoemulsion production.

3.7 NANOEMULSIONS IN DRUG DELIVERY

3.7.1 ORAL DRUG DELIVERY SYSTEMS

One of the main hurdle in the growth of an effective oral drug delivery system is the hydrophobic character and potential for degradation of most of the drug molecules in the GIT, since these are associated to absorption and bioavailability limitations. It is therefore very important to have an understanding of GIT and other oral components for an enhanced knowledge of many vital processes throughout oral drug administration (Silva et al., 2012). Regardless of the latest scientific and technological growth in the field of novel drug delivery systems, the incessant finding and therapeutic necessities of vigorous pharmaceutical moieties require an endless update and development approaches. Therefore, two main aspects should be kept in mind, targeted delivery system means the finest formulation methodology to provide the drug to the site of action and the selection of the most feasible and potent route of administration. In this scenario, the oral route of drug administration because of its various advantages such as ease of administration, nonpainful and economical production, remains the most favorite route (Desai et al., 2012; Patel et al., 2011).

The beneficial outcomes of nanoemulsions as colloidal drug delivery systems have been demonstrated in several studies such as: the use of biocompatible and biodegradable pharmaceutical ingredients leading to low or no toxicity, higher drug entrapment efficiency, drug protection against degradation, and at the end, ease of large-scale production. Despite of some limitations such as lack of labile molecules protection and their immediate release, nanoemulsions are becoming a stable alternative for liposomes (Bali et al., 2010; Sharma et al., 2010). These systems have proved to enhance the bioavailability of many lipophilic drugs by improving their absorption via GIT. The characteristic droplet size of the nanoemulsion and large surface area increases lipid digestion consequently leading to faster drug incorporation into the bile salts mixed micelles (Silva et al., 2012).

In a study, the impact of carrier (oil) used in the formulation of nanoemulsions-based drug delivery systems on bioaccessibility of oil soluble vitamin D3 was determined. It was found that the carrier oil nature was having a major role in the bioaccessibility of vitamin D3 using simulated GIT model. A comparatively higher bioaccessibility was found for the nanoemulsions developed from indigestible oil carriers that could be due to the existence of small vitamin-containing lipid droplets in the micelles. It was proposed that these minor lipid droplets could be adsorbed more rapidly by the human body and subsequently a higher vitamin bioavailability (Ozturk et al., 2015). In a similar study, Salvia-Trujillo and coresearchers (2015) observed the effect of carrier oil type on the functionality of lipophilic nutraceuticals (fucoxanthin) encapsulated within nanoemulsion. Nanoemulsions containing long-chain triacylglycerols and medium-chain triacylglycerols were entirely digested whereas formulations having indigestible oils were not digested in simulated GIT conditions. The animal studies presented that the fucoxanthin exhibited similar absorption behavior in the intestinal epithelial cells as was witnessed in the *in vitro* studies. The study highlighted the importance of the distinct *in vitro/in vivo* studies for the assessment of biological destiny of bioactive ingredients when incorporated into nanoemulsions.

3.7.2 PARENTERAL NANOEMULSION DRUG DELIVERY SYSTEM

The main objective of investigation in this particular area is to gain an enhanced drug delivery system with better drug targeting achieved by the modification in old and development in new emulsions. Parenteral route of drug administration is one of the most common and effective route which is accepted for drugs having lower bioavailability and with a narrow therapeutic window. For over five decades many formulations, mainly intravenous fats (oil/lipid) emulsions, have been in therapeutic use (e.g., Intralipid, approved in 1962 in Europe). Critically ill patients who cannot be orally feed are supplied with parenteral nutrition emulsions. These emulsions generally have triglycerides which upon absorption into the systemic circulation are than metabolized into vital fatty acids in the human body (Hörmann and Zimmer, 2016). One of the advantage of these systems is the recognized regulatory position of the raw materials used and their safe metabolism inside human body. Another big advantage of nanoemulsions is their capability of carrying nonwater-soluble drug moieties. As of today, many new medicinal compounds have a very poor aqueous solubility which can be greatly overcome by encompassing them into nanoemulsions. Moreover, the dosage and injection frequency can be greatly minimized throughout the therapy as the nanoemulsions can assure the drug release in a sustained or controlled mode (Lovelyn and Attama, 2011). During a study, thalidomide-loaded nanoemulsions were developed where the therapeutic effect could be achieved by a small dose of only 25 mg, after a storage duration of 2 months, he observed a substantial decline in drug content of the drug formulation at 0.01% that could be overwhelmed by the adding Polysorbate 80 (Araújo et al., 2011). Alayoubi et al. (2013) prepared a stable parenteral lipid nanoemulsions-based tocotrienol-rich fraction (TRF) and Simvastatin novel formulation for combination chemotherapy and evaluated its antiproliferative activities against MCF-7 and MDA-MB-231 human memory tumor cells. Nanoemulsions were fabricated encompassing high-pressure homogenization technique, TRF viscous 70/30 blend and medium-chain triglycerides as oily phase and simvastatin were dissolved at 9% (w/w) loading. The produced nanoemulsion was having 200 nm droplet size with a surface potential of -45 mV. During dissolution studies, after 24 h in sink conditions, approximately 20% simvastatin was released. The IC_{50} values 14 and 7 µm of TRF nanoemulsion against MCF-7 and MDA-MB-231 were decreased to 10 and 4.8 µm, respectively, upon the addition of simvastatin (Fig. 3.5). He found that parenteral lipidic nanoemulsions are a practicable delivery platform in anticancer chemotherapy.

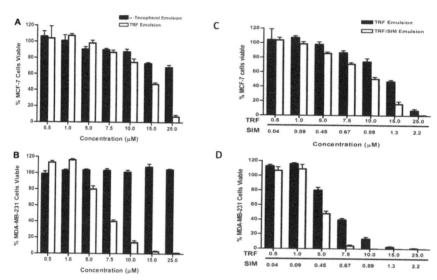

FIGURE 3.5 Left panel depicts antiproliferative α-tocopherol and TRF nanoemulsion activity against the estrogen receptors, positive (ER) MCF-7 (A) and ER negative MDA-MB-231 (B) cells, Right panel shows antiproliferative TRF lipid nanoemulsion activity with or without simvastatin (SIM) against estrogen receptor positive (ER) MCF-7 (C) and ER negative MDA-MB-231 (D) cells. (Reprinted with permission from Alayoubi, A. Y.; Anderson, J. F.; Satyanarayanajois, S. D.; Sylvester, P. W.; Nazzal, S. Concurrent Delivery of Tocotrienols and Simvastatin by Lipid Nanoemulsions Potentiates Their Antitumor Activity Against Human Mammary Adenocarcenoma Cells. *Eur. J. Pharm. Sci.* **2013,** *48* (3), 385–392. © 2013 Elsevier)

3.7.3 TRANSDERMAL ROUTE

The topical application of any drug moiety to the skin surface for the treatment of local skin diseases is termed as dermal drug delivery, while as transdermal drug delivery is basically the drugs absorption into systemic circulation via intact skin to cure many chronic illnesses. Oral route of drug delivery is associated with various clinical adverse effects which can be significantly overcome by using transdermal or dermal drug delivery systems. Transdermal route is advantageous over parenteral route because of its steady state controlled drug delivery over a prolonged time duration as well as self-administration is also thinkable. The patient can remove the transdermal patch any time in order to eliminate the drug input. The fluidity and transparent nature of nanoemulsions provide a pleasant feel on the skin. Only a few nanoemulsion-based formulations are available in the market till now which are approved for therapeutic purposes in comparison to other nanotechnology-based products (Table 3.2). There is no formulation of the marketed nanoemulsions that has been approved for transdermal delivery of hydrophobic compounds but either as topical or oral. Substantial efforts have been made and are in progress to overwhelm the impermeability issues through the human skin that has limited the transdermal application of a variety of drugs. Nanoemulsion is one of the most effective among these strategies proven by many *in vitro/in vivo* studies as a potential carrier for many hydrophobic compounds (Riehemann et al., 2009; Shakeel et al., 2012).

TABLE 3.2   List of Marketed Parenteral Nanoemulsion-based Products.

Sr. no.	Drug/bioactive	Marketed name	Manufacturer	Indications
1	Alprostadil palmitate	Liple[R]	Mitsubishi Pharmaceuticals, Japan	Platelet inhibitor, vasodilator
2	Clevidipine	Cleviprex[R]	The Medicine Company, USA	Calcium channel blocker
3	Cyclosporine	Restatis[R]	Allergan Pharmaceuticals	Ophthalmic emulsion
4	Dexamethasone	Limethasone[R]	Mitsubishi Pharmaceuticals, Japan	Steroid
5	Diazepam	Diazemuls[R]	Actavis Nordic, Europe	Sedative
6	Diclofenac Sodium	Oxalginnanogel[R]	Zyduscadila Pharmaceuticals	Analgesic, Anti-inflammatory
7	Etomidate	Etomidat-Lipuro[R]	Braun Melsungen, Europe	Anesthetic
8	Flurbiprofen	Lipfen[R]	Green Cross, Japan	NSAID
9	Saquinavir	Fortovase[R]	Roche Pharmaceuticals	Antiviral
10	Vitamins A, D, E, K	Vitalipid[R]	Fresenius Kabi, Europe	Parenteral nutrition

NSAID, nonsteroidal anti-inflammatory drug.

In addition, while conducting a study and his coresearchers developed and evaluated 3,5-dihydroxy-4-iso-propylstilbene (DHPS) nanoemulsions using low energy emulsification method encompassing isopropyl myristate as oil phase, polyoxyethylenated castor oil (EL-40) as a surfactant, and utilizing ethanol as a cosurfactant. The developed nanoemulsion was having uniform distribution, spherical droplets, and a low viscosity. An enhanced improvement in transdermal permeation of DHPS was observed with the nanoemulsion and could lead to a favorable new dosage form for the drug DHPS (Zhang et al., 2011). Shakeel and his fellow workers developed aceclofenac nanoemulsion in order to increase its skin permeation and anti-inflammatory effects encompassing Triacetin and Labrafil as oil phase, Tween 80 as surfactant, and Transcutol-P as a cosurfactant. Spontaneous emulsification technique was used to prepare the emulsion and was characterized for droplet size, viscosity, surface morphology, and refractive index. Aceclofenac transdermal flux through rat abdominal skin was studied by the use of Franz diffusion cell. Transdermal skin penetration profiles of optimized nanoemulsion and a conventional aceclofenac gel were compared, where the optimized nanoemulsion was found to have enhanced permeability parameters in comparison to the conventional and simple nanoemulsion gel. A significant increase in percentage inhibition was detected for optimized aceclofenac nanoemulsion after 24 h application in comparison to conventional aceclofenac gel and nanoemulsion gel. Intracellular transport was concluded to be the possible mechanism for permeation enhancement (Shakeel et al., 2007). Rachmawati et al. prepared and evaluated curcumin transdermal nanoemulsion in order to improve its permeability and stability. Curcumin nanoemulsion was developed using self-nanoemulsification technique incorporating glycerol monooleate as oil phase, Cremophor RH40, and Polyethylene glycol 400. The nanoemulsion was evaluated for *in vitro* permeation using vertical diffusion-cell and shaded snake skin of python reticulatus. He found that the nanoemulsion enhanced the permeation flux from hydrophilic matrix gel and was also protected against degradation (Rachmawati et al., 2015).

In an *in vivo* study domperidone nanoemulsion was developed to achieve enhanced percutaneous penetration. For the optimization of the surfactant, cosurfactant, and surfactant: cosurfactant and pseudoternary phase-diagrams were developed. The formulations were having smaller droplet size of <90 nm, uniform particle size distribution, and lower viscosity and were tested for their ex vivo permeation studies consuming rat skin and *in vivo* studies. The optimized formulation having oleic acid, polysorbate, diethylene-glycol-monoethylether, and water depicted a significant increase in the transdermal flux. While conducting *in vivo* studies, the optimized nanoemulsion formulation enhanced the relative bioavailability up to 3.5-folds when applied through transdermal route in comparison to oral administration of the drug suspension. Furthermore, after the transdermal application of domperidone nanoemulsion, the effective drug plasma concentration was maintained for an extended time period of 16 h indicating that the nanoemulsion system could be a favorable transdermal carrier of the drug for extended period.

3.8 CONCLUSION

Nanoemulsion-based formulations have been progressively used as nanocarriers both for diagnostic and therapeutic purposes in the field of biosciences. Their importance is depicted from their promising compatibility and capability to solubilize, carry, and target the encapsulated bioactive compounds to their active site of action. These colloidal systems can be applied to almost all the drug delivery routes and consequently holding the most potential for various fields such as cosmetology, therapeutics, and biotechnology. More and more research is required in this field in order to translate the wide range of nanoemulsions and excipient systems previously been developed in the labs to commercial products with established effectiveness. To validate the safety and efficacy of these systems more animal and human studies are required to ensure their reliable production with consistent properties in a commercially practical method.

KEYWORDS

- nanoemulsion
- routes of administration
- drug delivery
- encapsulated bioactive compounds
- advantages in modern health science

REFERENCES

Aboalnaja, K. O.; Yaghmoor, S.; Kumosani, T. A.; McClements, D. J. Utilization of Nano-
emulsions to Enhance Bioactivity of Pharmaceuticals, Supplements, and Nutraceuticals:
Nanoemulsion Delivery Systems and Nanoemulsion Excipient Systems. *Expert Opin.
Drug Deliv.* **2016,** *13* (9), 1–10.

Aboofazeli, R. Nanometric-scaled Emulsions (Nanoemulsions). *Iran J. Pharm. Res.* **2010,**
325–326.

Alayoubi, A. Y.; Anderson, J. F.; Satyanarayanajois, S. D.; Sylvester, P. W.; Nazzal, S. Concurrent Delivery of Tocotrienols and Simvastatin by Lipid Nanoemulsions Potentiates Their
Antitumor Activity Against Human Mammary Adenocarcenoma Cells. *Eur. J. Pharm. Sci.*
2013, *48* (3), 385–392.

Anton, N.; Vandamme, T. F. Nano-emulsions and Micro-emulsions: Clarifications of the
Critical Differences. *Pharm. Res.* **2011,** *28* (5), 978–985.

Araújo, F.; Kelmann, R.; Araujo, B.; Finatto, R., Teixeira, H., Koester, L. Development and
Characterization of Parenteral Nanoemulsions Containing Thalidomide. *Eur. J. Pharm. Sci.*
2011, *42* (3), 238–245.

Bali, V.; Ali, M.; Ali, J. Study of Surfactant Combinations and Development of a Novel Nanoemulsion for Minimising Variations in Bioavailability of Ezetimibe. *Colloids Surf. B Biointerfaces* **2010,** *76* (2), 410–420.

Brüsewitz, C.; Schendler, A.; Funke, A.; Wagner, T.; Lipp, R. Novel Poloxamer-based Nanoemulsions to Enhance the Intestinal Absorption of Active Compounds. *Int. J. Pharm.* **2007,**
329 (1), 173–181.

Černpjak, K.; Zvonar, A.; Gašperlin, M.; Vrečer, F. Lipid-based Systems as a Promising
Approach for Enhancing the Bioavailability of Poorly Water-soluble Drugs. *Acta pharmaceutica* **2013,** *63* (4), 427–445.

Chime, S.; Kenechukwu, F.; Attama, A. Nanoemulsions—Advances in Formulation, Characterization and Applications in Drug Delivery. In *Application of Nanotechnology in Drug
Delivery*; Ali, D. S., Ed.; In Tech: Croatia, 2014; pp 77–111.

Date, A. A.; Desai, N.; Dixit, R.; Nagarsenker, M. Self-nanoemulsifying Drug Delivery
Systems: Formulation Insights, Applications and Advances. *Nanomedicine* **2010,** *5* (10),
1595–1616.

dE Araujo, S. C.; dE Mattos, A. C. A.; Teixeira, H. F.; Coelho, P. M. Z.; Nelson, D. L.; dE
Oliveira, M. C. Improvement of *In vitro* Efficacy of a Novel Schistosomicidal Drug by
Incorporation into Nanoemulsions. *Int. J. Pharm.* **2007,** *337* (1–2), 307–315.

Desai, P. P.; Date, A. A.; Patravale, V. B. Overcoming Poor Oral Bioavailability Using
Nanoparticle Formulations—Opportunities and Limitations. *Drug Discov. Today Technol.*
2012, *9* (2), e87–e95.

Fasolo, D.; Schwingel, L.; Holzschuh, M.; Bassani, V.; Teixeira, H. Validation of an Isocratic
LC Method for Determination of Quercetin and Methylquercetin in Topical Nanoemulsions. *J. Pharma. Biomed. Anal.* **2007,** *44* (5), 1174–1177.

Hadinoto, K.; Sundaresan, A.; Cheow, W. S. Lipid–Polymer Hybrid Nanoparticles as a New
Generation Therapeutic Delivery Platform: A Review. *Eur. J. Pharm. Biopharm.* **2013,** *85*
(3), 427–443.

Hörmann, K.; Zimmer, A. Drug Delivery and Drug Targeting with Parenteral Lipid Nanoemulsions—A Review. *J. Control. Release* **2016,** *223*, 85–98.

Jain, S.; Patel, N.; Lin, S. Solubility and Dissolution Enhancement Strategies: Current Understanding and Recent Trends. *Drug Dev. Ind. Pharm.* **2015,** *41* (6), 875–887.

Kentish, S.; Wooster, T.; Ashokkumar, M.; Balachandran, S.; Mawson, R.; Simons, L. The Use of Ultrasonics for Nanoemulsion Preparation. *Innov. Food Sci. Emerg. Technol.* **2008,** *9* (2), 170–175.

Kong, M.; Chen, X. G.; Kweon, D. K.; Park, H. J. Investigations on Skin Permeation of Hyaluronic Acid Based Nanoemulsion as Transdermal Carrier. *Carbohydr. Polym.* **2011,** *86* (2), 837–843.

Koroleva, M. Y.; Yurtov, E. V. Nanoemulsions: The Properties, Methods of Preparation and Promising Applications. *Russ. Chem. Rev.* **2012,** *81* (1), 21–43.

Lovelyn, C.; Attama, A. A. Current State of Nanoemulsions in Drug Delivery. *J. Biomater. Nanobiotechnol.* **2011,** *2* (05), 626.

McClements, D. J. Nanoemulsions Versus Microemulsions: Terminology, Differences, and Similarities. *Soft Matter* **2012,** *8* (6), 1719–1729.

McClements, D. J.; Li, F.; Xiao, H. The Nutraceutical Bioavailability Classification Scheme: Classifying Nutraceuticals According to Factors Limiting Their Oral Bioavailability. *Annu. Rev. Food Sci. Technol.* **2015,** *6,* 299–327.

Mohsin, K.; Shahba, A.; Alanazi, F. Lipid Based Self Emulsifying Formulations for Poorly Water Soluble Drugs—An Excellent Opportunity. *Indian J. Pharm. Educ.* **2012,** *46* (2), 88–96.

Ozturk, B.; Argin, S.; Ozilgen, M.; McClements, D. J. Nanoemulsion Delivery Systems for Oil-soluble Vitamins: Influence of Carrier Oil Type on Lipid Digestion and Vitamin D 3 Bioaccessibility. *Food Chem.* **2015,** *187,* 499–506.

Patel, V. F.; Liu, F.; Brown, M. B. Advances in Oral Transmucosal Drug Delivery. *J. Control. Release* **2011,** *153* (2), 106–116.

Pathak, K.; Raghuvanshi, S. Oral Bioavailability: Issues and Solutions via Nanoformulations. *Clin. Pharmacokinet.* **2015,** *54* (4), 325–357.

Perazzo, A.; Preziosi, V.; Guido, S. Phase Inversion Emulsification: Current Understanding and Applications. *Adv. Colloid Interface Sci.* **2015,** *222,* 581–599.

Porter, C. J.; Trevaskis, N. L.; Charman, W. N. Lipids and Lipid-based Formulations: Optimizing the Oral Delivery of Lipophilic Drugs. *Nat. Rev. Drug Discov.* **2007,** *6* (3), 231–248.

Puri, A.; Loomis, K.; Smith, B.; Lee, J.-H.; Yavlovich, A.; Heldman, E.; Blumenthal, R. Lipid-based Nanoparticles as Pharmaceutical Drug Carriers: From Concepts to Clinic. *Crit. Rev. Ther. Drug Carrier Syst.* **2009,** *26* (6), 523–580.

Rachmawati, H.; Budiputra, D. K.; Mauludin, R. Curcumin Nanoemulsion for Transdermal Application: Formulation and Evaluation. *Drug Dev. Ind. Pharm.* **2015,** *41* (4), 560–566.

Rao, J.; Mcclements, D. J. Stabilization of Phase Inversion Temperature Nanoemulsions by Surfactant Displacement. *J. Agric. Food Chem.* **2010,** *58* (11), 7059–7066.

Rehman, F. U.; Shah, K. U.; Shah, S. U.; Khan, I. U.; Khan, G. M.; Khan, A. From Nanoemulsions to Self-nanoemulsions, with Recent Advances in Self-nanoemulsifying Drug Delivery Systems (SNEDDS). *Expert Opin. Drug Deliv.* **2017,** *14* (11), 1325–1340

Ren, S.; Park, M.-J.; Kim, A.; Lee, B.-J. *In vitro* Metabolic Stability of Moisture-sensitive Rabeprazole in Human Liver Microsomes and Its Modulation by Pharmaceutical Excipients. *Arch Pharm. Res.* **2008,** *31* (3), 406–413.

Riehemann, K.; Schneider, S. W.; Luger, T. A.; Godin, B.; Ferrari, M.; Fuchs, H. Nanomedicine—Challenge and Perspectives. *Angew Chem. Int. Ed.* **2009,** *48* (5), 872–897.

Salvia-Trujillo, L.; Sun, Q.; Um, B.; Park, Y.; McClements, D. *In vitro* and *In vivo* Study of Fucoxanthin Bioavailability from Nanoemulsion-based Delivery Systems: Impact of Lipid Carrier Type. *J. Funct. Foods* **2015**, *17*, 293–304.

Shakeel, F.; Baboota, S.; Ahuja, A.; Ali, J.; Aqil, M.; Shafiq, S. Nanoemulsions as Vehicles for Transdermal Delivery of Aceclofenac. *AAPS Pharm. Sci. Tech.* **2007**, *8* (4), 191–199.

Shakeel, F.; Shafiq, S.; Haq, N.; Alanazi, F. K.; Alsarra, I. A. Nanoemulsions as Potential Vehicles for Transdermal and Dermal Delivery of Hydrophobic Compounds: An Overview. *Expert Opin. Drug Deliv.* **2012**, *9* (8), 953–974.

Sharma, N.; Bansal, M.; Visht, S.; Sharma, P.; Kulkarni, G. Nanoemulsion: A New Concept of Delivery System. *Chron. Young Sci.* **2010**, *1* (2), 2.

Silva, H. D.; Cerqueira, M. A.; Souza, B. W.; Ribeiro, C.; Anides, M. C.; Quintas, M. A.; Coimbra, J. S.; Carneiro-DA-Cunha, M. G.; Vicente, A. A. Nanoemulsions of β-carotene Using a High-energy Emulsification–Evaporation Technique. *J. Food Eng.* **2011**, *102* (2), 130–135.

Silva, A. C.; Santos, D.; Ferreira, D.; Lopes, C. M. Lipid-based Nanocarriers as an Alternative for Oral Delivery of Poorly Water-soluble Drugs: Peroral and Mucosal Routes. *Curr. Med. Chem.* **2012**, *19* (26), 4495–4510.

Singh, S.; Pathak, K.; Bali, V. Product Development Studies on Surface-adsorbed Nanoemulsion of Olmesartan Medoxomil as a Capsular Dosage Form. *AAPS Pharm. Sci. Tech.* **2012**, *13* (4), 1212–1221.

Solans, C.; Izquierdo, P.; Nolla, J.; Azemar, N.; Garcia-Celma, M. Nano-emulsions. *Curr. Opin. Colloid Interface Sci.* **2005**, *10* (3), 102–110.

Tadros, T.; Izquierdo, P.; Esquena, J.; Solans, C. Formation and Stability of Nano-emulsions. *Adv. Colloid Interface Sci.* **2004**, *108*, 303–318.

Tan, C. P.; Nakajima, M. β-Carotene Nanodispersions: Preparation, Characterization and Stability Evaluation. *Food Chem.* **2005**, *92*, 661–671.

Wang, X.; Jiang, Y.; Wang, Y.-W.; Huang, M.-T.; HO, C.-T.; Huang, Q. Enhancing Anti-Inflammation Activity of Curcumin Through O/W Nanoemulsions. *Food Chem.* **2008**, *108* (2), 419–424.

Williams, H. D.; Trevaskis, N. L.; Yeap, Y. Y.; Anby, M. U.; Pouton, C. W.; Porter, C. J. Lipid-based Formulations and Drug Supersaturation: Harnessing the Unique Benefits of the Lipid Digestion/Absorption Pathway. *Pharm Res.* **2013**, *30* (12), 2976–2992.

Zhang, Y.; Gao, J.; Zheng, H.; Zhang, R.; Han, Y. The Preparation of 3,5-dihydroxy-4-isopropylstilbene Nanoemulsion and *In vitro* Release. *Int. J. Nanomed.* **2011**, *6*, 649.

CHAPTER 4

IMMUNOMODULATORY AND ANTIMICROBIAL EFFECTS OF NANOEMULSIONS: PHARMACEUTICAL DEVELOPMENT ASPECTS AND PERSPECTIVES ON CLINICAL TREATMENTS

MARCUS VINÍCIUS DIAS-SOUZA[1,3*], ARISTIDES ÁVILO NASCIMENTO[2,3], and THALLES YURI LOIOLA VASCONCELOS[3,4]

[1]Health Sciences Faculty, University Vale do Rio Doce, Minas Gerais, Brazil

[2]Head, Pharmaceutical School Internship Unit, Institute of Applied Theology (INTA), Ceará, Brazil

[3]Integrated Pharmacology and Drug Interactions Research Group (GPqFAR), Brazil

[4]Pharmacist of School Internship Unit, Institute of Applied Theology (INTA), Ceará, Brazil

*Corresponding author. E-mail: souzamv@mail.com

ABSTRACT

Nanoemulsions can be defined as oil-in-water or water-in-oil emulsions, which are prepared with the aid of surfactants and present droplets below a micron in size. In the pharmaceutical context, nanoemulsions are isotropic and kinetically stable dosage forms in which both hydrophobic and hydrophilic compounds can be effectively incorporated and explored for the treatment of different diseases. Most of the current pharmaceutical research in nanoemulsions has been directed for the treatment of inflammatory and infectious diseases. Different formulations have been explored both as vehicles for drugs already available at the pharmaceutical market, and as adjuvants for the delivery of antigens to mucosal tissues. Here, we present recent advances on the pharmaceutical development of nanoemulsions for treating inflammatory and infectious diseases.

4.1   INTRODUCTION

Nanoemulsions (NEs) can be defined as oil-in-water (O/W) or water-in-oil emulsions, which are prepared with the aid of surfactants and present droplets below a micron in size. Several approaches such as environmental and pharmaceutical have been proposed for the study of NEs. Regarding the environmental approach, unfortunately, some NEs have raised health concerns when released (or spontaneously formed, as a result of the presence of surfactants) in soil or water, for instance, due to the potential toxicity of the surfactants used in NEs manufacturing, and also due to their ability to load and transport toxic byproducts in the environment. Both situations represent risks for humans, animals, and plants. A complete separation of NEs for the clearance of toxic compounds is still technically difficult when dealing with such problems.

In the pharmaceutical context, the focus of this chapter, NEs are isotropic and kinetically stable dosage forms in which both hydrophobic and hydrophilic compounds can be effectively incorporated. The nanosized particles present a large surface area, which gives the formulation a low surface tension. Thus, they have been explored as multifunctional vehicles for the treatment of different diseases. Due to these characteristics, NEs are promising alternatives for the transdermal administration of drugs and biological compounds that are currently administered in solid dosage forms of oral intake, or in injectable dosage forms, such as tablets and vaccines, respectively (Sarker et al., 2005; Saberi et al., 2013). In this perspective, NEs become even more relevant not only because several hydrophobic and hydrophilic compounds of pharmaceutical interest are often difficult to be formulated in conventional dosage forms, but may also increase patient compliance to drug therapy, given that the administration of NEs in the skin or mucosal tissues can be more comfortable than using oral dosage forms with unmasked or poorly masked taste, and also injectable formulations (Saberi et al., 2013).

The development of NEs can be a challenge. Some properties of the formulations such as appearance, sensorial texture, color, and rheology, are influenced by the average droplet size of the emulsion. Techniques such as extrusion and ultrasound cavitation have been used in their development. NEs are generally transparent or milky, when droplet diameter is below 200 nm or near 500 nm, respectively. The average droplet size of O/W NEs, which is the most used type for pharmaceutical purposes, generally ranges from 100 to 500 nm (Saberi et al., 2013; Hörmann and Zimmer, 2016).

Most of the research in NEs for therapeutic use has been directed for the treatment of inflammatory and infectious diseases. Different formulations have been explored both as vehicles for drugs already available at the pharmaceutical market, and as adjuvants for vaccines of intranasal administration, aiming the penetration of antigens in the mucosa. In this chapter, we review in brief important advances on the pharmaceutical development of NEs for the treatment of inflammatory and infectious diseases. Moreover, perspectives on their clinical use are discussed.

4.2 IMMUNOREGULATORY POTENTIAL OF NANOEMULSIONS

The pharmacological regulation of the immune response is central for the treatment of diseases that involve inflammatory mechanisms. Formulations for downregulating the immune system are generally developed with drugs, whereas upregulation is usually achieved by the use of vaccine antigens. Hamouda et al. (2011) developed a soybean O/W NE adjuvant with a commercial influenza intramuscular vaccine with three viral strains. The formulation was administered intranasally in one dose to naïve ferrets. The NE induced higher titers at 1/50 of the standard intramuscular vaccine, produced significant cross immunity to five other influenza virus strains that were not included in the formulation and induced sterile immunity after challenge with homologous live virus. Seroconversion rates higher than 60% were achieved against the viral strains in the formulation.

Sharif et al. (2012) developed Ibuprofen-loaded (5%) NEs and investigated their permeability through the skin, using an *in vitro* model with the abdomen skin of Wistar rats. Ibuprofen was added to the NE both after it was completely formed and ready by water titration, and in the oil phase of the formulation, which contained the surfactants and cosurfactants. Skin samples of Wistar rats were hydrated from approximately 10–20% in water, in order to simulate the behavior of a healthy skin. The system was monitored by a gravimetric method. The NEs increased the permeation rate of ibuprofen up to 5.72–30 times when compared to a traditional formulation, and thus, were able to accelerate drug onset of action. The authors provided evidence that this effect was observed mostly due to the lower amount of surfactant and cosurfactant in the NE when compared to traditional topical formulations, and consequently, due to the lack of inflammatory effects induced by such molecules.

Myc et al. (2013) investigated the use of a NE adjuvant to enhance the immune response indirectly by phagocytosis of antigen-primed, apoptotic, epithelial cells, beyond direct uptake of antigen by dendritic cells. The authors used an *in vitro* model, and exposed TC-1 epithelial cells and JAWS II or bone-marrow-derived dendritic cells cultures to the formulation, using R-PE or DQ-OVA in antigen uptake assays. After the treatment, TC-1 cells underwent G2/M cell cycle arrest and apoptosis. TC-1 cells were rapidly engulfed when cocultured with JAWS II or bone-marrow-derived dendritic cells, and the expression of CD86 was upregulated. Moreover, the authors conducted *in vivo* assays. Intranasal instillations of the NEs with OVA were administered to female C57BL/6 mice in order to evaluate the potential use of the formulation as a mucosal vaccine adjuvant. A significant titer of IgG was observed 14 days after only one immunization of the formulation loaded with OVA, which increased in subsequent immunizations.

Eid et al. (2014) explored self-nanoemulsifying systems using Swietenia oil, which presents an anti-inflammatory potential comparable to acetyl-salicylic acid, beyond antimicrobial, antimutagenic, and anticancer properties. The formulation was developed using nonionic surfactants (Labrasol®, Polisorbate 20, Capmul®, and Labrafil®), and their effect on the formulation was investigated. The anti-inflammatory effect of the formulation was confirmed using the carrageenan-induced rat paw edema assay, and was then compared to the oil solution. The level of anti-inflammatory activity of the oil was superior to 70%, at the concentration of 4 mL/kg, followed by a 2 mL/kg dose, also superior to 70%. The inhibition activity of the NE, on the other hand, was of 64.5% and 68.9% for 0.5 mL/kg and 1 mL/kg dosages, respectively.

Orzechowska et al. (2015) investigated the apoptosis induced by a NE in human epithelial cells using the Annexin V assay. The authors observed that treated cells could die by pathways associated with apoptosis, necrosis, and/or autophagy, which is of interest for the development of vaccines. These pathways have implications on antigen uptake, processing, presentation, and generation of danger signals, due to the immunogenicity of the antigens in the formulation. The authors reported that the exposure of human epithelial cells to the formulation resulted in activation of caspases 1, 3, 6, 7, 8, and 9 and in increased expression of BIRC2 and GADD45A genes, involved in apoptosis pathway.

Bielinska et al. (2007) proposed the use of a NE as a mucosal adjuvant for a human needle-free anthrax vaccine with less side effects and requiring less administrations than the currently available vaccine dose. The authors conducted an intranasal immunization of mice and guinea pigs with the formulation loaded with rPA (recombinant *Bacillus anthracis* protective antigen) as an adjuvant, which penetrated the mucosal surface and loaded the antigen to dendritic cells. The immunization induced serum IgG and bronchial IgA and IgG against rPA after up to two administrations, and increased the expression of cytokines associated to the TH1 immune response. Also, the immunization resulted in 70% survival rate against intranasal challenge with $10\times$ LD50 and 40% survival rate against intranasal challenge with $100\times$ LD50.

This same research group provided evidence that the immunization using the NE could also induce a TH17 response in mice (Bielinska et al., 2010). A role for TLR2, TLR4, CD80/CD86, CD40, and IL-6 in the activation of this pathway was suggested: mice deficient in such molecules failed to produce TH17 immunity after being exposed to the formulation, and induction of TH17 response was enhanced by the presence of TLR2 and TLR4 receptors. More recently, the authors explored the NE to restore TH1/TH2 balance of immune responses after a TH2-polarized immunity picture was established by intramuscular immunization of mice with an alum-adjuvanted hepatitis B surface antigen (Bielinska et al., 2016). After an intranasal immunization with the formulation, the production of TH2 cytokines like IL-4 and IL-5 decreased, IL-10 expression was upregulated. IL-10 depletion suppressed the production of IL-4, IL-5, and IL-13, but interestingly, the expression of IFN-γ was not altered. After 6 weeks of immunization, the frequency of Tregs was elevated and the antigen-specific cytokine expression analysis in splenic lymphocytes showed induction of IFN-γ, TNF-α, and IL-17, with decreased levels of IL-4 and IL-5.

Alam et al. (2016) developed NEs for the treatment of contact dermatitis. The disease was induced in Wistar rats by the topical administration of 5% nickel sulfate in solid Vaseline. The authors explored the delivery of clobetasol propionate that like several drugs, presents low permeability across the skin. The preparation of different NEs formulations was conducted by aqueous phase titration, using algal oil as the oil phase, which is rich in ω-3 fatty acids, Tween 20, PEG 200 and water and varied surfactants (polysorbate 20, 60, and 80, and Labrasol) and cosurfactants (ethanol, Transcutol-P, Plurol oleique, PEG 200 and 400). The addition of carbopol 971 converted the NE into a thick hydrogel system. The final formulation droplets showed a small average diameter and the NE presented long-term stability. Using the carrageenan-induced hind paw methodology in Wistar rats, the authors observed inhibition of inflammation in 84.55% for the drug-loaded NE.

Dal Mas et al. (2016) developed an anti-inflammatory NE loaded with the stem bark extract of *Rapanea ferruginea*. The extract was incorporated on the oil phase, by phase inversion at low energy. The NE presented a moderate degree of skin irritation in the agarose overlay assay *in vitro*, and a pseudoplastic and thixotropic rheological behavior was observed. The anti-inflammatory potential of the formulation was confirmed by using the croton oil-induced edema ear model, in which the authors observed a decrease in TNF levels and in myeloperoxidase activity. The NE was significantly more efficient than a conventional emulsion containing 0.13% of the extract.

Mello et al. (2016) tested a methotrexate-loaded lipid NE (0.5 μmol/kg) in an *in vivo* model of antigen-induced arthritis, and compared the effectiveness of their formulation with the results of the use of a commercial formulation of the drug, prepared and administered at 0.5 μmol/kg. Arthritis was induced with methylated bovine serum albumin injected intra-articularly in New Zealand rabbits. The NE was then intravenously injected, and in order to determine the plasma decaying curves and the biodistribution of the drug, the authors used a 3H-cholesteryl ether labeled methotrexate, and monitored it by radioactive counting. The rabbits were sacrificed 24 h after the administration of the NE, and synovial fluid was collected. Protein leakage and cell content analyses were conducted, and histopathological analysis was performed with synovial membranes. The NE uptake was observed mainly in the liver, and the uptake by arthritic joints was two-fold higher when compared to the commercial formulation or to saline solution. Leukocyte infiltration into the synovial fluid was reduced by nearly 65%, and mononuclear and polymorphonuclear cells influx was reduced by 47% and 72%, respectively. On the other hand, the commercial formulation of methotrexate had no effect on cellular infiltration. Protein leakage on the joints of the rabbits was lower with the methotrexate NE treatment, when compared to the use of the commercial methotrexate formulation.

4.3   ANTIMICROBIAL NANOEMULSIONS

The efficacy of many drugs is limited by the low oral bioavailability. To overcome this disadvantage, effective strategies are needed to increase the bioavailability of these drugs (Aboalnaja et al., 2016). As exposed, a NE consists of a heterogeneous colloidal dispersion, developed to increase the solubility and bioavailability of hydrophobic compounds. The average size of the internal droplets is generally between 1 and 500 nm, and may reach up to 1000 nm. This type of formulation has an extensive contact surface where the droplets of the internal or dispersed phase are finely dispersed in the outer phase or dispersant droplets by means of surfactants (emulsifiers) and/ or cosurfactants to stabilize the dispersed system. By the process of encapsulation, NEs can protect drugs against degradation in the gastrointestinal tract (GIT) and modify pharmacokinetic parameters (Mason et al., 2006; Divsalar et al., 2012; Aboalnaja et al., 2016; Aulton, 2016; Campani et al., 2016; Piętka-Ottlik et al., 2016).

One important feature that distinguishes NEs from other emulsified systems is the different pattern in the physical and rheological properties with decreasing particle size. Major distinguishing advantages of a NE include its stability and easy penetration into biological membranes. Due to the very fine particle size and lower surface tension between the oil and water molecules, it almost has no tendency to agglomerate or precipitate, which reduces the possibility of creaming, coalescence, flocculation, phase reversal, or sedimentation. As a result, a NE is more stable than other conventional emulsified systems, being more translucent in comparison to the macroemulsions besides reducing the creaming and coalescence phenomena, very common in the latter (Santos-Magalhães et al., 2000; Bouchemak et al.; Azeem et al., 2009, Devarajan and Ravichandran, 2011, Prakash and Thiagarajan, 2011, Moghimi et al., 2016).

In nanotechnology, micelle systems are widely used to formulate drugs of low solubility in colloidal dispersions. The micelles are formed due to the surfactant power of the surfactant molecules to self-organize in an aqueous environment. As a consequence of the micellar structure, which provides a hydrophobic core and a hydrophilic surface, the micelles are usually used as solubilizing agents. NEs can also be obtained by liposomal technology. Liposomes are closed spherical vesicles consisting of an aqueous nucleus surrounded by one or more two-layer membranes (lamellae), which alternate with aqueous compartments. These double-layer membranes are mainly composed of phospholipids, but may also present other elements, such as cholesterol and antioxidants such as tocopherol, which are commonly employed in liposome formulations. Under a suitable lipid–water and temperature relationship, the lipids will organize into multilamellar vesicles. Due to their nature, the liposomes are capable of carrying both hydrophilic moieties and lipophilic moieties. The water-soluble drugs are incorporated into the aqueous compartments and the lipophilic drugs incorporated within the lipid bilayer. Liposomal formulations have the advantage of being able to control the release of the encapsulated drug (Aulton, 2016).

Antimicrobial NEs generally have particles ranging in size from 100 to 800 nm and should be bioactive at nontoxic concentrations in mammals. The selection of emulsifiers is extremely important in controlling the colloidal stability of NEs, and decisive in the possible interactions with the microbial cell membrane. The droplet size of the internal oil phase influences the antimicrobial activity of the NEs. The reduced size of NEs has a low surface tension, which fuses strongly with the membrane of bacterial, viral, and fungal cells, causing subsequent disruption of the cellular membrane of these microorganisms. This property does not apply to higher organism eukaryotic cells (Moghimi et al., 2016).

This activity of NEs is not specific, unlike traditional antimicrobials, thus allowing a broad-spectrum activity, limiting the generation capacity of resistance (Hwang et al., 2013; Moghimi et al., 2016). The incorporation of antimicrobials into NEs is a promising strategy to overcome the current challenges associated with antimicrobial therapy due to the unique physicochemical properties of NE. These include their large surface area, small size, and unique interactions with microorganisms and host cells, as well as their ability to be structurally and functionally modifiable (Zhang et al., 2007; Zhang et al., 2010).

The advantages of a nanoparticulate transport system containing an antimicrobial include targeted organ transport, relatively uniform tissue distribution, cellular intimality and improved solubility, prolonged drug release, minimized side effects, and increased patient compliance (Mansour et al., 2009; Sosnik et al., 2010). Other advantages of nanotechnology systems are: improved solubility of the drug, improved bioavailability and increased plasma half-life, protection against enzymatic degradation and humoral activity following oral administration, reduction of drug particle aggregation, low immunogenicity and antigenicity of the drug and targeting to specific organs, increased solubility of drugs, small internal droplet size, physical stability, lower viscosity, and ease of preparation (Shakeel et al., 2012; Aulton, 2016). Another advantage in the use of NE is the occurrence of increased retention time of the drug in the target region, therefore, occurrence of fewer side effects or decreased toxicity, because the drug will not act on undesired areas or sites. A lower amount of the drug is required because of better penetration, increased bioavailability, increased retention time, and less active substance loss (Sharma et al., 2010; Devarajan and Ravichandran, 2011).

In addition, the nanosystems themselves are able to overcome mechanisms of resistance to the specific drugs existing in certain microorganisms. Thus, the cotransport of multiple antimicrobials in these nanosystems is capable of having increased antimicrobial activity, acting in a synergistic way, surpassing the mechanisms of resistance of the microorganisms. These advantages are known as the main contributors to overcome the bacterial and fungal resistance associated with poor distribution of these antimicrobials (Zhang et al., 2010; Pelgrift and Friedman, 2013).

The antimicrobial activity and the mechanism of action of NEs are related to their ability to fuse with the outer membrane of microorganisms, to the electrostatic interaction between the cationic charge of the nanoparticles and to the anionic charge of the microorganisms, destabilizing the lipid bilayers of the bacterial membrane, leading to the rupture of the lipid membrane of the microorganism, increasing its cellular permeability. The antimicrobial activity of NEs is nonspecific, thus allowing broad spectrum activity, limiting the capacity for resistance generation. These and other features mentioned above make NEs a useful option for incorporation of antimicrobial drugs (Denyer and Stewart, 1998; Hamouda and Baker, 2000; Hamouda et al., 2001; Hemmila et al., 2010).

Fugizone® is a mixed micellar formulation, which is used to solubilize amphotericin B, an antifungal used to treat invasive fungal infections such as systemic candidiasis and histoplasmosis (Aulton, 2016). The organoselenic compounds, particularly 1,2-benzisoselanzol-3 (2H)-one and 2-phenyl-benzisoselanzol-3 (2H)-one, constitute an important class of multifunctional synthetic products with anti-inflammatory, antioxidant, cytoprotective, antinociceptive, antidepressant, and recently, antimicrobial activity has been described, particularly against Gram-positive bacteria and against yeasts (Piętka-Ottlik et al., 2016).

NEs with different particle sizes containing as antimicrobial agent lemon blueberry oil and soybean oil were developed using polysorbate 80 as an emulsifier. These NEs were tested against the standard strains of *Escherichia coli*, *Listeria monocytogenes*, *Salmonella typhimurium*, *Pseudomonas aeruginosa*, and *Bacillus cereus*. It has been observed that the smaller the droplet size of the oil phase of the NEs, containing the lemon blueberry oil, the greater the inhibition zone formed. For NEs containing soybean oil, a lower antimicrobial activity was observed when compared to the NE used with lemon blueberry oil, due to the larger internal phase droplet size. The preparation mode and the number of NE components may influence the antimicrobial properties, due to the fact that the low surface tension of the NE droplets can influence the bactericidal activity of the extracts used, acting synergistically. However, the antimicrobial effect of the NEs was due to the incorporated extracts of lemon blueberry oil and soybean oil, but was not related to the properties of the isolated NE (Buranasuksombat et al., 2011). A summary of NEs used against specific microorganisms is presented in Table 4.1.

TABLE 4.1 Antimicrobial Nanoemulsions with Potential Clinical Relevance.

Reference work	Antimicrobial spectrum	Possible clinical use
Buranasuksombat et al. (2011)	*Escherichia coli, Listeria monocytogenes, Salmonella typhimurium*	Gastroenteritis, listeriosis, salmonellosis
Lin et al. (2012)	*Helicobacter pylori*	Gastritis, duodenitis, peptic ulcer
Thakkar et al. (2015)	*Aspergillus niger*	Mycoses
Daood et al. (2015)	*Haemophilus influenzae, Streptococcus pneumoniae*	Meningitis
Yang et al. (2015)	*Candidatus liberibacter*	Huanglongbing (HLB)
Almadiy et al. (2016)	*E. coli, Pseudomonas aeruginosa, Staphylococcus aureus*	Gastroenteritis, acnes, boils, pneumonia, meningitis.
Ma et al. (2016)	*E. coli, L. monocytogenes, S. enteritidis*	Gastroenteritis, listeriosis, salmonellosis

One way in which the antimicrobial activity of NEs can be increased is with the addition of a cationic component such as cetylpyridinium chloride, which provides a positive surface charge on the NE particle and ultimately attracts efficiently to the negatively charged bacterial surface. In addition, cetylpyridinium chloride may act as a cosurfactant, thereby assisting in destabilization of the bacterial cell membrane. The electrostatic interaction of the inner phase of the positively charged NEs with negatively charged microbial walls increases the concentration of EO at the site of action. NEs are potentially a suitable alternative to traditional antimicrobial and antimicrobial agents for the treatment of bacteria resistant to conventional antimicrobials and those that form impenetrable biofilms (Hwang et al., 2013; Aulton, 2016; Donsì and Ferrari, 2016).

In order to increase the efficacy of azithromycin in the treatment of bacterial meningitis, a nanoemulsified system containing azithromycin has been developed to protect the drug from hepatic enzymatic metabolism, thereby increasing the circulation time of the drug in the bloodstream and, consequently, increasing the likelihood of their arrival in the brain. Because of the nanoemulsified system, drug permeability at the blood–brain barrier is increased, allowing the drug to reach the meninges in adequate amounts to produce its therapeutic effects against meningitis (McIntyre et al., 2012; Daood et al., 2015).

Another example of low-bioavailability drug due to poor solubility in water is itraconazole, with its absolute bioavailability after being administrated orally is only 55% due to its poor solubility and incomplete and erratic absorption. Thus, a NE was formulated with the aim of improving oral bioavailability, increasing the solubilization of the antifungal in the gastrointestinal tract. The NE was prepared by the ultrasonic emulsification method, with oil droplet size below 150 nm. The small droplet size resulted in a better penetration of the formulation into the intestinal membrane. The efficacy of itraconazole in the formulation was confirmed by the antimycotic study which showed greater zone of inhibition compared to the free drug (Thakkar et al., 2015).

A chitosan/heparin NE was developed with amoxicillin for the treatment of *Helicobacter pylori* infections. In the *in vitro* analysis of drug release of the NE particles, the system can control the release of amoxicillin in a mock gastrointestinal dissolution medium. Amoxycillin-loaded chitosan/heparin NE particles could be located in the intercellular spaces or cytoplasm of the cell, the site of *H. pylori* infection, and significantly increased inhibition of *H. pylori* growth compared to free amoxicillin (Lin et al., 2012).

The VivaGel® formulation, developed by Starpharma, is a gel that uses highly branched dentimer consisting of a central core, an inner dentifrice, and an outer surface of the dentifrice. In this formulation, the dendrimer is the active component, rather than being used as the release system. The dentifrice formulation has antiviral properties, due to its ability to bind to the virus and prevent a possible infection of the cells (Aulton, 2016). Other anti-retroviral NEs approved for clinical use or in clinical trials include Norvir® (ritonavir), delivered in semisolid hard gelatin capsules, Fortovase® (saquinavir), and Agenarase® (amprenavir), both delivered in soft gelatin capsules in liquid form (Shakeel et al., 2012).

Oral NEs may also be formulated for flavor masking. A study was carried out incorporating the isolated compounds thymol and eugenol, which when orally administered in conventional emulsions, have undesirable tastes, in addition to being weakly soluble in aqueous medium, and were formulated in an O/W NE using as emulsifier lauric arginate, a cationic substance with antimicrobial properties known to be safe (GRAS), associated with soy lecithin. The formulations were tested against *L. monocytogenes*, *E. coli*, and *Salmonella enteritidis*. The physical properties of the formulations were also evaluated in terms of size, physical stability, zeta potential, and morphology of the emulsion droplets, as well as release kinetics of lauric arginate lamellae *in vitro*. In this study, an *in vitro* improvement of the physical properties of the NEs formed was observed, as well as an increased antimicrobial activity *in vitro* against *L. monocytogenes*, *E. coli*, and *S. enteritidis* (Ma et al., 2016).

In the study conducted by Almadiy et al. (2016), NEs containing essential oils of four species of *Achillea* and their major fractions have been developed and tested against species of Gram-positive and Gram-negative food, such as *Staphylococcus aureus*, *E. coli*, and *P. aeruginosa*, respectively. The result of this research was in the development of NEs with adequate physical stability, without creaming, and with mean particle diameter between 100 and 300 nm. By incorporating the essential oils and fractions derived from four species of *Achillea* (*Achillea biebersteinii*, *Achillea fragrantissima*, *Achillea santolina*, and *Achillea millefolium*), with known antibacterial activity, a pronounced increase of the antibacterial activity of the essential oils and their predominant fractions was observed. O/W NEs are important vehicles for the delivery of lipophilic antimicrobials, such as *Achillea* spp. essential oils, incorporated into food products, nutraceuticals, and cosmetics. NEs prepared with essential oils and their fractions with antimicrobial activities showed an increase in antibacterial activity against foodborne bacteria (Almadiy et al., 2016).

Essential oils incorporated into NEs have further additional advantages such as minimizing the impact on the organoleptic properties of food products, increasing bioactivity, and better diffusion through cell membranes. As the droplet size of the internal oily phase decreases, the biological activity of lipophilic compounds used as antimicrobials increases due to the increased transport of these active molecules across the cell membranes (Donsi and Ferrari, 2016). Products containing essential oils as antimicrobial agents are promising formulations and should be tested in future research in combination with conventional antimicrobial therapy or even in combination among themselves (Bhargava et al., 2015).

4.4 CONCLUSIONS

NEs are suitable carriers for the delivery of anti-inflammatory and antimicrobial drugs through transdermal administration, representing an interesting approach to increase the bioavailability of hydrophobic drugs, and natural products as well. The spherical nanosized particles of such formulations explain their high permeation across the skin until the blood stream that is not reached by most of the currently available conventional topical emulsions. The clinical use of NEs open may increase the compliance to pharmacological treatments, as the topical use of medications can be not only more comfortable but also safer when compared to both orally administrated and injectable formulations.

ACKNOWLEDGMENTS

MVDS is head of GPqFAR and is currently supported by grants from Fundação de Amparo à Pesquisa do Estado de Minas Gerais (FAPEMIG). AAN and TYLV are thankful to INTA College.

KEYWORDS

- **nanoemulsions**
- **antimicrobial**
- **immunomodulation**
- **pharmaceutical**
- **dosage form**
- **treatment**

REFERENCES

Aboalnaja, K.; Yaghmoor, S.; Kumosani, T.; McClements, D. Utilization of Nanoemulsions to Enhance Bioactivity of Pharmaceuticals, Supplements, and Nutraceuticals: Nanoemulsion Delivery Systems and Nanoemulsion Excipient Systems. *Expert Opin. Drug Deliv.* **2016,** *13,* 1327–1336.

Alam, M. S.; Ali, M. S.; Zakir, F.; et al. Enhancement of Anti-dermatitis Potential of Clobetasol Propionate by DHA (Docosahexaenoic Acid) Rich Algal Oil Nanoemulsion Gel. *Iranian J. Pharm. Res.* **2016,** *15* (1), 35–52.

Almadiy, A.; Nenaah, G.; Al Assiuty, B.; Moussa, E.; Mira, N. Chemical Composition and Antibacterial Activity of Essential Oils and Major Fractions of Four *Achillea* Species and Their Nanoemulsions Against Foodborne Bacteria. *LWT-Food Sci. Technol.* **2016,** *69,* 529–537.

Araújo, F.; Kelmann, R.; Araújo, B.; Finatto, R.; Teixeira, H.; Koester, L. Development and Characterization of Parenteral Nanoemulsions Containing Thalidomide. *Eur. J. Pharm. Sci.* **2011,** *42,* 238–245.

Aulton, M. *Delineamento de Formas Farmacêuticas*, 4th ed.; Elsevier: Rio de Janeiro, 2016.

Azeem, A.; Khan, Z.; Aqil, M.; Ahmad, F.; Khar, R.; Talegaonkar, S. Microemulsions as a Surrogate Carrier for Dermal Drug Delivery. *Drug Dev. Ind. Pharm.* **2009**, *35*, 525–547.

Bhargava, K.; Conti, D.; da Rocha, S.; Zhang, Y. Application of an Oregano Oil Nanoemulsion to the Control of Foodborne Bacteria on Fresh Lettuce. *Food Microbiol.* **2015**, *47*, 69–73.

Bielinska, A. U.; Janczak, K. W.; Landers, J. J.; et al. Mucosal Immunization with a Novel Nanoemulsion-based Recombinant Anthrax Protective Antigen Vaccine Protects Against *Bacillus anthracis* Spore Challenge. *Infect. Immun.* **2007**, *75* (8), 4020–4029.

Bielinska, A. U.; Gerber, M.; Blanco, L. P.; et al. Induction of Th17 Cellular Immunity with a Novel Nanoemulsion Adjuvant. *Crit. Rev Immunol.* **2010**, *30* (2), 189–199.

Bielinska, A. U.; O'Konek, J. J.; Janczak, K. W.; Baker, J. R. Immunomodulation of TH2 Biased Immunity with Mucosal Administration of Nanoemulsion Adjuvant. *Vaccine* **2016**, *34* (34), 4017–4024.

Bouchemal, K.; Briançon, S.; Perrier, E.; Fessi, H. Nano-emulsion Formulation Using Spontaneous Emulsification: Solvent, Oil and Surfactant Optimisation. *Int. J. Pharm.* **2004**, *280*, 241–251.

Buranasuksombat, U.; Kwon, Y.; Turner, M.; Bhandari, B. Influence of Emulsion Droplet Size on Antimicrobial Properties. *Food Sci. Biotechnol.* **2011**, *20*, 793–800.

Campani, V.; Biondi, M.; Mayol, L.; Cilurzo, F.; Pitaro, M.; De Rosa, G. Development of Nanoemulsions for Topical Delivery of Vitamin K1. *Int. J. Pharm.* **2016**, *511*, 170–177.

Dal Mas J, Zermiani T, Thiesen LC, et al. Nanoemulsion as a Carrier to Improve the Topical Anti-inflammatory Activity of Stem Bark Extract of *Rapanea ferruginea. Int. J. Nanomed.* **2016**, *11*, 4495–4507.

Daood, G.; Basri, H.; Stanslas, J.; Fard Masoumi, H.; Basri, M. Predicting the Optimum Compositions of a Parenteral Nanoemulsion System Loaded with Azithromycin Antibiotic Utilizing the Artificial Neural Network Model. *RSC Adv.* **2015**, *5*, 82654–82665.

Daull, P.; Lallemand, F.; Garrigue, J. Benefits of Cetalkonium Chloride Cationic Oil-in-water Nanoemulsions for Topical Ophthalmic Drug Delivery. *J. Pharm. Pharmacol.* **2013**, *66*, 531–541.

Denyer, S.; Stewart, G. Mechanisms of Action of Disinfectants. *Int. Biodeter. Biodegr.* **1998**, *41*, 261–268.

Devarajan, V.; Ravichandran, V. Nanoemulsions: As Modified Drug Delivery Tool. *Int. J. Compr. Pharm.* **2011**, *2*, 1–5.

Divsalar, A.; Saboury, A.; Nabiuni, M.; Zare, Z.; Kefayati, M.; Seyedarabi, A. Characterization and Side Effect Analysis of a Newly Designed Nanoemulsion Targeting Human Serum Albumin for Drug Delivery. *Colloids Surf. B Biointerfaces* **2012**, *98*, 80–84.

Donsì, F.; Ferrari, G. Essential Oil Nanoemulsions as Antimicrobial Agents in Food. *J. Biotechnol.* **2016**, *233*, 106–120.

Eid, A. M.; El-Enshasy, H. A.; Aziz, R.; Elmarzugi, N. A. The Preparation and Evaluation of Self-nanoemulsifying Systems Containing Swietenia Oil and an Examination of Its Anti-inflammatory Effects. *Int. J. Nanomed.* **2014**, *9*, 4685–4695.

Hamouda, T.; Baker, J. R. Antimicrobial Mechanism of Action of Surfactant Lipid Preparations in Enteric Gram-negative bacilli. *J. Appl. Microbiol.* **2000**, *89*, 397–403.

Hamouda, T.; Myc, A.; Donovan, B.; Shih, A. Y.; Reuter, J. D.; Baker, J. R. A Novel Surfactant Nanoemulsion with a Unique Non-irritant Topical Antimicrobial Activity Against Bacteria, Enveloped Viruses and Fungi. *Microbiol. Res.* **2001**, *156*, 1–7.

Hamouda, T.; Sutcliffe, J. A.; Ciotti, S.; Baker, J. R. Intranasal Immunization of Ferrets with Commercial Trivalent Influenza Vaccines Formulated in a Nanoemulsion-based Adjuvant. *Clin. Vaccine Immunol.* **2011,** *18* (7), 1167–1175.

Hemmila, M.; Mattar, A.; Taddonio, M.; Arbabi, S.; Hamouda, T.; Ward, P.; Wang, S.; Baker, J. Topical Nanoemulsion Therapy Reduces Bacterial Wound Infection and Inflammation After Burn Injury. *Surgery* **2010,** *148*, 499–509.

Hörmann K1, Zimmer A2. Drug Delivery and Drug Targeting with Parenteral Lipid Nano-emulsions: A Review. *J Control. Release* **2016,** *223*, 85–98.

Hwang, Y.; Ramalingam, K.; Bienek, D.; Lee, V.; You, T.; Alvarez, R. Antimicrobial Activity of Nanoemulsion in Combination with Cetylpyridinium Chloride in Multidrug-resistant *Acinetobacter baumannii. Antimicrob. Agents Chemother.* **2013,** *57*, 3568–3575.

Kelmann, R.; Kuminek, G.; Teixeira, H.; Koester, L. Carbamazepine Parenteral Nanoemul-sions Prepared by Spontaneous Emulsification Process. *Int. J. Pharm.* **2007,** *342*, 231–239.

Lin, Y.; Chiou, S.; Lai, C.; Tsai, S.; Chou, C.; Peng, S.; He, Z. Formulation and Evaluation of Water-in-oil Amoxicillin-loaded Nanoemulsions Using for *Helicobacter pylori* Eradica-tion. *Process Biochem.* **2012,** *47*, 1469–1478.

Ma, Q.; Davidson, P.; Zhong, Q. Nanoemulsions of Thymol and Eugenol Co-emulsified by Lauric Arginate and Lecithin. *Food Chem.* **2016,** *206*, 167–173.

Mansour, H. M.; Rhee, Y. S.; Wu, X. Nanomedicine in Pulmonary Delivery. *J. Nanomed.* **2009,** *4*, 299–319.

Mason, T.; Wilking, J.; Meleson, K.; Chang, C.; Graves, S. Nanoemulsions: Formation, Structure, and Physical Properties. *J. Phys. Condens. Matter* **2006,** *18*, R635–R666.

McIntyre, P.; O'Brien, K.; Greenwood, B.; van de Beek, D. Effect of Vaccines on Bacterial Meningitis Worldwide. *Lancet* **2012,** *380*, 1703–1711.

Mello, S. B.; Tavares, E. R.; Bulgarelli, A.; Bonfá, E.; Maranhão, R. C. Intra-articular Meth-otrexate Associated to Lipid Nanoemulsions: Anti-inflammatory Effect Upon Antigen-induced Arthritis. *Int. J. Nanomed.* **2013,** *8*, 443–449.

Mello, S. B.; Tavares, E. R.; Bulgarelli, A.; Bonfá, E.; Maranhão, R. C. Anti-inflammatory Effects of Intravenous Methotrexate Associated with Lipid Nanoemulsions on Antigen-induced Arthritis. *Clinics.* **2016,** *71* (1), 54–58.

Moghimi, R.; Aliahmadi, A.; McClements, D.; Rafati, H. Investigations of the Effectiveness of Nanoemulsions from Sage Oil as Antibacterial Agents on some Food Borne Pathogens. *LWT-Food Sci. Technol.* **2016,** *71*, 69–76.

Musa, S.; Basri, M.; Masoumi, H.; Karjiban, R.; Malek, E.; Basri, H.; Shamsuddin, A. Formu-lation Optimization of Palm Kernel Oil Esters Nanoemulsion-loaded with Chloramphen-icol Suitable for Meningitis Treatment. *Colloids Surf. B Biointerfaces.* **2013,** *112*, 113–119.

Myc, A.; Kukowska-Latallo, J. F.; Smith, D. M.; et al. Nanoemulsion Nasal Adjuvant W805EC Induces Dendritic Cell Engulfment of Antigen-primed Epithelial Cells. *Vaccine* **2013,** *31* (7), 1072–1079.

Orzechowska, B. U.; Kukowska-Latallo, J. F.; Coulter, A. D.; Szabo, Z.; Gamian, A.; Myc, A. Nanoemulsion-based Mucosal Adjuvant Induces Apoptosis in Human Epithelial Cells. *Vaccine* **2015,** *33* (19), 2289–2296.

Pelgrift, R.; Friedman, A. Nanotechnology as a Therapeutic Tool to Combat Microbial Resis-tance. *Adv. Drug Deliv. Rev.* **2013,** *65*, 1803–1815.

Piętka-Ottlik, M.; Lewińska, A.; Jaromin, A.; Krasowska, A.; Wilk, K. Antifungal Organose-lenium Compound Loaded Nanoemulsions Stabilized by Bifunctional Cationic Surfac-tants. *Colloids Surf. A Physicochem. Eng. Asp.* **2016,** *510*, 53–62.

Prakash, U. R. T.; Thiagarajan, P. Nanoemulsions for Drug Delivery Through Different Routes. *Res. Biotecnhol.* **2011,** *2,* 1–13.

Saberi, A. H.; Fang, Y.; McClements, D. J. Fabrication of Vitamin E-enriched Nanoemulsions: Factors Affecting Particle Size Using Spontaneous Emulsification. *J. Colloid Interface Sci.* **2013,** *391,* 95–102.

Santos-Magalhães, N.; Pontes, A.; Pereira, V.; Caetano, M. Colloidal Carriers for Benzathine Penicillin G: Nanoemulsions and Nanocapsules. *Int. J. Pharm.* **2000,** *208,* 71–80.

Sarfaraz Alam, M.; Ali, M. S.; Zakir, F.; et al. Enhancement of Anti-dermatitis Potential of Clobetasol Propionate by DHA (Docosahexaenoic Acid) Rich Algal Oil Nanoemulsion Gel. *Iranian J. Pharm. Res.* **2016,** *15* (1), 35–52.

Sarker, D. K. Engineering of Nanoemulsions for Drug Delivery. *Curr. Drug. Deliv.* **2005,** *2* (4), 297–310.

Shakeel, F.; Shafiq, S.; Haq, N.; Alanazi, F.; Alsarra, I. Nanoemulsions as Potential Vehicles for Transdermal and Dermal Delivery of Hydrophobic Compounds: An Overview. *Expert Opin. Drug Deliv.* **2012,** *9,* 953–974.

Sharif, B. M.; Torabi, S.; Azarpanah, A. Optimization of Ibuprofen Delivery Through Rat Skin from Traditional and Novel Nanoemulsion Formulations. *Iranian J. Pharm. Res.* **2012,** *11* (1), 47–58.

Sharma, N.; Bansal, M.; Visht, S.; Sharma, P. K.; Kulkarni, G. T. Nanoemulsion: A New Concept of Delivery System. *Chron. Young Sci.* **2010,** *1,* 2–6.

Sosnik, A.; Carcaboso, Á.; Glisoni, R.; Moretton, M.; Chiappetta, D. New Old Challenges in Tuberculosis: Potentially Effective Nanotechnologies in Drug Delivery. *Adv. Drug. Deliv. Rev.* **2010,** *62,* 547–559.

Sutradhar, K.; Amin, M. Nanoemulsions: Increasing Possibilities in Drug Delivery. *Eur. J. Nanomed.* **2013,** *5,* 97–110.

Thakkar, H.; Khunt, A.; Dhande, R.; Patel, A. Formulation and Evaluation of Itraconazole Nanoemulsion for Enhanced Oral Bioavailability. *J. Microencapsul.* **2015,** *32,* 559–569.

Yang, C.; Powell, C.; Duan, Y.; Shatters, R.; Zhang, M. Antimicrobial Nanoemulsion Formulation with Improved Penetration of Foliar Spray Through Citrus Leaf Cuticles to Control Citrus Huanglongbing. *PLoS One* **2015,** *10,* e0133826.

Zhang, L.; Gu, F.; Chan, J.; Wang, A.; Langer, R.; Farokhzad, O. Nanoparticles in Medicine: Therapeutic Applications and Developments. *Clin. Pharmacol. Ther.* **2007,** *83,* 761–769.

Zhang, L.; Pornpattananangkul, D.; Hu, C.; Huang, C. Development of Nanoparticles for Antimicrobial Drug Delivery. *Curr. Med. Chem.* **2010,** *17,* 585–594.

SECTION II
NANOSUSPENSIONS

CHAPTER 5

NANOSUSPENSIONS: FORMULATION, CHARACTERIZATION, AND APPLICATIONS

M. M. CHOGALE[1], A. A. DATE[2], V. N. GHODAKE[1], S. A. PAYGHAN[3], J. I. DISOUZA[3], and VANDANA B. PATRAVALE[1*]

[1]*Department of Pharmaceutical Sciences and Technology, Institute of Chemical Technology, Nathalal Parekh Marg, Matunga (E), Mumbai 400019, Maharashtra, India*

[2]*The Center for Nanomedicine, Wilmer Eye Institute, Johns Hopkins School of Medicine, Baltimore, MD 21205, USA*

[3]*Department of Pharmaceutics, Tatyasaheb Kore College of Pharmacy, Warananagar, Panhala, Kolhapur, Maharashtra, India*

[*]*Corresponding author. E-mail: vbp_muict@yahoo.co.in*

ABSTRACT

Nanosuspensions are composed of nanosized crystals of pure drug dispersed in an aqueous or nonaqueous system containing a stabilizer/surfactant and/ or cosurfactant. Nanosuspensions have evolved as the delivery system of choice for formulation of most of Biopharmaceutics Classification System Class II and IV drugs. Ease of manufacture, amenability to be formulated for different drugs and feasibility for large-scale manufacture further present nanosuspensions as a popular approach for formulation of water-insoluble drugs. This chapter focuses on the various aspects of nanosuspensions, including their formulation components, preparation methods, unique features, methods of characterization, and applications in various routes of administration. The chapter concludes with a case study that illustrates the development of a nanosuspension formulation based on the principles of Quality by Design and optimization by Design of Experiments.

5.1   INTRODUCTION

The water solubility of a drug is of paramount importance in the process of drug and formulation development as it significantly impacts the pharmacokinetic and pharmacodynamic parameters (Rodriguez-Aller et al., 2015). The Biopharmaceutics Classification System (BCS) categorizes poorly water-soluble active pharmaceutical ingredients (APIs) into BCS Class II (low solubility–high permeability) and IV (low solubility–low permeability). Historical evidence suggests that drugs belonging to BCS Class II and IV show dissolution rate-limited absorption which leads to poor and/or highly variable bioavailability. Due to these reasons, poorly water-soluble drugs have always been viewed as high-risk developmental candidates. Ironically, around 40% of the top 200 oral drugs in the global pharmaceutical market have poor water solubility (Takagi et al., 2006). Moreover, 90% of the new chemical entities and 75% of the drug candidates under development are also poorly water-soluble (Rodriguez-Aller et al., 2015). Considering the current landscape of high-throughput drug discovery screening programs and the perpetual need for development of drug candidates with higher potency, the rate of generation of poorly water-soluble compounds is unlikely to diminish (Merisko-Liversidge and Liversidge, 2011). Hence, there will always be a need to develop formulation strategies for efficient delivery of poorly water-soluble APIs.

Over the years, significant efforts have been made to develop approaches for successful formulation of poorly water-soluble drugs. Strategies such as salt formation of ionizable drugs, prodrug synthesis, solubilization using water-miscible cosolvents or oily vehicles, solubilization in surfactant micelles, solid dispersions, formation of inclusion complexes with cyclodextrins, formulation of emulsions or microemulsions, and formulation into liposomes are routinely used to improve delivery of poorly water-soluble drugs (Merisko-Liversidge and Liversidge, 2011; Patravale et al., 2004; Rodriguez-Aller et al., 2015). However, the aforementioned strategies have been successful in limited cases and none of them are capable of tackling delivery challenges associated with poorly soluble molecules that are insoluble in water, oil, and most of the organic solvents (Merisko-Liversidge and Liversidge, 2011; Patravale et al., 2004; Rabinow, 2004). Furthermore, these strategies involve use of excipients at a significantly higher ratio compared to the drug and they are usually unable to achieve high drug loading. Hence, the poorly water-soluble drugs with high therapeutic dose are unlikely to be suitable candidates for the above formulation strategies.

For delivery of drugs with high therapeutic dose and/or poor solubility in water, oils, and organic solvents, reduction of particle size to improve dissolution rate and surface area for absorption could be an attractive approach. For many years, micronization has been used to improve dissolution rate and bioavailability of certain drugs. However, for drugs with extremely low solubility, micronization is not a suitable approach (Merisko-Liversidge and Liversidge, 2011; Patravale et al., 2004; Rabinow, 2004). Hence, transformation of poorly water-soluble drugs into "nanosuspensions" or nano-sized drug particles has emerged as a promising (and potentially universal) approach in the drug delivery field. Over the last 15 years, the utility, versatility, translational capability, and commercial viability of nanosuspensions have been demonstrated via extensive investigations from academia and pharmaceutical industry.

Nanosuspensions can be defined as a colloidal dispersion of nanoscale drug particles in liquid media, with typical size range of 200–600 nm and a suitable stabilizer (or mixture of stabilizers). The nanosuspensions can be generated in aqueous or nonaqueous [polyethylene glycol (PEG) or oils] vehicles depending upon end application (Chin et al., 2014). Nanosuspensions represent a nanotechnology that has been rapidly translated from laboratory to commercial scale, and several pharmaceutical products based on nanosuspensions are commercially available (Table 5.1).

TABLE 5.1 List of Pharmaceutical Products Based on Nanosuspensions.

Trade name	Drug	Approval year	Final dosage form	Technology used for nanosizing	Route of administration
Rapamune®	Rapamycin/ sirolimus	2000	Tablet	NanoCrystal®	Oral
Emend®	Aprepitant	2003	Capsules	NanoCrystal®	Oral
Tricor®	Fenofibrate	2004	Tablet	NanoCrystal®	Oral
Triglide®	Fenofibrate	2005	Tablet	IDD-P®	Oral
Megace® ES	Megestrol acetate	2005	Liquid Nano-suspension	NanoCrystal®	Oral
Invega Sustenna®	Paliperidone palmitate	2009	Liquid Nano-suspension	NanoCrystal®	Parenteral (intramuscular)

5.2 MERITS OF NANOSUSPENSIONS

Nanosuspensions have emerged as a dosage form of choice for the formulation of high dose, water-insoluble drugs owing to their advantages over other nanocarriers. These benefits of nanosuspensions are enlisted and briefly described in the following sections.

5.2.1 ENHANCEMENT OF DISSOLUTION VELOCITY

The size reduction of a drug leads to an enhanced surface area, and thus, according to the Noyes–Whitney equation, an improved dissolution velocity (Noyes and Whitney, 1897). Therefore, in the case of drugs for which the dissolution velocity is the rate-limiting step, micronization maybe a suitable approach to successfully augment the bioavailability. By progressing from micronization further down to nanonization, the particle surface and thereby the dissolution velocity is further enhanced.

5.2.2 INCREASE IN SATURATION SOLUBILITY

The universal notion is that saturation solubility is a constant value and function of the compound, the dissolution medium, and temperature conditions. This concept is relevant for powders with size in the micrometer range or more. However, below a critical size of 1–2 μm, the saturation solubility is also dependent on the crystalline structure and particle size. It increases with a reduction in particle size below 1000 nm (Dressman et al., 1998). For drug nanosuspensions, the enhancement in saturation solubility (C) and surface area (A) leads to a rise in the dissolution velocity (dX/dt) according to the Noyes–Whitney equation.

where, dX/dt is the dissolution velocity, D is the diffusion coefficient, A is the surface area, h is the diffusional distance, Cs is the saturation solubility, and Ct is the concentration of the drug around the particles (Fig. 5.1).

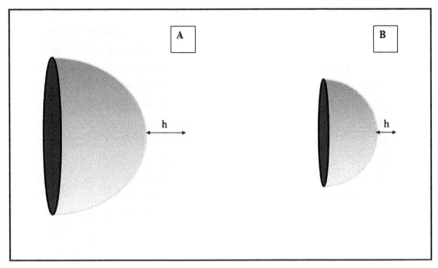

FIGURE 5.1   Comparison of diffusional distance of a (A) microcrystal and a (B) nanocrystal. (Reprinted from Chogale, M. M.; Ghodake V. N.; Patravale, V. B. Performance Parameters and Characterizations of Nanocrystals: A Brief Review. *Pharmaceutics* **2016,** *8* (3). https://www.ncbi.nlm.nih.gov/pubmed/27589788. https://creativecommons.org/licenses/by/4.0/)

An imperative factor herein is the diffusional distance "*h*," as a part of the hydrodynamic boundary layer h_H, which is also strongly reliant on the particle size, as the Prandtl equation shows (Mosharraf and Nystrom, 1995):

$$h_H = k \, (L^{1/2}/V^{1/2})$$

where, h_H is the hydrodynamic boundary layer, k is a constant, L is the length of the particle surface, and V is the relative velocity of the flowing liquid surrounding the particle.

According to the Prandtl equation, decrease in particle size leads to a reduced diffusional distance "*h*" and subsequently an enhanced dissolution velocity, as described by the Noyes–Whitney equation. Due to the enhanced saturation solubility, the concentration gradient between gut lumen and blood augments and accordingly absorption by passive diffusion rises. Hence, a decrease in the particle size of the drug to the nanometer range leads to an enhancement in solubility and consequently the dissolution velocity. Both, solubility and dissolution velocity are very critical factors in terms of improvement of the bioavailability of poorly soluble drugs.

Furthermore, according to the Kelvin equation, the dissolution pressure increases with enhanced curvature, which means reduced particle size (Fig. 5.2).

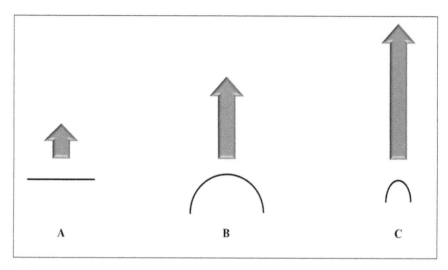

FIGURE 5.2 Increasing dissolution pressure over (A) a flat surface, (B) curvature of a microparticle, and (C) curvature of a nanocrystal. (Reprinted from Chogale, M. M.; Ghodake V. N.; Patravale, V. B. Performance Parameters and Characterizations of Nanocrystals: A Brief Review. *Pharmaceutics* **2016,** *8* (3). https://www.ncbi.nlm.nih.gov/pubmed/27589788. https://creativecommons.org/licenses/by/4.0/)

Particles in the nanometer range offer a huge curvature; consequently a higher dissolution pressure is achieved driving the equilibrium toward dissolution. The Ostwald–Freundlich equation correlates saturation solubility of the drug and the particle size:

where, Cs is the saturation solubility, $C\alpha$ is the solubility of the solid consisting of large particles, r is the interfacial tension of substance, V is the molar volume of the particle material, R is the gas constant, T is the absolute temperature, is the density of the solid, and r is the radius.

It is apparent that the saturation solubility (Cs) of the drug increases with a decrease of particle size (r). However, this effect is pronounced for materials that have mean particle size of less than 2 μm.

5.2.3 INCREASED PERMEABILITY AND ADHESIVENESS

Another feature of nanosuspensions (nanocrystals) is a marked enhancement in their adhesiveness compared to microparticles. The increased adhesiveness is responsible for improving the oral absorption of poorly soluble drugs besides the increased saturation solubility and dissolution velocity.

Nanosuspensions also offer the advantages of enhanced permeability across the skin and mucosal membranes. Transdermal delivery is feasible only for a few drug molecules. This is mainly owing to the constraint in permeation of clinically effective concentrations of drugs through the skin barrier. This constraint is more marked in case of poorly water-soluble drugs. In spite of the enhanced permeation properties of water insoluble drugs due to their lipophilicity, rate of drug release is the rate-limiting step for these drugs. Enhanced permeability of drug nanosuspensions is a size-dependant property; wherein particles with size less than 40 nm were found to permeate the skin via the follicular route while larger sized particles showed limited permeation due to the tight network of epidermal Langerhan's cells. Another study reveals that when the particle size was higher than 5 μm, almost negligible permeation was seen through the stratum corneum, while particles in the range of 500–750 nm manifested better permeation into the hair follicle of the human skin (Ponchel et al., 1997).

5.2.4   LONG-TERM STABILITY

Nanosuspensions are endowed with long-term stability, which is also one of the requirements for the successful regulatory approval of a pharmaceutical formulation. The marked physical stability of nanosuspension is mainly owing to lack of aggregation and absence of "Ostwald ripening" phenomenon. The lack of Ostwald ripening phenomenon is primarily owing to the narrow size distribution of nanosuspensions (Mantzaris, 2005). It was reported that the particles in highly dispersed systems tend to grow, due to the differences in saturation solubility in the milieu of variably sized particles. This phenomenon is referred to as Ostwald ripening (Jacobs and Muller, 2000). The solute concentration is higher in the region of the smaller particles in contrast to that of the larger particles due to the greater saturation solubility of the smaller ones. The molecules therefore will diffuse from the area of the small particles to the area of the large particles driven by the concentration gradient and recrystallize on the surface of the larger particles. The perpetual dissolution of the small particles and recrystallization on the surface of the larger ones usher the formation of microparticles. Moreover, nanoparticles dispersed in a medium tend to aggregate. The physical instability of the system is due to the tendency of the nanoparticles to reduce the high surface energy created by the large interface between the solid and the liquid medium (Drews and Tsapatsis, 2007). This can be averted by incorporation of either of the multiple kinds of stabilizers such as the ionic surfactants, nonionic surfactants, and amphiphilic copolymers (Kocbek et al., 2006).

5.3 FORMULATION COMPONENTS OF NANOSUSPENSIONS

Nanosuspensions are composed of nanosized crystals of pure drug in an aqueous or organic solvent system stabilized by a suitable surfactant and/or cosurfactant/stabilizer. Each of the formulation components of nanosuspensions are described in detail in the following sections.

5.3.1 ACTIVE PHARMACEUTICAL INGREDIENTS

As mentioned in Section 5.1, nanosuspensions are used as a formulation strategy for BCS Class II and IV drugs (poorly water soluble). Currently, poorly soluble drugs makeup 1/3rd of United States Pharmacopeia-recognized drugs. Lipophilic compounds have poor aqueous solubility and imperfect dissolution profile which leads to their low bioavailability. Therefore, formulating new poorly water-soluble molecules to obtain an adequate bioavailability has become a serious and challenging scientific, industrial, and medical issue. "Grease ball" and "brick dust" molecules are two types of poorly soluble drug compounds (Muller et al., 1995). Grease ball molecules are highly lipophilic with high log P due to no interactions with water. Brick dust molecules have melting point above 200°C and low log P. Their poor solubility in water is caused by the strong intermolecular bonding and high lattice energy in solid state (Muller et al., 1995; Muller et al., 1999). Hence, such poorly water-soluble molecules with bioavailability concerns are usually formulated as nanosuspensions. Moreover, versatility of nanosuspensions to be converted into a range of dosage forms further widens the scope of their application.

5.3.2 STABILIZERS

Like with all other nanosized systems, long-term stability is a true challenge with drug nanosuspensions. The greatest stability problem is aggregation (described in Section 5.2.4), and here the role of stabilizer is crucial. The preparation of nanosuspensions generates an enhancement in the particle surface. Due to the increased Gibbs free energy in the system associated with the formation of additional interfaces with the micronization/nanon-ization methods, the nanosuspensions exhibit thermodynamically unstable states and have a tendency to minimize the total free energy by agglom-eration/aggregation (Wu et al., 2011). Similarly, the spontaneous nucleation and growth events in precipitation techniques tend to facilitate a decrease in the free energy. Both these aggregation and growth behaviors related to the change of free energy result in thermodynamic instability of the nanocrystal suspensions. Unlike particle growth during precipitation, which depends on many factors, particle aggregation mostly depends on the stabilizer, its type, and concentration used (Peltonen and Hirvonen, 2010).

Kinetically, the aggregation process depends on its activation energy, which can be increased by including stabilizers to the system. The appro-priate stabilizer thus provides a barrier to aggregation. Theoretically, the prin-ciple of an energetic barrier can be explained by the DLVO theory (named after Boris Derjaguin, Lev Landau, Evert Verwey, and Theodoor Overbeek), which basically describes the interaction of solid particles in a liquid medium in terms of attractive and repulsive interactions between the electric double layers surrounding the particles in solution (Van Eerdenbrugh et al., 2008a, 2008b). The balance between the attractive and repulsive forces is adjusted by the stabilizers in the nanocrystalline systems. Additionally, the kinetic stability of a suspension may also be controlled by some surface forces such as hydrophobic force. The stabilization is based on the stabilizer adsorption on the particle surfaces in order to decrease the free energy of the system and interfacial tension of the particles. The principle mechanisms of stabiliza-tion can be classified into electrostatic and steric stabilization. Electrostatic stabilization involves charged or ionic surfactants, which preserve an elec-trostatic repulsion among particles and steric stabilization relates to the use of nonionic polymers, which offer a steric repulsion between particles. In nanocrystals, the stabilizing effect is mainly achieved via steric stabilization. The polymers are adsorbed physically onto the drug surfaces, and strong adsorption, full surface coverage, and slow desorption associated with high steric repulsion are requirements for efficient stabilization. The thickness of the stabilizer layer with polymeric stabilizers has been calculated to be approximately 1.5–16 nm and the surface coverage 0.15–1.6 g/cm^2, which is a high enough value for efficient steric stabilization (Lee et al., 2003).

Irrespective of the preparation method used, the selection of a suitable type of stabilizer and its optimal concentration are very important for stabi-lizing the smaller sized particles and for maintaining the physical stability during the shelf life of the final product (Van Eerdenbrugh et al., 2009). The stabilizer used may be natural polymers (celluloses, vegetables, starches, and sugars), semisynthetic nonionic polymers (hydroxypropyl methylcellulose, methylcellulose, hydroxyethyl cellulose, and hydroxypropyl cellulose), semisynthetic ionic polymers (sodium carboxymethylcellulose and sodium alginate), synthetic linear polymers (polyvinyl pyrrolidone, polyvinyl alcohol, and derivatives), synthetic copolymers (poloxamers and polyvinyl alcohol–PEG graft copolymers), or surfactants of ionic type (sodium dodecyl sulfate, sodium docusate, and sodium deoxycholate) or nonionic type (polysorbates and sorbitan esters). Additionally, an application of a combination of ionic surfactants with polymeric stabilizers may provide an enhanced stabilization which combines the advantages of both electrostatic and steric stabilization (Peltonen et al., 2016). In general, when media milling is considered, high stabilizer concentrations (in some cases even up to 100% relative to the drug weight) ease both the nanosuspension production and improve the stability (Van Eerdenbrugh et al., 2008a, 2008b).

Cellulose derivatives are efficient to increase viscosity of the nanosuspension to reduce particle mobility and to minimize direct contact of the nanocrystals in the dried formulation by shielding their surface. At the same time, redispersability will be hampered due to the slow dissolution and the possible entanglement of these polymers. Substitution degree as well as molecular weight, which are closely related to viscosity, are important parameters to take into account when selecting the suitable polymer. Also, in pharmaceutical applications, GRAS (generally regarded as safe) status or prior approval as a pharmaceutical excipient by the drug authorities within the drug delivery route in question, or the monograph in a pharmacopoeia, are important selection criteria, too. The semisynthetic polymers may show poor stabilizing behavior, partially because of the viscosity-limited concentration in which they can be prepared. High viscosity of polymers may decrease the diffusion velocity of polymer molecules, slowing down the movement of milling pearls and hindering the energy delivery, and thus delaying the preparation procedure (Van Eerdenbrugh et al., 2008a, 2008b). The linear synthetic polymers show a better stabilizing potential when applied at higher concentrations, an effect that was even more pronounced for some of the synthetic copolymers. Practically, the physicochemical properties of the stabilizer such as molecular weight, melting point, *log* P, aqueous solubility, and density may offer, in most cases, an explanation for a certain behavior. For instance, a stabilizer with a low melting point causes problems in the manufacturing process at high temperatures.

When considering the bottom-up approaches, the particle growth drives the stabilizer adsorption, which is influenced by both the thermodynamic and kinetics aspects on the interfacial surfaces in order to lower the interfacial energy and increase the nucleation rate (Van Eerdenbrugh et al., 2009). The interactions between the drug, the solvent, and the antisolvent are the factors that control the drug diffusion rate from the solvent phase to the boundary region of the antisolvent phases, affinity of the stabilizer molecule and the kinetics of adsorption, and thereby, considerably alter the particle size (Peltonen and Hirvonen, 2010; Sinha et al., 2013). Slower diffusion of the drug molecules results in fast adsorption of the stabilizer, inhibits the growth, and increases the nucleation rate. Faster diffusion of the drug molecules and increased hydrophobic interactions result in aggregation of the particles and increase the particle size. The affinity of stabilizer for the particle surface regulates its adsorption kinetics. If the affinity is higher, adsorption is faster, and hence, a smaller particle size is obtained. Increasing the surfactant concentration helps in faster adsorption because of a higher concentration gradient thereby resulting in a lesser particle size. Though, above the critical micelle concentration (CMC), the surfactant adsorption on particle surfaces is reduced and aggregation is increased, which leads to larger particle sizes. Above the CMC, the micellar solubilization effect may enhance the particle growth by facilitating the Ostwald ripening. An example of this is a study with Pluronic F127 as a stabilizer for paclitaxel (PTX) nanocrystals (Deng et al., 2010). Below CMC, the stabilizer had high affinity on drug surfaces and stable nanocrystals were formed. However, with concentrations above the CMC, affinity was lost and the nanosuspensions became unstable. This was explained by the fact that above CMC, the monomers had a higher affinity toward micelle formation than drug surfaces, hence lowering the nanocrystal stability. Evidently, the concentration of the surfactant required for better stabilization also depends on the molecular structure. Lower molar concentration is required for nonionic surfactants that have a longer hydrophobic chain and a bigger hydrophilic head group, which provides effective steric hindrance and thus reduce the aggregation tendency (Sinha et al., 2013).

Polymers help to stabilize the particles essentially in the same way as surfactants, that is, adsorption at the solid–liquid interface and minimization of the surface tension, resulting in an increased rate of nucleation. Adsorption of polymers provides steric hindrance. Some ionic polymers (e.g., Polymer JR 400) are able to facilitate both steric and electrostatic stabilizations. Use of a polymer above its critical flocculation concentration (CFC) increases the particle size, essentially in the same way as CMC in the case of surfactants. Also, it has been observed that polymers with higher molecular weights provide better steric stabilization. The viscosity of a polymer also plays a crucial role in controlling the particle size. Highly viscous polymers, for example, hydroxyl propyl methyl cellulose (HPMC), reduce the mobility of the nuclei/particles, thus reducing both the appearance time and the collision frequency of nuclei, and hence reduce the final particle size. The stabilizer selection may also be based on their surface tension, since the stabilizers with lower surface tensions have been detected as more efficient compared to the ones with higher surface tension (Sinha et al., 2013). Finally, the placement of a stabilizer in an organic phase has been reported to reduce the required amount of stabilizer, resulting in an increased nucleation rate. The drug-to-stabilizer ratio in the formulation may vary from 1:20 to 20:1 and should be investigated for a specific case (Liversidge et al., 1992).

Table 5.2 lists some examples of drug–stabilizer combinations.

TABLE 5.2   Overview of Stabilizing Excipients Used in Nanosuspension Formulation.

Stabilizer	API	References
Multiple stabilizer combinations		
Surfactant combinations: lecithin (20%)–sodium cholic acid (16.7%)	Prednisolone	Muller et al., 1998
Lecithin (20%)–tyloxapol (20%)	Budesonide	Jacobs and Muller, 2002
Poloxamer 188 (100%)–lecithin (50%)	Buparvaquone	Muller et al., 2002
Tween 80 (20%)–lecithin (10%)	Azithromycin	Zhang et al., 2007
HPC–sodium lauryl sulfate (1.6%)	Cilostazol	Liversidge et al., 1992
PVA (100%)–poloxamer 188 (200%)	Budesonide	Muller et al., 2002
Single stabilizer systems		
Lecithin	RMKP 22	Peters et al., 1999
Poloxamer 407	Itraconazole	Mouton et al., 2006
Tween 80	Albendazole	Kumar and Rao, 2008
Tyloxapol	Budesonide	Jacobs and Muller , 2002
Polymers		
Acacia gum	UCB-35440-3	Hecq et al., 2006
Polyvinyl alcohol (30–70 kDa)	Beclomethasone	Muller et al., 1995
Povidone K1	Danazol	Liversidge et al., 1992
HPMC, PVP, lecithin, pluronic F68	Deacetymyco-epoxydiene	Wang et al., 2011
PVP, HPMC	Ibuprofen	Plakkot et al., 2011
Vitamin E TPGS	NVS-102	Ghosh and Schenck, 2012
Pluronic F127, HPMC	Nimodipine	Fu et al., 2013
HPMC, Tween 80	Naproxen	Kumar and Burgess, 2014

| Tween 20, HPMC, Vitamin E TPGS | Cinnarizine, naproxen | Kayaert and Anné, 2011 |
| HPMC, PVP, HPC, Vitamin E TPGS | Naproxen | Ghosh et al., 2013 |

HPMC, hydroxyl propyl methyl cellulose; TPGS, tocopheryl polyethylene glycol succinate.

Thus, surfactants and/or stabilizers are often used to enhance the wettability of the nanosuspension and thereby improve dispersion and dissolution rate of the nanocrystals (Dolenc et al., 2009; Van Eerdenbrugh et al., 2008a, 2008b). Moreover, the quantity of these surfactants has to be screened carefully to avoid solubilization of the API during water removal.

5.3.3 ORGANIC SOLVENTS

The parameters to be considered when selecting an organic solvent for making nanoparticles include physicochemical properties of the solvents and their ability to dissolve the API.

As these techniques are still in their infancy, elaborate information on formulation considerations is not available. The acceptability of organic solvents in the pharmaceutical arena, their toxicity potential, and their ease of removal from the formulation need to be considered when formulating nanosuspensions. The pharmaceutically acceptable and less hazardous water-miscible solvents, such as ethanol and isopropanol, and partially water-miscible solvents, such as ethyl acetate, ethyl formate, butyl lactate, triacetin, propylene carbonate, and benzyl alcohol, are preferred in the formulation over the conventional hazardous solvents such as dichloromethane. Furthermore, partially water miscible organic solvents can be used as the internal phase of the microemulsion when the nanosuspensions are to be produced using a microemulsion as a template (Gassmann et al., 1994).

5.3.4 COSURFACTANTS

The choice of cosurfactant is critical when using microemulsion templates to formulate nanosuspensions. Since cosurfactants can drastically impact phase behavior, the effect of cosurfactant on uptake of the internal phase for selected microemulsion composition and on drug loading should be investigated. Although the literature describes the use of bile salts and dipotassium glycyrrhizinate as cosurfactants, various solubilizers, such as Transcutol, glycofurol, ethanol, and isopropanol can be safely used as cosurfactants in the formulation of microemulsions (Muller et al., 1995).

5.3.5 OTHER ADDITIVES

Poorly water-soluble molecules are usually formulated using excipients that aid in improving dissolution rate and storage stability. Excipients increase drug dissolution rate by increasing active drug surface area in contact with the dissolution medium (Sanganwar et al., 2010). Examples of excipients include:

1. cosolvents such as PEG-400 (Zhang et al., 2007),
2. wetting agents such as sorbitan ester derivatives (Hiemenz and Rajagopalan, 1997),
3. disintegrant such as croscarmellose sodium (Peters and Muller, 1999),
4. cyclodextrins such as β-cyclodextrins (Kayser, 2000), and
5. micelles and lipid-based systems such as emulsions (Trotta et al., 2003), microemulsions, liposomes, solid lipid nanoparticles, and solid dispersions.

Nanosuspensions may also contain additives such as buffers, salts, polyols, osmogent, and cryoprotectant, depending on either the route of administration or the properties of the drug moiety (Trotta et al., 2001).

5.4 METHODS OF PREPARATION

Various techniques that have been developed for the development of nanosuspensions can be broadly categorized as (1) top-down methods, (2) bottom-up methods, and (3) methods employing combinatorial approaches (Gao et al., 2014; Möschwitzer, 2013; Scholz and Keck, 2015; Wais et al., 2016). The top-down methods involve transformation of micron-size or larger drug particles into nanoscale drug particles via different particle reduction technologies and most of the commercially available nanosuspensions-based products utilize the top-down methods for generating nanosuspensions. Bottom-up methods involve dissolution of the drug in a suitable solvent or supercritical fluids, followed by controlled precipitation in the presence of stabilizers to yield stabilized drug nanosuspensions.

5.4.1 TOP-DOWN METHODS

Depending upon the principle and mechanism behind particle size reduction, top-down methods can be further divided into (1) wet media milling and (2) high-pressure homogenization (HPH). Both the top-down methodologies usually result in significant heat generation during the manufacturing of nanosuspensions which can be detrimental to heat-sensitive drugs (Scholz and Keck, 2015; Wais et al., 2016). However, technological advances have been made to develop nanosuspensions by top-down methods without causing any heat generation during the manufacturing process.

5.4.1.1 WET MEDIA MILLING TECHNOLOGY

Wet media milling technology has been adapted from media milling methodologies used in paint, photographic, and magnetic industries. The patented process of generating drug nanosuspensions using wet media milling method is termed as NanoCrystal® technology and is currently owned by Elan Drug Technologies (Möschwitzer, 2013). The media mill used for the generation of drug nanosuspensions consists of milling chamber, milling shaft, and recirculation chamber as depicted in Figure 5.3. The process of wet media milling involves introduction of poorly water-soluble drug suspended in a stabilizer solution into the milling chamber containing suitable milling media and the mixture is agitated for an appropriate period of time to generate drug nanosuspensions (Merisko-Liversidge and Liversidge, 2011; Möschwitzer, 2013; Patravale et al., 2004; Wais et al., 2016). The high energy and shear forces generated by the collision among milling media, drug particles, and milling chamber wall provide the necessary force to disintegrate large drug particles into drug nanosuspensions. Due to the simplicity of the wet media milling technology, it is possible to develop drug nanosuspensions at laboratory scale as well as large scale (500–1000 kg of drug nanosuspension per batch) (Merisko-Liversidge and Liversidge, 2011).

Wet media milling technology generally yields smaller and fairly monodisperse drug nanosuspensions without the need to convert large drug particles into smaller particles using micronization or precipitation. The processing time can vary from 30 min to several hours depending upon the physicochemical properties and the concentration of the drug, type of the milling media, and desired particle size of nanosuspensions (Merisko-Liversidge and Liversidge, 2011; Möschwitzer, 2013; Patravale et al., 2004; Wais et al., 2016). The milling media used for generation of nanosuspensions can be made from materials such as highly cross-linked polystyrene media, ultradense ceramic, yttrium-stabilized zirconia, zirconium oxide, and glass beads. The material and size of the milling media can have impact on the processing time and particle size of the nanosuspensions (Merisko-Liversidge and Liversidge, 2011; Möschwitzer, 2013; Patravale et al., 2004; Scholz and Keck, 2015; Wais et al., 2016).

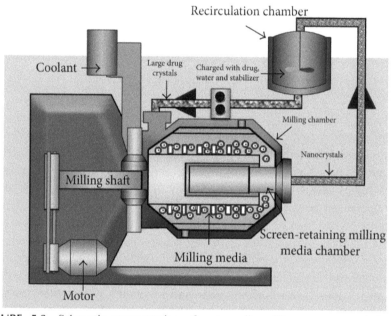

FIGURE 5.3 Schematic representation of wet media milling process used for the development of nanosuspensions. (Reprinted from Sutradhar, K. B.; Khatun, S.; Luna, I. P. Increasing Possibilities of Nanosuspension. *J. Nanotechnol.* **2013**).

Product contamination due to potential erosion of the milling media material has been one of the main concerns associated with wet media milling technology. However, highly cross-linked polystyrene beads (PolyMill milling media) have been demonstrated to result in very minimal trace impurities in the final nanosuspension after optimization of media milling process parameters (Merisko-Liversidge and Liversidge, 2011; Möschwitzer, 2013; Patravale et al., 2004; Scholz and Keck, 2015; Wais et al., 2016). In fact, the products developed using PolyMill milling media can be successfully used for development of parenteral nanosuspensions. To date, wet media milling (NanoCrystal®) technology has been successfully used for manufacturing of several nanosuspension-based pharmaceutical products in the United States, Canada, European Union (EU), and Japan (Merisko-Liversidge and Liversidge, 2011).

5.4.1.2   HPH TECHNOLOGY

Depending on the operating method used to achieve homogenization, HPH technology can be categorized into microfluidization (jet-stream homogenization) and piston-gap homogenization. Unlike wet media milling approach, HPH methods typically require conversion of large drug particles into micron-sized particles.

5.4.1.2.1   Microfluidization method

Microfluidization method involves use of particle collision and shear forces to disintegrate drug particles into nanosuspensions. The process of generating nanosuspensions using this method is also called as insoluble drug-delivery particles (IDD-P™) technology and it is currently owned by Skyepharma PLC, Canada (Junghanns and Müller, 2008; Sinha, 2015; Wais et al., 2016). Micron-sized drug particles suspended in a stabilizer solution are transported through a designed interaction chamber at a very high speed. The geometry of the interaction chamber design dictates the process of particle disintegration. In the "Z"-shaped chamber, drug suspension under high velocity suddenly undergoes changes in the flow direction causing collision between particles and chamber wall leading, whereas in the "Y"-shaped chamber, two high-velocity streams of drug suspension collide on each other causing particle disintegration (Sinha, 2015; Wais et al., 2016). The parameters that affect the process of nanosuspensions development are drug density, drug concentration, and number of microfluidization cycles. IDDP technology has been used for the development of commercially available fenofibrate nanosuspensions (Triglide®) (Junghanns and Müller, 2008).

5.4.1.2.2   Piston-gap Homogenization Method

Nanosuspensions developed using piston-gap homogenization method are termed as either DissoCubes or NanoPure technology (PharmaSol GmBH, Germany), depending upon the vehicle used for their fabrication. DissoCube technology uses aqueous vehicles for the fabrication of nanosuspensions, whereas NanoPure technology uses nonaqueous vehicles (PEG or oils) for the generation of nanosuspensions. The schematics of the piston-gap homogenization process are shown in the Figure 5.4.

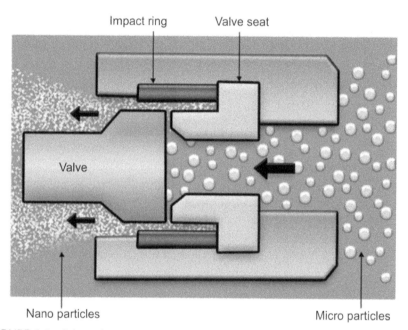

FIGURE 5.4   Schematic representation of piston-gap homogenization process. (Reprinted from Sutradhar, K.; Amin, M. Nanoemulsions: Increasing Possibilities in Drug Delivery. *Eur. J. Nanomed.* **2013,** *5,* 97–110. https://www.degruyter.com/downloadpdf/j/ejnm.2013.5.issue-2/ejnm-2013-0001/ejnm-2013-0001.pdf)

As demonstrated in Figure 5.4, micronized drug particles suspended in a stabilizer solution are forced to pass through a very narrow homogenization gap at extremely high velocity in the homogenizer. As per Bernoulli's equation, the dramatic increase in the velocity causes corresponding decrease in the static pressure below the boiling point of the vehicle at the room temperature. This leads to generation of gas bubbles which implode when the drug suspension leaves the homogenization gap (cavitation). The cavitation phenomenon, particle collision, shear forces, and turbulent flow generated during the homogenization process result in the disintegration of drug particles leading to formation of nanosuspensions (Junghanns and Müller, 2008; Möschwitzer, 2013; Patravale et al., 2004; Scholz and Keck, 2015; Wais et al., 2016). As nonaqueous vehicles have low vapor pressure, in case of nanosuspensions generated using NanoPure technology, cavitation phenomenon does not contribute to the particle size reduction (Junghanns and Müller, 2008; Möschwitzer, 2013; Scholz and Keck, 2015). Piston-gap homogenizers can be operated at homogenization pressure between 100 and 2000 bars and depending upon the instrument, they can homogenize as small as 3 mL to as high as thousand liters of drug suspension. HPH can be carried out in continuous or discontinuous mode. Various studies have shown that parameters such as homogenization pressure, number of homogenization cycles, drug concentration, and stabilizer concentration affect the process of nanosuspensions development (Junghanns and Müller, 2008; Möschwitzer, 2013; Patravale et al., 2004; Scholz and Keck, 2015; Wais et al., 2016).

5.4.2   BOTTOM-UP TECHNIQUES

As mentioned earlier, bottom-up technologies involve solubilization of poorly water-soluble molecules in a suitable solvent followed by controlled precipitation/crystallization of drug in the presence of an appropriate stabilizer to yield nanosuspensions. By optimizing the conditions for drug precipitation/crystallization, it is possible to generate nanosuspensions varying in particle size depending upon the end application. Based on the techniques used for solubilization and/or crystallization of drugs, the bottom-up methods can be further classified into (1) solvent–antisolvent methods, (2) supercritical fluid-based methods, and (3) cryogenic methods (Wais et al., 2016).

5.4.2.1   SOLVENT–ANTISOLVENT METHODS

Solvent–antisolvent methods have been used for many decades for development of a variety of nanocarriers including nanosuspensions. List and Sucker (1988) first reported the preparation of nanosuspensions using solvent-antisolvent or "hydrosol" method and the intellectual property was owned by Novartis (previously Sandoz) (Chan and Kwok, 2011; Scholz and Keck, 2015). Solvent–antisolvent methods utilize organic solvents (belonging to Class III or IV industrial organic solvents) to dissolve the drug, and the solvent is rapidly added to the antisolvent (typically water)-containing stabilizer to generate nanoscale drug particles. The formation of nanoscale particles is governed by Marangoni effect and parameters such as interfacial turbulence, flow diffusion, surface tension, solvent to antisolvent ratio, dielectric constant of solvent and antisolvent, speed of mixing, stabilizer type and concentration, and temperature have a significant impact on the physicochemical properties of nanosuspensions (Wais et al., 2016). Although hydrosol method is very simple, controlled nucleation and crystal growth of drug is difficult to achieve. Hence, hydrosol method often results in polydisperse and usually metastable amorphous particles that rapidly transform into larger crystalline particles.

Over the years, efforts have been made to properly control the rate and kinetics of drug nucleation and/or to stabilize amorphous drug nanoparticles. Researchers from BASF, Germany developed solvent–antisolvent method utilizing organic solvent and digestible oils to generate stable amorphous nanoparticles of nutraceuticals such as carotenoids (Junghanns and Müller, 2008; Möschwitzer, 2013; Scholz and Keck, 2015). This technology was later acquired by Soliqs/Abbott and it is available under the trade name of Nanomorph®. Researchers have also been working on approaches such as sonoprecipitation (use of ultrasonic energy), flash nanoprecipitation (use of impinging jet mixers or multi-inlet vortex mixers), and high-gravity reactive precipitation (use of rotating packed bed) to obtain nanosuspensions with desired particle size, stability, and narrow size distribution (Chan and Kwok, 2011; Patravale et al., 2004; Du et al., 2015; Sinha et al., 2013; Wais et al., 2016). These approaches have been developed to improve micromixing between solvent and antisolvent that yields rapid drug nucleation and drug supersaturation with much greater degree of control over formation of drug nanoparticles.

Solvent–antisolvent methods are not useful for the drugs that exhibit poor solubility in water, oils, and organic solvents. It is also important to note that the organic solvents used for solvent–antisolvent methods should be present in amounts less than toxicologically acceptable levels in the end product.

5.4.2.2 SUPERCRITICAL FLUID-BASED METHODS

Supercritical fluid-based methodologies have shown a great utility in engineering of particulate carriers such as polymeric microparticles. In last 15 years, applications of supercritical fluid-based methodologies have also been established for development of nanosuspensions. Unlike solvent–anti-solvent methods, supercritical fluid-based methods usually do not result in any residual solvent in the end product. In addition, high diffusivity, low viscosity, low surface tension, and higher solubilization power of supercritical fluids are also advantageous for particle engineering (Chan and Kwok, 2011; Patravale et al., 2004; Girotra et al., 2013; Wais et al., 2016). Furthermore, supercritical fluids can be easily removed from the end product by simple depressurization. Supercritical CO_2 is most commonly used for particle engineering due to nontoxic, inert, and nonflammable nature of CO_2. Supercritical fluid-based methodologies can be further classified as rapid expansion of supercritical solution (RESS), rapid expansion of supercritical solutions into a liquid solvent (RESOLV), and supercritical antisolvent precipitation (SAS).

In the RESS and RESOLV methods, poorly water-soluble drug is dissolved in the supercritical fluid. In the RESS method, the supercritical fluid containing dissolved drug passes through a small nozzle at supersonic speed and undergoes decompression. The decompression process leads to rapid expansion of supercritical fluid which is followed by high supersaturation of the drug in the fluid droplets; finally resulting in generation of uniform supercritical fluid nucleation of the drug. The physicochemical properties of the generated drug particles depend on the drug properties and concentration, pre- and post-expansion temperature and pressure, nozzle design, drug–supercritical fluid interaction, and the solubility of drug in the supercritical fluid. Studies have shown that RESS method yields nanoscale drug particles occasionally and micron-sized particles are obtained in many cases (Patravale et al., 2004; Wais et al., 2016). To overcome this issue, RESOLV process was developed. In the RESOLV method, supercritical fluid containing solubilized drug is passed through a nozzle and expanded into a chamber containing liquid to prevent particle aggregation occurring in the expansion process. The liquid that can act as antisolvent for the drug is selected and it may also contain suitable stabilizers to generate nanosuspensions with desired physicochemical characteristics (Chan and Kwok, 2011; Patravale et al., 2004; Girotra et al., 2013; Wais et al., 2016).

SAS method has been adapted for the drugs that do not show good solubility in the supercritical CO_2. In the SAS method, supercritical fluid acts as an antisolvent for the drug. Typically, drug of interest is dissolved into the organic solvent that has good miscibility with the supercritical fluid. The organic solvent containing dissolved drug is sprayed into an excess flowing continuum of supercritical fluid which leads to bidirectional interfacial diffusion and mixing between drug solution and supercritical antisolvent, thus resulting in rapid supersaturation and precipitation of drug into nanoscale particles (Chan and Kwok, 2011; Patravale et al., 2004; Girotra et al., 2013; Wais et al., 2016).

5.4.2.3   CRYOGENIC METHODS

Cryogenic methods such as freeze drying and spray freeze drying are routinely used for transformation of nanosuspensions (developed using top-down methods) into solid drug nanoparticles (Van Eerdenbrugh et al., 2008a, 2008b; Wais et al., 2016). However, recently, emulsion freeze drying method has been developed to directly develop solid drug nanoparticles. Emulsion freeze drying involves solubilization of drug into suitable organic solvents such as dichloromethane, chloroform, cyclohexane, and o-xylene. The organic solvent is emulsified into aqueous solution containing suitable stabilizer. The nanoscale emulsion droplets are rapidly frozen using liquid nitrogen to yield a frozen mass. The frozen emulsion is lyophilized using suitable equipment to remove water as well as organic solvent, resulting in formation of solid drug nanoparticles in the stabilizer matrix. The emulsion freeze drying process typically results in amorphous and porous particles. The physicochemical properties of the solid drug nanoparticles depend on parameters such as emulsion droplet size, type and concentration of stabilizer, dielectric constant of organic phase, phase–volume ratio, drug concentration, freezing rate, and freeze drying parameters (McDonald et al., 2014; Wais et al., 2016). Cryogenic methods can be divided into spray freeze drying and emulsion freeze drying. To date, only few solvents have been explored for the emulsion freeze drying process and drugs that are not soluble in these solvents cannot be utilized for emulsion freeze drying method. Furthermore, given the fact that conventional freeze dryers are not equipped to remove organic solvents, suitable modifications in the instrument would be required while scaling-up this method from lab to commercial scale.

5.4.3   COMBINATION METHODS

Combination methods have been mainly utilized to overcome disadvantages associated with top-down processes such as long milling time or multiple homogenization cycles. In addition, optimally designed combination methods can also yield smaller size nanosuspensions. The first combination method, Nanoedge® was developed by Baxter and it uses a combination of microprecipitation (solvent–antisolvent method) and HPH to generate nano-suspensions (Junghanns and Müller, 2008; Möschwitzer, 2013; Rabinow, 2004; Scholz and Keck, 2015). The microprecipitation step induces defects in the drug crystals and reduces the particle size. However, this process also leads to generation of amorphous particles. Subsequent HPH of micropre-cipitated drug further reduces particle size of the drug and also transforms metastable amorphous drug particles into stable crystalline nanosuspensions (Rabinow, 2004) as shown in Figure 5.5.

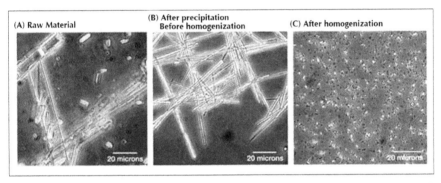

FIGURE 5.5 Nanosuspensions engineered using Nanoedge technology; (A) drug crystals are too large and hard to run through a homogenizer, (B) after microprecipitation, relatively uniform needle shape drug crystals are obtained that are amenable to homogenization, and (C) homogenization of easily fracturable crystals yields nanosuspensions. (Reprinted with permission from Rabinow, B. E. Nanosuspensions in Drug Delivery. *Nat. Rev. Drug. Discov.* **2004,** *3,* 785–796. © 2004 Nature Publishing Group)

After homogenization, organic solvent is removed from the nanosus-pension to yield final product. The other combination methods involve use of spray drying–HPH (H42 technology) (Scholz and Keck, 2015), freeze drying–HPH (H96 technology) (Scholz and Keck, 2015), high-speed stir-ring–HPH (ARTcrystal® technology), and wet media milling–HPH (smart-Crystal® technology) (Al Shaal et al., 2010; Romero et al., 2015).

5.5 CHARACTERIZATION OF NANOSUSPENSIONS

For the successful fabrication of a nanocrystal formulation, determination of suitable parameters for characterization is as important as selection of the appropriate excipients, to ensure that the unique properties respon-sible for the performance of nanocrystals are within the desired limits. The forthcoming sections discuss in detail the various characterization tests for the evaluation of nanocrystals.

5.5.1 PARTICLE SIZE AND POLYDISPERSITY INDEX

The mean particle size and width of particle size distribution (polydispersity index; PI) are the most important parameters for a nanosuspension, since they are directly linked with the physicochemical properties such as saturation solubility, dissolution velocity, physical stability, and biological performance. The physical stability of nanosuspensions is related to the PI and should be as low as possible for long-term stability of nanosuspensions. The PI value of 0.1–0.25 shows a fairly narrow size distribution, while a PI value more than 0.5 indicates a very broad distribution. The most commonly used techniques for particle size measurements of nanosized systems are dynamic light-scattering techniques, static light-scattering techniques, and microscopy. The mean particle size of nanosuspensions is ideally analyzed by dynamic light scattering also known as photon correlation spectroscopy (Keck and Muller, 2005). However, the measurement of particle size above 6 µm by using this technique is not feasible. Hence, optical microscopy and low-angle static light scattering (laser light diffraction) are used for the detection of larger particles, especially for the nanosuspensions that are meant for parenteral and pulmonary delivery. The advantage of light microscopy is the visible nature of analysis which therefore yields a doubt-free result; however, a major drawback is lack of any statistical significance because it is not possible or very time consuming to analyze 10,000 particles or more, necessary for a valid analysis. Laser diffractometry (LD) is a robust technique and has the advantage over all the other techniques to be able to analyze large particles, small nanoparticles, and mixtures of small and large particles within only a single measurement (Bott and Hart, 1990; Xu, 2005). A Coulter counter analysis is essential for nanosuspensions to be administrated intravenously. In contrast to the volume distribution of the LD analysis, the Coulter counter data give an absolute value that is the absolute number of particles per volume unit for the different size classes (Gao et al., 2008).

Apart from these, the other techniques used for the particle size analysis include confocal laser scanning microscopy, scanning probe microscopy, and scanning tunneling microscopy.

5.5.2 SURFACE CHARGE

The particle surface charge is ideally quantified in terms of the "zeta potential," which is measured via the electrophoretic mobility of the particles in an electric field. Zeta potential determines the physical stability of nanosuspension and can be used to predict long-term stability. According to the literature, a zeta potential of at least −30 mV for electrostatic and −20 mV for sterically stabilized systems is desired to obtain a physically stable nanocrystal suspension. The upper limit for the zeta potential is +30 mV. The zeta potential is determined by measuring the electrophoretic particle velocity in an electrical field. During the particle movement, the diffuse layer is shed off, hence the particle acquires a charge due to the loss of the counter ions in the diffuse layer. This charge at the border of the shedding is termed as the zeta potential (Li, 2006).

5.5.3 SOLID STATE PROPERTIES

The solid state properties (polymorphic crystal form, solvate/hydrate form, degree of crystallinity) influence the apparent solubility and thereby the dissolution rate. Ideally, the thermodynamically most stable crystalline form is desirable to prevent the peril of solid state transformations during production, storage, and/or administration. To increase the dissolution velocity and bioavailability of nanocrystals, it is more amenable to formulate the nanocrystals in a metastable crystalline form or even prepare the amorphous equivalent of nanocrystals. However, this is not being practiced commonly.

Different nanocrystal manufacturing conditions and procedures can have an impact on the ensuing solid state form. Furthermore, the environmental conditions affect the thermodynamically stable polymorphic form. For example, hydrate forms are generally more stable (and therefore less soluble) in aqueous media. Therefore, if the drug is susceptible to hydrate formation, then the potential or factors triggering the said conversion should be extensively investigated during stability studies in different conditions (Chogale et al., 2016). X-ray powder diffraction, thermal analytical techniques [differential scanning calorimetry (DSC), thermogravimetry, etc.)], and vibrational spectroscopy (infrared and Raman) are the most commonly used methods to determine and monitor the solid state form of nanocrystals.

X-ray diffraction (XRD) studies are usually performed for the confirmation of drug crystallinity following its conversion to a nanocrystal formulation. Every crystalline substance gives a specific pattern; the same substance always yields the same pattern; and in a mixture of substances each displays its pattern irrespective of the others. The XRD pattern of a substance therefore represents the unique fingerprint of the substance. XRD has been used to compare the impact of nanocrystal preparation process on the crystalline nature of the drug. Thermal techniques such as DSC and thermogravimetry and FTIR have also been employed to study modifications in the nature of the API during nanocrystal processing. Raman spectroscopy has been further used as an in-line quality control tool to study the impact of the various processing steps on the nature of nanocrystals and to identify the precise step which ascertains the properties of the final formulation (Chogale et al., 2016).

5.5.4 PARTICLE MORPHOLOGY

Ideally, the shape or morphology of the nanocrystals can be determined using a transmission electron microscope (TEM) and/or a scanning electron microscope (SEM). A wet sample of suitable concentration is needed for the TEM analysis. When the formulated nanosuspensions are to be converted into a dried powder (e.g., by spray drying or lyophilization), a SEM analysis is crucial to monitor alterations in the particle shape and size before and following the process of the water removal. Principally, agglomeration may be observed following water removal, leading to an increase in the particle size. Such changes and more can be observed through SEM. To play down the magnitude of the increase of particle size, some excipients are included as "protectants" during drying operation. Mannitol is usually used as a cryoprotectant in lyophilization, which recrystallizes around the nanocrystals during the water-removal operation, thus preventing particle interaction and agglomeration (Gao et al., 2008). Agglomeration within a certain limit is allowable when the final particle size is still in an acceptable range. Moreover, the dried powder should be easily redispersed into stable nanosuspensions. The shape of the drug crystals depends on its crystalline structure. Particle shape is of prime importance when the nanocrystals are to be formulated as dry powder inhalers (DPIs) for direct lung delivery of the drugs. Aerodynamic diameter is a critical parameter that determines the lung deposition of the DPIs. Different particle shapes yield different drag forces and particle terminal-settling velocities, which in turn influence the lung deposition of the DPIs. Enhancing particle surface roughness would ascribe to decreased aerodynamic diameter of the particles. This would in turn have a higher possibility of deep lung deposition as compared to spherical particles (Chogale et al., 2016). In another study, elongated particles were found to have better aerodynamic behavior as compared to spherical ones. Particle aerosolization is another factor responsible for deep lung delivery. Aerosolization is mainly influenced by the interparticle interaction and interaction between the wall of the inhaler and the particles. All types of particle interactions are linked to van der Waals forces which are the particle surface morphology, size, shape, electrostatic properties, and hygroscopicity. Particle shape characterized by low contact area and van der Waals force have a lower tendency to aggregate and hence can be readily dispersed in the air. Elongated particles are not ideal for aerosolization owing to their large attractive forces. Besides SEM and TEM, atomic force microscopy and surface plasmon resonance (SPR) have been used in specific cases to determine the various dimensions of particle shape.

5.5.5 SATURATION SOLUBILITY AND DISSOLUTION VELOCITY

The determination of the saturation solubility and dissolution velocity is a very critical step as these parameters together help foresee any alteration in the *in vivo* performance (blood profiles, plasma peaks, and bioavailability) of the drug. Since nanosuspensions are known to enhance the saturation solubility of the drug, determination of this parameter remains an important investigational tool. The saturation solubility of the drug in various physiological buffers as well as at different temperatures should be assessed using methods described in the pharmacopoeias. The determination of the dissolution velocity of nanosuspensions depicts their merits over conventional formulations, especially when designing the nanoparticulate-based sustained release dosage forms based on nanoparticulate drugs (Patravale et al., 2004).

5.5.6 IN VIVO PERFORMANCE

The establishment of an *in vitro/in vivo* correlation and the monitoring of the *in vivo* performance of the drug is absolutely essential irrespective of the route of administration or the final dosage form. It is of utmost importance in the case of intravenously injected nanosuspensions since the *in vivo* behavior of the drug is related to the organ distribution, which is a function of its surface properties such as surface hydrophobicity and interactions with plasma proteins (Blunk et al., 1993, 1996; Luck et al., 1997). In fact, the qualitative and quantitative composition of the protein absorption pattern observed after the intravenous injection of nanoparticles is recognized as the essential factor for organ distribution. Hence, suitable techniques have to be used in order to evaluate the surface properties and protein interactions to get an idea of *in vivo* behavior. Techniques such as hydrophobic interaction chromatography can be used to determine surface hydrophobicity, whereas 2D polyacrylamide gel electrophoresis (2D PAGE) can be employed for the quantitative and qualitative estimation of protein adsorption following intravenous injection of drug nanosuspensions in animals (Wallis and Muller, 1993).

5.6   APPLICATIONS OF NANOSUSPENSIONS

5.6.1   *NANOSUSPENSIONS FOR PARENTERAL DELIVERY*

Injectables are an integral part of pharmaceutical formulations. Despite its invasive nature, parenteral route is still preferred when rapid delivery to the target site and rapid onset of action is required. The parenteral route can also enable delivery of drugs that are not absorbed through gastrointestinal tract or undergo extensive first-pass metabolism. Finally, parenteral route is also used for administration of long-acting depot formulations and vaccines. The development of parenteral formulations of poorly water-soluble drugs represents a great challenge to formulation scientists. Conventional approaches for developing parenteral formulation of poorly water-soluble drug include use of extreme pH conditions, high amounts of cosolvents, surfactants, or macrocyclic solubilizers (Chin et al., 2014; Merisko-Liversidge and Liversidge, 2011; Patravale et al., 2004). In case of lipid-soluble drugs, oily depots or parenteral emulsions can be developed depending upon end application. However, these approaches are often associated with problems such as anaphylactic reaction, solvent-related toxicity issues, pain on injection, and precipitation of drug at the injection site. Furthermore, drugs that are not soluble in water, oil, and cosolvents or drugs with higher therapeutic dose cannot be easily administered by parenteral route. Nanosuspensions can enable parenteral delivery of such drugs. Nanosuspensions do not use any cosolvents, toxic solvents, or ingredients and they can also achieve very high loading of drug (up to 400 mg/mL in certain cases) without affecting syringeability (Chin et al., 2014; Merisko-Liversidge and Liversidge, 2011; Patravale et al., 2004; Rabinow, 2004). Hence, nanosuspensions meet almost all criteria of an ideal parenteral drug delivery system. Nanosuspensions can also enable preclinical evaluation of poorly water-soluble molecules generated in the drug discovery pipeline. In this section, we discuss key examples that demonstrate potential of nanosuspensions for the parenteral delivery.

It is important to note that formulation of poorly water-soluble drug into nanosuspension may alter its pharmacokinetic parameters compared to the same drug formulated into solution or other conventional approaches. It is well known that particle size, shape, and surface properties of nanocarriers significantly influence their biodistribution and pharmacokinetics and nanosuspensions are no exception to this (Chin et al., 2014; Merisko-Liversidge and Liversidge, 2011; Patravale et al., 2004; Rabinow, 2004). However, as nanosuspensions contain pure drug, the solubility of poorly water-soluble drug in the blood pool also significantly affects the pharmacokinetics and biodistribution. If the drug is readily soluble in the blood, then nanosuspensions will have pharmacokinetic and biodistribution profile similar to that of solution formulations. This was nicely demonstrated in the case of flurbiprofen. Flurbiprofen injectable solution has high pH and it causes significant irritation to veins. Formulation of flurbiprofen into nanosuspensions overcame drawbacks of solution formulation. Interestingly, flurbiprofen nanosuspensions and flurbiprofen solution showed similar pharmacokinetic and biodistribution profile due to high solubility of flurbiprofen in the blood (Merisko-Liversidge and Liversidge, 2011; Pace et al., 1999).

Nanosuspensions of poorly water-soluble drugs that are not easily soluble in the blood pool show significantly altered pharmacokinetics and biodistribution. Like most of the nanocarriers, these nanosuspensions are rapidly cleared by macrophages of mononuclear phagocytic system (Kupffer cells in the liver and spleen). The phagocytosed drug nanosuspensions, depending upon the pH-solubility profile of the drug, will dissolve in the acidic pH of phagolysosome, slowly release the drug, or exit the phagolysosome. The nanosuspensions that undergo phagolysosomal escape will end up in cytoplasm and then into the extracellular space. Such nanosuspensions can show significantly lower C_{max} but significantly higher mean residence time (MRT). This altered pharmacokinetic and biodistribution profile can be very useful for the certain therapeutic agents (triazole antifungals) with high C_{max} associated toxicity and longer residence time-associated efficacy (Rabinow, 2004).

Currently available itraconazole solution formulation (Sporanox®; Janssen Pharmaceutica) contains 30% hydroxypropyl-β-cyclodextrin (HPBCD) to solubilize itraconazole. This formulation can be used for parenteral as well as oral delivery. While injectable itraconazole solution is a great option for the treatment of systemic candidiasis, the product label indicates that the formulation should be used with caution for patients with renal and/or heart conditions. These warnings are due to renal toxicity issues associated with the use of high amounts of HPBCD and due to negative inotropic issues associated with the use of itraconazole. In order to overcome these issues, itraconazole was reformulated into nanosuspensions. The pharmacokinetics and biodistribution of intravenously administered itraconazole nanosuspension and Sporanox® solution were studied in the rats (Rabinow et al., 2007). Interestingly, itraconazole nanosuspensions showed reduced C_{max} but significantly higher half-life and MRT compared to Sporanox®. This behavior was due to sequestration of the nanosuspensions by the organs of mononuclear phagocytic system, followed by slow release of the drug. Unlike Sporanox® solution, it was also possible to administer higher doses of itraconazole nanosuspensions without any signs of toxicity. Finally, itraconazole nanosuspensions showed significantly higher efficacy in immunocompromised rats, challenged with a lethal dose of either itraconazole-sensitive or itraconazole-resistant *Candida albicans*. McKee and coworkers evaluated cardiac toxicity potential of itraconazole nanosuspensions and Sporanox® in dogs (McKee et al., 2010). As expected, injection of Sporanox solution caused significant reduction in myocardial contractility (negative inotropic effect), whereas itraconazole nanosuspensions, at twofold higher dose and infusion rate, did not show any changes in myocardial contractility. Thus, itraconazole nanosuspensions appear to have clear advantage over Sporanox® solution with respect to tolerability as well as antifungal efficacy.

Like other nanocarriers, it is also possible to improve *in vivo* circulation of nanosuspensions by modifying their size, shape, surface properties, and plasma protein adsorption patterns. Use of PEG-containing (PEGylated) stabilizers such as PEGylated phospholipids or pluronics will help to achieve longer *in vivo* circulation. Gao and colleagues have shown that coating of PTX nanocrystals with PEGylated stabilizer Pluronic F68 led to significantly increased *in vivo* circulation and significantly reduced liver uptake compared to PTX nanocrystals without Pluronic F68 coating (Gao et al., 2016). Danhier and coworkers showed that Pluronic F108-coated nanosuspensions of MTKi-327 (kinase inhibitor with anticancer activity) have significantly higher *in vivo* circulation compared to sulfobutyl ether cyclodextrin-solubilized MTKi-327 (Danhier et al., 2014). Furthermore, MTKi-327 nanosuspensions showed significantly higher anticancer activity compared to MTKi-327 solution, MTKi-327-loaded long circulating polymeric nanoparticles, and polymeric micelles. This investigation clearly showed the potential of nanosuspensions for improved parenteral delivery. Shegokar et al. focused on development of nevirapine (anti-HIV drug) nanosuspensions with different surface coating leading to different protein adsorption patterns. The effect of coating nevirapine nanosuspensions with bovine serum albumin, dextran, and PEG 1000 was studied on *in vitro* protein adsorption and *in vivo* biodistribution including HIV reservoirs such as brain. Interestingly, the material used for coating of nanosuspensions had a significant impact on protein adsorption and *in vivo* biodistribution and bovine serum albumin-coated nanosuspensions were able to show significant accumulation in the brain compared to nevirapine nanosuspensions coated with other materials (Shegokar and Singh, 2011).

Nanosuspensions can also be developed using stabilizers such as D-α-tocopherol-PEG succinate (TPGS) that have inherent ability to inhibit P-glycoprotein (P-gp)-mediated efflux of drugs. These modifications could be useful for improving delivery of anticancer agents as most of the anticancer agents are P-gp substrates. Liu and coworkers developed nanosuspensions of PTX (a known P-gp substrate) coated with TPGS. *In vitro* and *in vivo* activity of TPGS-coated PTX nanosuspensions was tested against resistant human ovarian cancer cells (NCI/ADR-RES). Interestingly, TPGS-coated PTX nanosuspensions showed significantly higher *in vitro* and *in vivo* anticancer activity compared to commercially available PTX injectable formulation (Taxol). Similar *in vitro* and *in vivo* observations were noted in another investigation against lung cancer cells (Liu et al., 2010).

In recent years, a great emphasis has been given to develop long-acting, intramuscular injectable formulations based on nanosuspensions. As nanosuspensions can be formulated even at a very high concentration of drug (as high as 400 mg/mL) without affecting their syringeability, they can offer a better alternative to the conventional injectable depots. Paliperidone palmitate (atypical antipsychotic) is the first molecule that became commercially available as long-acting injectable nanosuspension. It is currently being marketed under trade name Invega Sustenna®. It is available as lipid dispersion in the prefilled syringes at a concentration of 156 mg/mL of paliperidone palmitate and can be injected into deltoid or gluteal muscle. Invega Sustenna® is able to offer at least 1 month long release of paliperidone palmitate and is recommended for patients suffering from compliance issues (Merisko-Liversidge and Liversidge, 2011). Interestingly, paliperidone palmitate nanosuspension can be stored up to 2 years without affecting its physical and chemical stability. This reiterates the advantages associated with formulating drugs as nanosuspensions.

Long-acting injectable nanosuspensions have also been evaluated for long-term preexposure prophylaxis from HIV infections (Landowitz et al., 2016). Currently, nanosuspensions of two anti-HIV drugs, rilpivirine (a poorly water- and oil-soluble non-nucleoside reverse transcriptase inhibitor) and cabotegravir (a poorly water-soluble HIV-1 integrase inhibitor), are actively being evaluated in preclinical and clinical trials. Both the drugs can be formulated into nanosuspensions at a concentration as high as 200 mg/mL. Intramuscular administration of rilpivirine and cabotegravir nanosuspensions have successfully shown ability to prevent repeated intravaginal HIV infection in humanized mice and/or macaques from 1 to 2 months depending upon the study (Baert et al., 2009; Kovarova et al., 2015; Radzio et al., 2015; van 't Klooster et al., 2010). This clearly indicated the ability of intramuscular long-acting injectable nanoformulations to maintain protective drug levels in the vaginal tissue. Recently, multiple-compartment pharmacokinetics of long-acting rilpivirine nanosuspension after single intramuscular dose were evaluated in six HIV-negative volunteers. Rilpivirine concentrations in the plasma and genital tissues peaked at 6–8 days and were detectable up to 84 days. After intramuscular dose of 1200 mg rilpivirine, cervicovaginal fluids showed considerable anti-HIV activity at day 28 and 56 indicating that long-acting rilpivirine injectable formulation could be useful for long-term HIV preexposure prophylaxis (Jackson et al., 2014).

5.6.2 NANOSUSPENSIONS FOR ORAL DELIVERY

Oral route is the most convenient route for drug administration due to simplicity and high degree of patient compliance. Oral medications face much lesser hurdles from regulatory agencies for their approval compared to products intended for parenteral or other routes of administration. The oral drug delivery segment of pharmaceutical market was reported to be $64.3 billion in 2013 and is expected to rise to $100 billion by 2018 (Date et al., 2016). However, oral delivery of poorly water-soluble drugs (belonging to BCS Class II and IV) is quite challenging. Studies have shown that poorly water-soluble drugs often exhibit low oral bioavailability due to dissolution rate-limited absorption and/or intraindividual (fed and fasted state) and/or interindividual variability. These issues can be efficiently resolved by formulating poorly water-soluble drugs into nanosuspensions. However; nanosuspensions are unlikely to improve the oral delivery of poorly water-soluble drugs whose oral bioavailability is low due to low permeability or rapid presystemic and/or first-pass metabolism (Merisko-Liversidge and

Liversidge, 2011; Patravale et al., 2004; Rabinow, 2004). To date, numerous literature reports have demonstrated the utility of nanosuspensions in oral delivery. In this section, we focus on discussing some key examples reported in the literature and on the commercially available oral products based on nanosuspensions.

One of the classic examples that established potential of oral nanosuspensions was reported more than 20 years ago. Liversidge and coworkers observed that oral bioavailability of commercial suspension of a gonadotropin inhibitor, danazol (average particle size ~10 µm) was about 5.2% (Liversidge et al., 1992). Interestingly, formulating danazol into nanosuspension (average particle size ~169 nm) led to a dramatic improvement in pharmacokinetic parameters [(more than 10-fold improvement in C_{max} and area under the curve (AUC)] and oral bioavailability (82.3% oral bioavailability). The investigation also demonstrated that danazol nanosuspensions were able to reduce intra- and interindividual variability in absorption of danazol (Liversidge et al., 1992). The reduction in fed and fasted state variation in oral bioavailability has also been nicely demonstrated in case of posaconazole. Compared to marketed posaconazole suspension (Noxafil®; average particle size ~5 µm), posaconazole nanosuspensions (average size ~150 nm) were significantly able to reduce fed to fasted state differences in oral bioavailability (Lentz et al., 2007; Merisko-Liversidge and Liversidge, 2011).

Nanosuspensions can also help to achieve rapid onset of action and/or reduction of local side effects in case of certain drugs such as naproxen. Nonsteroidal anti-inflammatory drugs (NSAIDs) including naproxen often lead to gastric irritation at least in part due to formation of insoluble crystals in the gastric fluids. Formulation of naproxen into nanosuspensions (particle size reduction from 20–30 µm crystals to 270 nm nanosuspensions) was able to reduce gastric irritation of naproxen due to rapid absorption, improved solubility in gastric fluids, reduced crystallinity, and gastric residence time. Furthermore, the naproxen nanosuspension also significantly reduced indicating that rapid onset of action can also be achieved with nanosuspensions (Liversidge and Conzentino, 1995).

Rapamune® is the first commercial product that contained drug nanosuspensions. The product was launched in the year 2000 and it is available in the form of tablets that contain 1, 2, or 5 mg rapamycin. Rapamycin (sirolimus) is a poorly water-soluble (~2 µg/mL) immunosuppressant, which is used for preventing organ rejection. Rapamycin was originally available as a lipid-based liquid solution. Rapamycin lipid solution did not have acceptable taste and patient-friendly dispensing protocol. Furthermore, the lipid solution required storage in the refrigerator. Formulation of rapamycin into <200 nm nanosuspensions (using NanoCrystals® technology) and their subsequent conversion into tablet proved to be advantageous. The Rapamune tablets (containing rapamycin nanosuspension) showed at least 25% higher bioavailability compared to rapamycin liquid suspension (Junghanns and Müller, 2008; Merisko-Liversidge and Liversidge, 2011; Möschwitzer, 2013). Furthermore, Rapamune tablets do not require cold storage and they offer a high degree of patient compliance.

In the year 2001, the second nanosuspensions-containing product, Emend® was made commercially available. Emend® is a spray-coated capsule formulation that contains aprepitant nanosuspensions. Aprepitant is a poorly water-soluble crystalline base (water solubility: 3–7 μg/mL) with ability to inhibit human substance P/neurokinin 1 receptors. It functions as an anti-emetic and prevents chemotherapy-induced nausea and vomiting. The conventional aprepitant formulations were associated with significant increase in oral bioavailability in the fed state. The effect was more pronounced for 300 mg dose compared to 100 mg dose of aprepitant. The reduction of aprepitant oral bioavailability variation between fed and fasted state was essential for the patient acceptability and commercial success of the product. Wua and colleagues compared pharmacokinetics and bioavailability of orally delivered aprepitant suspension (particle size ~5 μm) and aprepitant nanosuspensions (particle size ~120 nm) in dogs. The study was performed to predict the performance of aprepitant nanosuspensions in humans. Interestingly, aprepitant nanosuspensions showed around fourfold improvement in fasted-state oral bioavailability and completely abolished the variation in fed- and fasted-state variations in bioavailability (Junghanns and Müller, 2008; Merisko-Liversidge and Liversidge, 2011; Möschwitzer, 2013). Aprepitant nanosuspensions were converted into a solid formulation that is filled into capsule for human use. Emend® is approved in United States, Europe, and Japan and available in three strengths containing 40, 80, and 125 mg of aprepitant.

Fenofibrate is a poorly water-soluble drug used for the treatment of hypercholesterolemia or mixed dyslipidemia. The oral bioavailability of fenofibrate is unknown due to its extreme insolubility in vehicles suitable for injectables. However, studies have shown that relative oral bioavailability of fenofibrate is significantly increased in fed state (up to 35% increase) due to its high lipid solubility (Junghanns and Müller, 2008). Hence, to reduce the variation in fed- and fasted-state oral bioavailability, fenofibrate was formulated into nanosuspensions and later converted into tablets. Two tablet products, Tricor® and Triglide®, containing fenofibrate nanosuspensions, are commercially available. Tricor (48 or 145 mg of fenofibrate) tablets contain fenofibrate nanosuspensions formulated using NanoCrystal technology, whereas Triglide (50 or 160 mg of fenofibrate) tablets contain fenofibrate nanosuspensions manufactured using IDD-P technology. The data filed with FDA indicates that high-dose fenofibrate nanosuspensions-containing tablets (145 or 160 mg of fenofibrate) are bioequivalent to 200 mg micronized fenofibrate capsule. Furthermore, these products have significantly minimized variation in the oral bioavailability during fed and fasted states.

Our group has been working on development of oral nanoformulations of a variety of antimalarial drugs with the objective to achieve reduction in dose and improvement in efficacy. We have developed nanosuspensions of atovaquone, a poorly water-soluble antimalarial drug Studies have shown that atovaquone has extremely low solubility throughout the pH range in the gastrointestinal tract and it has highly variable and low oral bioavailability. Atovaquone is available as oral tablets (Malarone®; 250 mg atovaquone) for the treatment of malaria and as oral suspension (Mepron®; 150 mg/mL atovaquone) for the treatment of *Pneumocystis carinii* pneumonia. We developed atovaquone nanosuspensions using Nanoedge technology and converted them into solid powder using freeze drying. Our studies showed that atovaquone nanosuspensions showed three- to four-fold improvement in oral bioavailability in rats compared to marketed Malarone® tablets. More importantly, atovaquone nanosuspensions, at the one-fourth dose compared to Malarone® tablets, showed significantly higher antimalarial activity and survival in *Plasmodium berghei*-infected mice (Borhade et al., 2014). Thus, rationally designed nanosuspensions of antimalarial drugs could bring a paradigm shift in the treatment of malarial infections.

5.6.3 NANOSUSPENSIONS FOR PULMONARY DELIVERY

Pulmonary drug delivery represents a noninvasive route for local and systemic therapies. Moreover, drugs administered through the pulmonary route avoid the first-pass metabolism of the gastrointestinal tract, and may eliminate the barrier to patient compliance caused by injections. However, the fate of the medication administered by this route is highly variable depending on parameters such as the particle size of the aerosol, particle morphology and geometry, surface adhesive properties, lung properties of the patient, and mechanism and rate of elimination from the respiratory tract (Yang et al., 2008). Among the properties of the drug formulation, the aerodynamic diameter of the aerosol describes its aerodynamic behavior with consideration of size, density, and shape, and its optimization is of primary importance for a successful pulmonary delivery (Beck-Broichsitter et al., 2012).

In fact, different mechanisms govern the pulmonary deposition of inhaled particles depending on the aerodynamic diameter, and values between 1 and 5 μm are required to target the lower respiratory region. Formulations containing nanoparticulates have demonstrated many advantages in pulmonary route, among them, drug nanosuspensions show a great potential for pulmonary delivery of poorly soluble drugs (Yang et al., 2008; Zhang et al., 2011). Two kinds of pulmonary formulations containing drug nanocrystals have been reported. One is aqueous nanosuspensions packaged and administered by a nebulizer. Drug nanocrystals can be retained and conveyed into the lungs by the minute aerosol droplets generated by the nebulizer. The other one is inhaled powders containing dried nanocrystals dispersed in some inhalable carriers (e.g., inhalable lactose) (Kayser et al., 2003). These powders can be inhaled by commercial DPIs. Compared to the conventional microparticles for pulmonary use, nanoparticles afford a more homogenous distribution of drugs in the nebulized droplet or among the inhalable carriers.

When drug nanocrystals in aerosol droplets deposit in the lung, they can spread more evenly on the lung surface, especially when coated with stabilizers having a good spreadability. The drug nanocrystals then dissolve rapidly in the fluid lining the lung leading to a high concentration owing to their nano-range size. This is very advantageous for the localized treatment or prophylaxis of respiratory diseases. In humans, orally inhaled aerosols greater than 5–10 μm are mostly deposited in the oropharynx and incapable of reaching the lung, while the finest aerosols (<1 μm) are mostly exhaled without deposition (Heyder et al., 1986). Many studies have demonstrated that pulmonary delivery of nanosuspensions is associated with higher drug concentrations in lung tissue and markedly raise the lung to serum ratio of drugs compared with other administration route (Edwards and Dunbar, 2002; Labiris and Dolovich, 2003; Mansour et al., 2009; Patton and Byron, 2007). Furthermore, this is also conducive to a rapid onset of systemic effect since high drug concentrations in the alveoli would raise the driving force for permeation through the alveolar membrane, and result in higher C_{max} and earlier T_{max} (Gill et al., 2007; Kraft et al., 2004; Yang et al., 2008). When nanosuspensions contain particles in amorphous form characterized by higher inner energy, this effect is more obvious (Kraft et al., 2004; Yang et al., 2010).

Nanosuspensions can be successfully applied to overcome the problems associated with the use of microparticles for inhalation (oropharyngeal deposition) and pressured metered dose inhalers (use of banned CFCs) (Azarmi et al., 2008; Ostrander et al., 1999). The nebulization of nanosuspensions generates aerosol droplets of the preferred size loaded with a large amount of drug nanocrystals. Using nebulized nanosuspensions, the respirable fraction is distinctly increased in comparison to conventional metered dose inhalers (Ostrander et al., 1999). The smaller the particle size of the drug nanocrystals, the higher the drug loading of the aerosol droplets (Jacobs and Muller , 2002; Wiedmann et al., 1997). Therefore, the required nebulization time is distinctly reduced (Kraft et al., 2004). Beside this, drug nanocrystals show an increased mucoadhesive behavior, leading to a prolonged residence time at the mucosal surface of the lung. Nanosuspensions are suitable to formulate poorly soluble drugs without the extensive use of organic cosolvents, thus reducing their *in vivo* interference and potential toxicological effects. Moreover, the concentration of the active principle in the nanosuspension is not limited by solubility in the vehicle, thus permitting to reach a wider dose range.

Compared to dry powder formulations, nanosuspensions consisting of prewetted particles may improve dissolution and consequent lung absorption, though maintaining the solubility-dependent prolonged release from nanocrystals (Yang et al., 2008). Disposition in the lungs can be controlled via the size distribution of the generated aerosol droplets. Compared with microcrystals, the drug is more evenly distributed in the droplets when using a nanosuspension. The numbers of crystals are higher; consequently, the possibility that one or more drug crystals are present in each droplet is higher (Hernandez-Trejo et al., 2004). Delivery of water-insoluble drugs to the respiratory tract is very important for the local or systemic treatment of diseases. Many important drugs for pulmonary delivery show poor solubility both in water and nonaqueous media, for example, corticosteroids such as budesonide or beclomethasone dipropionate.

As reported, nanocrystal suspensions have the merit of being devoid of any matrix and polymeric carrier usually prevalent in other nanocarriers. This is of highest importance in pulmonary drug delivery since the deposition of nanoparticles on the surface of the lungs have shown a variable toxicological potential, dependant on matrix degradability and chemical composition (Gill et al., 2007).

Nanosuspensions of fluticasone and budesonide prepared using a wet milling technique, showed prolonged residence time in the lungs as compared to solutions *in vivo*, with a difference between the two drugs which can be ascribed to their different solubility (Yang et al., 2008). Another study eliminated the possibility of administering fluticasone coarse suspensions due to unacceptable aerodynamic properties of particles. Conversely, efficient delivery and dose-dependent lung deposition were achieved after administration of nanosuspension aerosol to mice *in vivo* (Chiang et al., 2009).

Zhang et al. compared the bioavailability of baicalein in rats after pulmonary administration versus intravenous injection. A baicalein nanosuspension was formulated using a combination of antisolvent recrystallization and HPH methods. It was administered to the lungs and compared to an intravenous injection of a baicalein solution. The different formulations showed no significant difference in their pharmacokinetic parameters, thus providing an example of potential systemic administration through pulmonary route (Zhang et al., 2011).

Tolerability of nanosuspensions after repeated dose should also be assessed, especially for those therapeutic agents that are required for long-term therapies. Recently, Rundfeldt et al. described a nanosuspension of an itraconazole as potential treatment for bronchopulmonary aspergillosis. The nanosuspension was produced using a wet milling technique, and the final formulation was screened for the content of inorganic contamination potentially derived from the grinding media, since inorganic contaminations are not acceptable for inhaled preparations. The pulmonary administration of the nanosuspension resulted in a prolonged pulmonary exposure to the drug, with itraconazole lung concentration 28-fold higher than the MIC of *Aspergillus fumigatus* at 24 h following the inhalation. Moreover, repeated dosages of the nanosuspension (once daily for 7 days) showed no histological or behavioral changes in rats (Rundfeldt et al., 2013). Due to the promising results in preclinical animal studies, the pharmacokinetics of nanosuspensions after pulmonary administration was explored on human volunteers.

In a trial on healthy adult volunteers reported by Kraft et al., budesonide nanosuspension was compared to a commercial regular suspension of the same drug, reporting comparable plasma AUC but higher C_{max} and shorter T_{max}. These results are in accordance with the more rapid rate of solubility and subsequent absorption of nanocrystals compared to bigger particles (Kraft et al., 2004).

5.6.4 NANOSUSPENSIONS FOR OCULAR DELIVERY

Nanosuspensions can prove to be a boon for drugs that exhibit poor solubility in lachrymal fluids. Approaches such as suspensions and ointments have been recommended for the delivery of such drugs. Although suspensions offer advantages such as prolonged residence time in a cul-de-sac (desirable for effective treatment for most ocular diseases) and avoidance of the high tonicity created by water-soluble drugs, their performance depends on the intrinsic solubility of the drug in lachrymal fluids. Thus, the intrinsic dissolution rate of the drug in lachrymal fluid directs its ocular bioavailability. However, the intrinsic dissolution rate of the drug will vary because of the constant inflow and outflow of lachrymal fluids. Hence, suspensions may fail to give consistent performance. However, nanosuspensions, by their inherent ability to improve the saturation solubility of the drug, represent an ideal approach for ocular delivery of hydrophobic drugs (Patel et al., 2013). Moreover, the nanoparticulate nature of the drug allows its prolonged residence in the cul-de-sac, giving sustained release of the drug. These nanosuspensions also sustained drug release and were more effective over a longer duration. Nanosuspensions also impart stability to the drug in the formulation. It has been recommended that particles be less than 10 nm to minimize particle irritation to the eye, decrease tearing and drainage of instilled dose, and therefore increase the efficacy of an ocular treatment; a fact endorsed in various published articles (Hui and Robinson et al., 1986; Schoenwald and Stewart, 1980; Pignatello et al., 2002).

The efficacy of nanosuspensions in improving ocular bioavailability of three different types of glucocorticoids; hydrocortisone, prednisolone, and dexamethasone which are practically water insoluble has been demonstrated in several research studies (Ali et al., 2011; Armaly , 1986; Kassem et al., 2007). They represent the three classes of short-, medium-, and long-acting steroids, respectively. They are widely used topically for the treatment of inflammatory conditions of the conjunctiva and anterior segment of the eye. The present therapy with these drugs mostly requires frequent instillation in the conjunctival sac, which, besides leading to poor patient compliance, may result in administration of a large dose which, in turn, may induce glaucoma, damage optic nerve, and result in cataract formation (Armaly, 1986).

Kassem and colleagues compared ocular bioavailability of various glucocorticoids (prednisolone, dexamethasone, and hydrocortisone) from nanosuspensions, solutions, and microcrystalline suspensions (Kassem et al., 2007). The formulations were instilled into the lower cul-de-sac of the rabbit eye and intraocular pressure (IOP) was measured at frequent time intervals up to 12 h. The area under percentage increase in IOP *versus* AUC values for all the suspensions were higher than that for the respective drug solutions. Moreover, higher extent of drug absorption and more intense drug effects were observed for nanosuspensions of all the steroids compared to the solutions.

In another study, Ali compared ocular bioavailability of hydrocortisone nanosuspensions prepared by precipitation and milling method with hydrocortisone solution in rabbits post-topical instillation (Ali et al., 2011). A sustained drug action which was represented in terms of changes in intraocular pressure was maintained up to 9 h for the nanosuspensions compared to 5 h for the drug solution. It can be concluded that nanosuspensions could be an efficient ophthalmic drug delivery system for delivery of poorly soluble drugs. In addition, nanosuspension can also be incorporated into hydrogels or ocular inserts for achieving sustained drug release for stipulated time period (Patel et al., 2013).

5.6.5 NANOSUSPENSIONS FOR TOPICAL/MUCOSAL DELIVERY

Among the various strategies for enhancing dermal application, nanocrystals can be considered a rather new but highly interesting approach (Shegokar and Muller , 2010). In spite of introduction of the concept of nanosuspensions over 30 years ago, very few attempts have been made to study nanosuspensions as a tool for skin permeation. However, a few published results clearly indicate that this technology could be very effective for improving dermal bioavailability of actives with poor solubility. Moreover, besides the increased saturation solubility and dissolution rate, nanocrystals also exhibit the property of increased adhesiveness to the skin thus facilitating the dermal delivery. The interest toward this technology for dermal application has been rising since 2007, when the first antiaging and skin-protective cosmetic products based on nanosuspensions of poorly soluble antioxidants rutin and hesperidin were introduced to the market (Petersen, 2008).

Rutin nanocrystals were produced by a combination process of low-energy pearl milling followed by HPH. When applied to the skin of human volunteers, a nanosuspension with 5% rutin as nondissolved nanocrystals showed a 500-fold higher antioxidant activity compared with a 5% solution of a water-soluble rutin-glycoside derivative. Due to the higher solubility of active rutin as a nanocrystal, there is an increased concentration gradient between dermal formulation and skin, which leads to a higher penetration when compared to large drug crystals. Rutin penetrating into the skin is immediately replaced by fast-dissolving molecules from the nanocrystals. Moreover, compared to the hydrophilic rutin-glucoside, the original lipophilic molecule rutin can better penetrate the skin where it is believed to have a higher affinity to the sites of action than the synthetically modified derivative (Petersen, 2008). As a consequence of these positive outcomes, several researches on nanosuspensions of antioxidant molecules have been reported in literature since 2007 (Kobierski et al., 2009; Mishra et al., 2009; Mitri et al., 2011; Romero et al., 2015).

First, Mishra et al. produced nanocrystals of hesperetin, the aglycone of hesperidin, which not only has antioxidative but also anti-inflammatory properties and, therefore, is thought to be a very effective antiaging compound. Hence, in this study nanosuspensions of hesperetin were produced by HPH employing four different stabilizers suitable for dermal use such as Poloxamer 188, Inutec SP1, Tween 80, and Plantacare 2000, possessing different mechanisms of stabilization. Interestingly, as predicted from the zeta potential measurements, nanosuspensions stabilized by Inutec and Plantac were stable with no change in mean diameter size. Slight increases in size were found for nanosuspensions stabilized by Poloxamer and Tween, which is not considered to impair their use in dermal formulations (Mishra et al., 2009).

In 2011, Kobierski et al. developed nanocrystals of antioxidant resveratrol to be developed as dermal cosmetic formulations such as creams, lotions, and gels. Since preservation is essential for dermal formulations, resveratrol nanosuspensions were produced by a HPH process and the effects of preservatives on physical stability were monitored as a function of cycle numbers (Kobierski et al., 2009).

Mitri et al. prepared a lutein nanosuspension by means of HPH to enhance dissolution velocity and saturation solubility, which are major factors determining skin penetration. Lutein is a well-known antioxidant and anti-free radical used in cosmetic, nutraceutical industry with potential application in pharmaceutics as supportive antioxidant. The main results from this study indicated that lutein nanocrystals significantly increased drug saturation solubility as well as permeation through cellulose nitrate membranes, compared to coarse powder. Moreover, no permeation through pig ear skin occurred, which supports the fact that lutein rather penetrates the skin and retains its antioxidant activity (Mitri et al., 2011).

A solid in oil nanosuspension of diclofenac sodium was the first pharmaceutical formulations in which nanosuspension technology was applied for topical delivery of a poorly soluble therapeutic agent (Piao et al., 2008). In this study, Piao et al. successfully dispersed diclofenac sodium into isopropyl myristate (oil phase) as a nanosized suspension via complex formation with sucrose ester surfactant. The drug flux across the Yucatan micro pig skin model was increased up to approximately fourfold compared with a coarse oil suspension used as control.

Nanosuspension gel formulations were also studied as suitable formulation approach for improving the skin permeability of a poorly soluble molecule such as ibuprofen for the treatment of acute and chronic arthritic conditions (Ghosh et al., 2013). Since the ibuprofen oral administration results in gastric mucosal damage subsequently leading to ulceration, its topical delivery can overcome many side effects such as gastric complications. In this study, tocopheryl polyethylene glycol 1000 succinate and HPMC were used as the basic components in nanoparticle ibuprofen formulations. TPGS was used to enhance the drug permeability (tested in both porcine and human skin) and to stabilize the system by hydrophobic interactions, while HPMC was used as a steric stabilizer to inhibit crystal growth of the drug in the formulation. Results of experiments clearly indicated that the overall skin permeation enhancement process strongly depended on the solubilizer and the particle size of the drug crystals. These factors resulted in higher drug release due to the formation of supersaturated solution around the crystals, and thus a high concentration gradient was produced between the drug crystals and skin surface.

Lai et al. tested nanosuspensions prepared by simple precipitation method as a tool to improve cutaneous targeting and photo stability of tretinoin (TRA), poor water-soluble and unstable molecule, commercially formulated in cream and gel forms for the treatment of acne vulgaris (Lai et al., 2013). Drug skin permeation and deposition were studied *in vitro* by diffusion experiments through newborn pig skin, while TRA photo stability was investigated by irradiating samples with UV light. As an appropriate comparison, an oil in water (o/w) nanoemulsion was also prepared and tested. Indeed, nanoemulsions have shown to be particularly useful as vehicle for dermal and transdermal delivery of hydrophobic compounds for pharmaceutical, cosmetic, and chemical industry applications. During 8-h percutaneous experiments, the nanosuspension was able to localize the drug into the pig skin with a very low transdermal drug delivery (which is responsible for the systemic side effects of this drug), whereas the nanoemulsion greatly improved drug permeation.

Therefore, this work demonstrated the high potentiality of nanosuspensions in dermal drug delivery. It is worth noticing that nanosuspensions are almost exclusively composed of the drug nanoparticles with small amounts of biocompatible and safe surfactants such as the soy lecithin used in this work. Although only a few toxicological data are available at the moment, no side effects are known or expected from nanocrystal topically applied. Indeed, as suggested by Muller et al., each solid macro/microparticle applied on the skin will convert into nanocrystal during its dissolution process and, up to now, no intolerability has been reported (Muller et al., 2008). As already reported above, also in this research paper authors justified the increased drug dermal delivery as a consequence of increased concentration gradient between the dermal formulation and the skin. More specifically, they claimed that finely divided and uniformly suspended TRA nanocrystals possess an increased dissolution rate due to their large surface area and increased saturation solubility. Solid drug dissolves in the vehicle, diffuses through the vehicle to the skin, establishes local phase equilibrium with the outer layer of skin, and finally the larger concentration gradient penetrates the skin forming a depot in the lipophilic stratum corneum from which it diffuses.

Furthermore, the topical application of TRA nanocrystals has the advantage of increasing the drug photo stability, in comparison with the control nanoemulsion. In the last 2 years, the application of nanocrystal technology has been extended to medium soluble compounds and a new mechanism by which nanocrystals can improve dermal delivery has been proposed, in addition to the simple increase of concentration gradient between formulation and skin (Romero et al., 2015). This second mechanism involves the hair follicles.

Nanocrystals with an appropriate size (approximately 700 nm) can accumulate into these shunts, which act as a depot from which the drug can diffuse into the surrounding cells for prolonged release (Lademann et al., 2007; Patzelt et al., 2011). In a novel interesting paper, Zhai et al. produced nanocrystals to enhance the skin penetration of a medium soluble active such as caffeine (Zhai et al., 2014). Moreover, they developed a specific production process, namely, a low-energy milling in low dielectric constant dispersion media and a selected stabilizer to overcome crystal growth and fiber formation caused by supersaturation and recrystallization effects, typical with medium soluble compounds.

The novel concept introduced by Zhai et al. consisted of formulating nanocrystals from medium soluble actives, and applying a dermal formulation containing additional nanocrystals, which should act as a fast-dissolving depot, increase saturation solubility, and accumulate in the hair follicles, to further enhance skin penetration. For this purpose, they produced caffeine nanocrystals with varied size ranging from 660 nm (optimal for hair follicle accumulation) to 250 nm (optimal for fast dissolution), by pearl milling process in ethanol/propylene glycol 3:7 with 2% carbopol (Zhai et al., 2014).

5.6.6 SURFACE-MODIFIED NANOSUSPENSIONS FOR TARGETED DELIVERY

By modifying the surface properties of nanosuspensions, their *in vivo* behavior/distribution can be altered and this could be useful to achieve targeted delivery. The release of drug from nanosuspensions is usually rapid or "burst release" which may cause toxicity and severe side effects. Therefore, modification of the surface is required in order to control drug release and/or prolonged residence at the site of action. For example, nanosuspensions-based formulations used for targeting the monocyte phagocytic system in the treatment of lymphatic-mediated diseases can cause toxicity due to accumulation of drug (Charmanand and Stella, 1992). It is notable that after intravenous administration of a nanosuspension, the drug nanoparticles are taken up by mononuclear phagocytic system (MPS) cells and Kupffer cells, as observed in the case of various other colloidal drug carriers. The particles are recognized by the host as foreign bodies and are phagocytosed by the macrophages mainly in the liver (60–90%), spleen (approximately 1–5%), and to a very small extent in the lungs. This natural targeting does not have an effect on the safety profile of the drug. In fact, it helps in increasing the drug tolerance as MPS cells or Kupffer cells can act as a controlled release vehicle for the drug, enabling its prolonged action, faster onset of action, and improved dose proportionality. Natural targeting can be used to treat mycobacterial, fungal, or leishmanial infections by targeting macrophages, if the infectious pathogens persist intracellularly; especially for leishmanicidal drugs such as amphotericin B or buparvaquone. This leads to safety improvement, reducing side effects on other organs, permitting higher dosing, and improved efficacy. Moreover, in order to achieve efficient macrophage targeting, the surface properties of nanosuspensions could be altered in a controlled way to modify the plasma protein adsorption pattern (Kayser, 2005).

The currently available antiretroviral (ARV) chemotherapy has lot of limitations which need to be overcome. As penetration of ARV drugs into the viral reservoir sites is constrained, a high dose has to be given with consequent intolerance and toxicity (Amiji et al., 2006). In addition, many drugs cannot effectively reach or reside in these sites in sufficient concentrations and for the necessary duration to exert the therapeutic response. Hence, studies involving ARV drug-loaded nanoparticles for targeting to the macrophages have consequently emerged. Shegokar et al. prepared surface-modified Nevirapine nanosuspensions for viral reservoir targeting. The nanosuspension was prepared by HPH technique and the surface was modified with surface modifiers such as serum albumin, PEG 1000, and dextran 60 by physical adsorption method. The improved accumulation of drug nanosuspensions in various organs of rat was observed when compared to the plain drug solution. The surface-coated nanosuspension also showed higher MRT values in brain, liver, and spleen as compared to pure drug solution which evidenced improved bioavailability and prolonged residence of the drug at the target site. Challenges in developing new therapeutic strategies include not only identifying novel active agents but also improving the delivery of a drug at the biologic level. The development of a new drug delivery strategy is important. In another case study, Yin et al. evaluated the biological effect of PEG-modified PTX nanosuspensions based on human serum albumin for tumor. PTX is widely used chemotherapeutic agent against ovarian cancer (Yin et al., 2016). Yet, PTX therapy is challenged with high hydrophobicity, nonselectivity, and rapid systemic clearance. The commercial formulation of PTX, Taxol®, also suffers from drawbacks such as nonselective biodistribution and undesirable pharmacokinetics. Taxol® is generally infused at 135–175 mg/m^2 within 3–24 h. However, the terminal half-life and C_{max} was found to be around 5 h and 59.37 μg/mL, respectively. The formulation shows rapid blood clearance followed by a high blood concentration, which when above the therapeutic window, could lead to low therapeutic indices and acute therapeutic toxicity. While high concentrations of free PTX are found in lung, heart, and kidney, insufficient drug concentration is detected in tumors, which contributes to severe peripheral toxicity and low therapeutic index. Abraxane™, the first nanotechnology-based formulation approved by FDA was a formulation of nab-PTX nanosuspensions which demonstrated high drug accumulation in tumors. Success is primarily attributed to a combination of passive (enhanced permeability and retention effect) and active (endothelial transcytosis through binding HSA to glycoprotein gp60 receptor) targeting methods. To overcome the disadvantages of Abraxane™, solid dispersion of PTX was formulated using PEG as the solubilizer. The solid dispersion of PTX–PEG was later mixed with HSA solution and homogenized under high pressure to achieve PTX–PEG–HSA nanosuspensions. As expected, PEG-modified PTX–PEG–HSA nanosuspensions achieved increased drug half-life in circulation and a subsequent increase in tumor accumulation over PTX–HSA. Prolonged circulation of PTX–PEG–HSA in blood was seen after *in vivo* pharmacokinetic studies. These studies also indicated that PTX–PEG–HSA achieved prolonged blood circulation, illustrated by an 8.8-fold and 4.8-fold increase in area under the curve (AUC) of PTX compared to Taxol® and PTX–HSA, while a 3.2-fold and 1.5-fold increase in the MRT of PTX in PTX–PEG–HSA was observed.

HSA-mediated active targeting further suppressed nonspecific distribution of PTX to normal tissues, which permitted enhanced antitumor efficacy in S180 mice over Taxol® and PTX–HSA. Safety of intravenously administered PTX–PEG–HSA was confirmed through lower hemolytic activity, a 2.2-fold and 1.2-fold increase in LD50 (113.4 mg/kg) over Taxol®, and PTX–HSA alongside the absence of local venous irritation.

Talekar et al. developed the folate-targeted PIK-75 nanosuspension for tumor-targeted delivery to improve therapeutic efficacy in human ovarian cancer model (Talekar et al., 2013). The nanosuspension was prepared by HPH technique with folic acid–Poloxamer (FA-P188) conjugate. *In vitro* studies in SKOV-3 cells indicated a two-fold improvement in drug uptake and 0.4-fold lowering in IC_{50} value of PIK-75 when treated with targeted nanosuspension compared to nontargeted nanosuspension. The improvement in cytotoxicity was attributed to an elevated caspase 3/7 and highly reactive oxygen species activity. A 5–10 fold enhanced PIK-75 accumulation in the tumor with both the nanosuspension formulations compared to PIK-75 suspension was observed in *in vivo* studies. The targeted nanosuspension showed an enhanced downregulation of pAkt compared to nontargeted formulation system. These results illustrate the opportunity to formulate PIK-75 as a targeted nanosuspension to enhance uptake and cytotoxicity of the drug in tumor.

5.7   QUALITY BY DESIGN-BASED DEVELOPMENT OF ATOVAQUONE NANOSUSPENSION FOR PARENTERAL DELIVERY

Pharmaceutical product development based on Quality by Design (QbD) approach as highlighted in ICH Q8 guideline outlines a systematic approach to design a quality product that begins with predetermined objectives and emphasizes on product and process understanding and determination of control strategy based on sound science and quality risk management (ICH, 2009). It is a concept which states that "quality should be built in by design."

In the current guideline setting, the concept of QbD is focused whose objectives are:

- To define and achieve meaningful product quality attributes which ensure clinical efficacy and safety for the patients consistently.
- To achieve enhanced product and process design and understanding so as to be able to put into place suitable controls to reduce product variability and hence reduce the risk to product quality.

- To achieve improved efficiencies in product development, registration, and manufacturing.
- To enhance capability to manage post-approval changes during the product life cycle and assist in continual improvement in process/ product.

The various stages elucidating product development using QbD include:

- Determination of target product profile (TPP) or quality target product profile (QTPP) that identifies the critical quality attributes (CQAs) of the drug product.
- Identification of the critical material attributes (CMAs) of the drug substance and other inactives/excipients which can have an impact on the CQAs.
- Identification of critical process parameters (CPPs) which can have an impact on the CQAs.
- An iterative risk assessment-based approach to achieve a thorough understanding of the product composition and process to minimize the risk to product quality and establish linkage of CMAs and CPPs to the CQAs.
- Empirical or enhanced (Design of Experiment) approach based on the criticality and complexity of the material and/ or process attributes to generate a design space.
- A control strategy based on design space that includes specifications for drug substance, excipients, and drug product as well as in-process controls during each step of the manufacturing process.
- Continual improvement and life cycle management.

This knowledge with respect to material and process understanding derived from pharmaceutical development studies provides a scientific understanding to establish the design space, specifications, and manufacturing controls.

The very important concept of design space herein refers to the allowable quantitative combination of material (composition) and process parameters that assures quality within the specified limit. This is a very important element in any product design and development as it determines the boundary within which the product can be manufactured within the specification limits with confidence. With defined design space and updated risk analysis that ensures the risk minimization within the acceptable limits, a control strategy is proposed. Further, any change in the control strategy desires an assessment and regulatory approval (Chogale et al., 2016).

Herein, to illustrate the application of QbD in the development of nano-suspensions, development of a fast-action intravenous atovaquone nanosus-pension based on the principle of QbD has been exemplified. Atovaquone nanosuspension is prescribed for the treatment of severe malaria.

5.7.1 ESTABLISHMENT OF QTPP

The QTPP for intravenous atovaquone nanosuspension is depicted in Table 5.3 incorporating the summary of quality characteristics of the formulation from a patient point of view.

TABLE 5.3   QTPP for Intravenous Atovaquone Nanosuspension.

QTPP element	Target	Justification
Dosage form	Dry powder for reconstitution with sterile water for injection (lyophilized nanosuspension)	Formulation as a dry powder would circumvent instability problems with aqueous formulation of drug
Route of administration	IV bolus following reconstitution	Intravenous route was proposed to ensure rapid onset of action and due to inability of severe malaria patients to take medications orally
Injection volume	2–10 mL	Since route of administration is an intravenous bolus, 2–10 mL was decided as the injection volume for clinical efficacy
Identity	Meets pharmacopoeial requirement	To ensure safety and efficacy of the product
Strength	75 mg/vial	Dose required for antimalarial activity
Assay	97.5–101.5% of label claim	To ensure safety and efficacy of product as per USP specifications
pH	5.5–7.5	This pH range is acceptable for parenteral administration to avoid pain and irritation
Reconstitution time	Easy reconstitution within 1–2 min	To ensure quick administration
Insoluble particulate matter	Meets pharmacopoeial requirements	To ensure safety and efficacy of product

TABLE 5.3 *(Continued)*

QTPP element	Target	Justification
Impurities and related substances	Meets ICH requirements	To ensure safety and efficacy of product
Content uniformity	Meets pharmacopoeial requirements	To ensure accurate dosing
In vitro release	Immediate release: greater than 80% release in 30 min	To ensure rapid onset of action
Bacterial endotoxin	Meets pharmacopoeial requirements	To ensure patient safety
Sterility	Meets pharmacopoeial requirements	To ensure patient safety
Excipients	Acceptable at levels used for the proposed market	To ensure patient safety
Shelf life	24 months	To maintain safety and efficacy of product
Container/ closure system	Single dose vial with rubber closure sealed with aluminum caps	To maintain the sterility and therapeutic potential of drug product during shelf life

5.7.2 PROPOSED FORMULATION AND MANUFACTURING PROCESS

The product contains two parts, namely, sterile lyophilized powder and Sterile Water for Injection. The manufacturing of Sterile Water for Injection was done in house by the routine process which was well controlled, and hence will not be discussed in the forthcoming sections.

Atovaquone is a lipophilic drug with extremely poor aqueous solubility. Intravenous formulation development for atovaquone is challenging due to its poor solubility in hydrophilic cosolvents, surfactants, oils, and lipids used/approved for intravenous administration (Scholer et al., 2001). Nano-suspensions provide a feasible alternative to formulate such lipophilic drugs having high melting point and low solubility in water and oils (Rabinow, 2004).

HPH was chosen as the manufacturing process owing to its amenability to scale up, faster production compared to media milling, and extensive experience and expertise with the technique (Patravale et al., 2004).

5.7.3   IDENTIFICATION OF CQA-BASED ON TPP AND PROPOSED MANUFACTURING PROCESS

Based on proposed formulation and manufacturing process, CQAs of atovaquone nanosuspension were identified and categorized as those of high, medium, and low risk as shown in Table 5.4.

TABLE 5.4   Critical Quality Attributes (CQAs) of Atovaquone Nanosuspension.

Quality attribute	Target/acceptance criteria	Criticality	Justification
Appearance	Yellow lyophilized powder/cake	Low	Not directly linked to safety and efficacy of product so low criticality
Identity	UV spectrum and chromatogram match the reference standard	Low	This will affect the clinical safety and efficacy but can be controlled with appropriate API source selection
Assay	97.5–101.5%	High	This will affect the safety and efficacy. Process variables may influence the assay and hence assay should be evaluated at each step which makes it critical
pH	5.5–7.5	Medium	This is required to avoid pain and irritation but can be controlled through preliminary evaluation and monitoring at product release thus, considered to be medium critical
Redispersibility of lyophilized powder/cake	Easy dispersion within 1–2 min	High	Is critical to obtain quick and complete redispersion and nanosuspension of the desired particle size and could be affected by material attributes and process parameters and hence would be studied in detail during manufacture

TABLE 5.4 *(Continued)*

Quality attribute	Target/acceptance criteria	Criticality	Justification
Insoluble particulate matter	Meets pharmacopoeial requirements	Medium	Critical to ensure safety and efficacy of product but would be controlled through proper selection of API, excipients, control of manufacturing facility, and monitored at product release
Impurities and related substances	Meets ICH requirements	High	To ensure safety and efficacy of product
Content uniformity	Meets pharmacopoeial requirements	Medium	Critical to ensure accurate dosing and would be monitored after each unit operation and at product release and hence medium critical
Bacterial endotoxin	Meets pharmacopoeial requirements	Medium	Critical to safety but would be controlled through proper control of raw materials used in the manufacture and stringent environmental control of the manufacturing facility
Sterility	Meets pharmacopoeial requirements	Medium	Critical to safety but would be controlled through proper control of raw materials used in the manufacture and stringent environmental control of the manufacturing facility
Excipients	Meets pharmacopoeial requirements	Medium	Critical but can be controlled through proper excipient source selection
Particle size of premilled suspension for HPH	Less than 25 μm	High	Critical to prevent clogging of HPH valve and needs to be monitored before proceeding for HPH
Particle size of nanosuspension	100–200 nm	High	Is critical since release profile is directly related to the particle size of the nanosuspension, and material attributes and process parameters can impact particle size, and hence will be monitored throughout the manufacture

TABLE 5.4 *(Continued)*

Quality attribute	Target/acceptance criteria	Criticality	Justification
Viscosity of nanosuspension	Acceptable syringeability and easy injectability	Medium	Critical to ensure syringeability and injectability but will be controlled and monitored at release hence medium criticality
Release profile	Greater than 80% release in 30 min	High	Any variation in release profile will lead to batch to batch variation and affect the pharmacokinetic profile and so is highly critical

The product attributes identified to be highly critical were assay, impurities, and particle size of premilled suspension for HPH, particle size of nanosuspension, redispersibility of the lyophilized powder/cake, and release profile. These were subjected to further risk analysis.

5.7.4   SCREENING OF SURFACTANT/STABILIZER

Various excipients were screened for their role as a wetting agent and stabilizer on the basis of the following criteria, namely, compendial status, acceptability for parenteral administration, Inactive Ingredients Database (IID) limits, and low microbial load. Final screening of surfactant and/ or stabilizer was done on basis of saturation solubility measurement and contact angle determination. On the basis of these studies, Tween 80 and Kollidon 12 PF were selected for further formulation optimization.

5.7.5   INITIAL RISK ASSESSMENT

Initial risk assessment for formulation of nanosuspension is shown in Table 5.5, wherein the CMAs and CPPs which were identified to be of high risk for formulation of nanosuspension were studied further during product development. Previous knowledge was utilized to identify the variables having a low potential for causing risk to product quality.

TABLE 5.5   Initial Risk Assessment for Formulation of Nanosuspension.

Product CQA	Critical material attributes				Critical processing parameters	
	API purity	API particle size	Conc. of Tween 80	Conc. of Kollidon® 12 PF	Homogenization pressure	Number of HPH cycles
Assay	Low	Low	Low	Low	Low	Low
Impurities	Low	Low	Low	Low	Low	Low
Particle size	Low	Low	High	High	High	High
Release profile	Low	Low	Medium	Medium	Medium	Medium

API obtained was of high purity and its purity verified by chromatography, hence it would be a low risk factor to the assay and impurities of the final product. Also, particle size of API was considered to be of low risk since an API having constant particle size $D_{0.9}$ of less than 100 μm was obtained from the API supplier. Similarly, based on previous experience with HPH technique for nanosuspension manufacture, the concentration of Tween 80 and Kollidon® 12 PF, homogenization pressure, and number of HPH cycles were reported to be of high risk to the particle size and hence were optimized further for product development. Drug–excipient compatibility studies revealed concentration of Tween 80 and Kollidon® 12 PF to be of low risk to the assay and impurities of the product. Similarly, the drug was found to be stable to HPH processing, and hence homogenization pressure and number of HPH cycles were expected to be of low risk to the assay and impurities of product. The concentration of Tween 80, Kollidon® 12 PF, homogenization pressure, and number of HPH cycles were identified to be of medium risk to the release profile since the excipients chosen were for promoting solubility, wettability, and the HPH process would result in particle size reduction. Also, the release profile of nanosuspension is directly linked to its particle size, and particle size and effect of all these variables on particle size was studied further by Design of Experiments.

5.7.6 EXPERIMENTAL DESIGN AND OPTIMIZATION

Design Expert was utilized to study the effect of Kollidon® 12 PF (at 0.5%, 0.75%, and 1% levels) and Tween 80 concentration (at 0.5%, 0.75%, and 1%

levels) on the particle size of nanosuspension and particle size after 24 h as shown in Table 5.6.

TABLE 5.6 Experimental Optimization Design for Atovaquone Nanosuspension.

Factor	Range investigated
Concentration of Tween 80	0.5–1
Concentration of Kollidon® 12 PF	0.5–1
Homogenization pressure	900–1500
Number of HPH cycles	20–30

Response	Goal	Acceptance
Particle size of nanosuspension	$D_{0.9}$ of 100–200 nm	$D_{0.9}$ of 100–200 nm
Particle size of nanosuspension after 24 h	$D_{0.9}$ of 100–200 nm	$D_{0.9}$ of 100–200 nm

Figure 5.6 depicts the impact of homogenization pressure and number of HPH cycles on the particle size of the nanosuspension and its PDI at a constant ratio of drug:stabilizer:wetting agent.

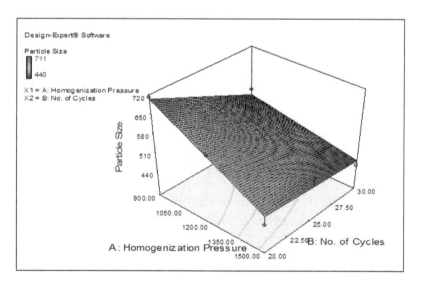

FIGURE 5.6 **(See color insert.)** Impact of homogenization pressure and number of cycles on particle size.

From Figure 5.6, it was observed that an increase in homogenization pressure and number of HPH cycles resulted in a decrease in the particle size. The design space obtained is depicted in Figure 5.7.

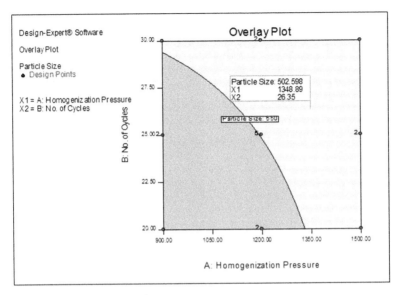

FIGURE 5.7   **(See color insert.)** Design space.

5.7.7   UPDATED RISK ASSESSMENT FOR FORMULATION OF NANOSUSPENSIONS AND THEIR JUSTIFICATION

After successful product and HPH process optimization, all the variables and unit operations were reassessed toward the risk and it was observed that the risk was minimized at each stage as depicted in Table 5.7.

TABLE 5.7   Updated Risk Assessment for Formulation of Atovaquone Nanosuspension.

Product CQA	Critical material attributes				Critical processing parameters	
	API purity	API particle size	Conc. of Tween 80	Conc. of Kollidon® 12 PF	Homogenization pressure	Number of HPH cycles
Assay	Low	Low	Low	Low	Low	Low
Impurities	Low	Low	Low	Low	Low	Low
Particle size	Low	Low	Low (0.7–0.8%)	Low (0.7–0.8%)	Low (1140–1260 bars)	Low (23–27 cycles)
Release profile	Low	Low	Low	Low	Low	Low

Based on experiments, it was concluded that by operating within the defined design space as shown in Figure 5.7, the product having the desired particle size and stability was obtained. Thus, the risk associated with the concentration of Kollidon® 12 PF, Tween 80, homogenization pressure, and number of HPH cycles on particle size was minimized. This in turn minimizes the risk associated with release profile since it is directly related to the particle size. The obtained nanosuspension would further be freeze dried to obtain the desired powder for reconstitution.

5.8   CONCLUSION

Nanosuspensions have interesting features such as ease of manufacture and scale-up, increased surface area of drug, increased dissolution velocity and saturation solubility, increased drug absorption, enhanced long-term physical stability, increased bioadhesivity, increased drug loading, and the ability to reduce variation in oral bioavailability in fed and fasted state. To date, particle size reduction or top-down technologies have dominated the large-scale production of drug nanosuspensions. Significant efforts are being directed to develop alternative technologies to enable or accelerate the development of nanosuspensions of hydrophobic drugs that are not suitable candidates for top-down methodologies. Furthermore, significant advances have been made in transforming nanosuspensions into various types of liquid, semisolid, and solid dosage forms. These features have widened the scope of nanosuspensions to various routes of administration. To date, extensive studies have been carried out to establish utility of nanosuspensions for oral and parenteral delivery. In recent years, significant efforts have been directed towards developing nanosuspensions for local delivery (pulmonary, dermal, and ocular). Thus far, nanosuspensions have been mainly used in the commercial oral products due to their ability to achieve quick and uniform absorption of drugs independent of fed and fasted state of the patients. However, in recent years, emphasis has also been given to develop long-acting injectable nanosuspensions for intramuscular administration leading to successful commercialization of two long-acting injectable nanosuspensions. In conclusion, formulation into nanosuspensions has proven to be a unique, versatile, and translational approach to overcome problems associated with delivery of drugs with poor solubility in water, oil, and organic solvents.

KEYWORDS

- **dissolution velocity**
- **nanocrystals**
- **nanosuspensions**
- **saturation solubility**
- **stabilizers**
- **quality by design**

REFERENCES

Al Shaal, L.; Müller, R. H.; Shegokar, R. smartCrystal Combination Technology: Scale Up from Lab to Pilot Scale and Long Term Stability. *Pharmazie* **2010,** *65,* 877–884.

Ali H. S.; York, P.; Ali, A. M.; Blagden, N. Hydrocortisone Nanosuspensions for Ophthalmic Delivery: A Comparative Study Between Microfluidic Nanoprecipitation and Wet Milling. *J. Control. Release* **2011,** *149,* 175–181.

Amiji, M.; Vyas, T.; Shah, L. Role of Nanotechnology in HIV/AIDS Treatment: Potential to Overcome the Viral Reservoir Challenge. *Discov. Med.* **2006,** *6,* 157–162.

Armaly, M. F. Corticosteroid Glaucoma. In *Glaucoma*; Cairns, J. E., Ed.; Grune & Stratton: London, 1986; pp 697–710.

Azarmi, S.; Roa, W. H.; Löbenberg, R. Targeted Delivery of Nanoparticles for the Treatment of Lung Diseases. *Adv. Drug Deliv. Rev.* **2008,** *60* (8), 863–875.

Baert, L.; van 't Klooster, G.; Dries, W.; François, M.; Wouters, A.; Basstanie, E.; Iterbeke, K.; Stappers, F.; Stevens, P.; Schueller, L.; Van Remoortere, P.; Kraus, G.; Wigerinck, P.; Rosier, J. Development of a Long-acting Injectable Formulation with Nanoparticles of Rilpivirine (TMC278) for HIV Treatment. *Eur. J. Pharm. Biopharm.* **2009,** *72,* 502–508.

Beck-Broichsitter, M.; Merkel, O. M.; Kissel, T. Controlled Pulmonary Drug and Gene Delivery Using Polymeric Nano-carriers. *J. Control. Release* **2012,** *161* (2), 214–224.

Blunk, T.; Hochstrasser, D. F.; Sanchez, J. C.; Muller, B. W. Colloidal Carriers for Intravenous Drug Targeting: Plasma Protein Adsorption Patterns on Surface-modified Latex Particles Evaluated by Two-dimensional Polyacrylamide Gel Electrophoresis. *Electrophoresis* **1993,** *14* (12), 1382–1387.

Blunk, T.; Hochstrasser, D. F.; Luck, M. A.; Calvor, A.; Muller, B. W.; Muller, R. H. Kinetics of Plasma Protein Adsorption on Model Particles for Controlled Drug Delivery and Drug Targeting. *Eur. J. Pharm. Biopharm.* **1996,** *42,* 262–268.

Borhade, V.; Pathak, S.; Sharma, S.; Patravale, V. Formulation and Characterization of Atovaquone Nanosuspension for Improved Oral Delivery in the Treatment of Malaria. *Nanomedicine (Lond.)* **2014,** *9,* 649–666.

Bott, S. E; Hart, W. H. Particle Size Analysis Utilizing Polarization Intensity Differential Scattering. U. S. Patent 4,953,978, April 14, 1992.

Chan, H. K.; Kwok, P. C. Production Methods for Nanodrug Particles Using the Bottom-up Approach. *Adv. Drug. Deliv. Rev.* **2011,** *63,* 406–416.

Charmanand, W. C. N.; Stella, V. J. *Lymphatic Transport of Drugs*; CRC Press: Boca Raton, FL, USA, 1992.

Chiang, P. C.; Alsup, J. W.; Lai, Y.; Hu, Y.; Heyde, B. R.; Tung, D. Evaluation of Aerosol Delivery of Nanosuspension for Pre-clinical Pulmonary Drug Delivery. *Nanoscale Res. Lett.* **2009,** *4,* 254–261.

Chin, W. W.; Parmentier, J.; Widzinski, M.; Tan, E. H.; Gokhale, R. A Brief Literature and Patent Review of Nanosuspensions to a Final Drug Product. *J. Pharm. Sci.* **2014,** *103,* 2980–2999.

Chogale, M. M.; Ghodake V. N.; Patravale, V. B. Performance Parameters and Characterizations of Nanocrystals: A Brief Review. *Pharmaceutics* **2016,** *8* (3). https://www.ncbi.nlm.nih.gov/pubmed/27589788 (accessed Nov 23, 2016).

Danhier, F.; Ucakar, B.; Vanderhaegen, M. L.; Brewster, M. E.; Arien, T.; Préat, V. Nanosuspension for the Delivery of a Poorly Soluble Anti-cancer Kinase Inhibitor. *Eur. J. Pharm. Biopharm.* **2014,** *88,* 252–260.

Deng J.; Huang, L.; Liu, F. Understanding the Structure and Stability of Paclitaxel Nanocrystals. *Int. J. Pharm.* **2010,** *390* (2), 242–249.

Dolenc, A.; Kristl, J.; Baumgartner, S.; Planinsek, O. Advantages of Celecoxib Nanosuspension Formulation and Transformation into Tablets. *Int. J. Pharm.* **2009,** *376* (2), 204–212.

Dressman, J.; Amidon, G.; Reppas, C. Dissolution Testing as a Prognostic Tool for Oral Drug Absorption: Immediate Release Dosage Forms. *Pharm. Res.* **1998,** *15,* 11–22.

Drews, T.; Tsapatsis, M. Model of the Evolution of Nanoparticles to Crystals via an Aggregative Growth Mechanism. *Micropor. Mesopor. Mat.* **2007,** *101,* 97–107.

Du, J.; Li, X.; Zhao, H.; Zhou, Y.; Wang, L.; Tian, S.; Wang, Y. Nanosuspensions of Poorly Water-soluble Drugs Prepared by Bottom-up Technologies. *Int. J. Pharm.* **2015,** *495,* 738–749.

Edwards, D. A.; Dunbar, C. Bioengineering of Therapeutic Aerosols. *Ann. Rev. Biomed. Eng.* **2002,** *4,* 93–107.

Fu, Q.; Sun, J.; Zhang, D.; Li, M.; Wang, Y.; Ling, G.; Liu, X.; Sun, Y.; Sui, X.; Luo, C.; Sun, L.; Han, X.; Lian, H.; Zhu, M.; Wang, S.; He, Z. Nimodipine Nanocrystals for Oral Bioavailability Improvement: Preparation, Characterization and Pharmacokinetic Studies. *Colloids Surf. B Bioint.* **2013,** *109,* 161–166.

Gassmann, P.; List, M.; Schweitzer, A.; Sucker, H. Hydrosols-alternatives for the Parenteral Application of Poorly Water Soluble Drugs. *Eur. J. Pharm. Biopharm.* **1994,** *40* (2), 64–72.

Gao, L.; Zhang, D.; Chen, M. Drug Nanocrystals for the Formulation of Poorly Soluble Drugs and Its Application as a Potential Drug Delivery System. *J. Nano. Res.* **2008,** *10* (5), 845–862.

Gao, L.; Liu, G.; Ma, J.; Wang, X.; Wang, F.; Wang, H.; Sun, J. Paclitaxel Nanosuspension Coated with P-gp Inhibitory Surfactants: II. Ability to Reverse the Drug-resistance of H460 Human Lung Cancer Cells. *Colloids. Surf. B Biointerfaces* **2014,** *117,* 122–127.

Gao, W.; Chen, Y.; Thompson, D. H.; Park, K.; Li, T. Impact of Surfactant Treatment of Paclitaxel Nanocrystals on Biodistribution and Tumor Accumulation in Tumor-bearing Mice. *J. Control Release* **2016,** *237,* 168–176.

Ghosh, I.; Schenck, D. Optimization of Formulation and Process Parameters for the Production of Nanosuspension by Wet Media Milling Technique: Effect of Vitamin E TPGS and Nanocrystal Particle Size on Oral Absorption. *Eur. J. Pharm. Sci.* **2012,** *47,* 718–728.

Ghosh, I.; Schenck, D.; Bose, S.; Liu, F.; Motto, M. Identification of Critical Process Parameters and Its Interplay with Nanosuspension Formulation Prepared by Top Down Media Milling Technology—A QbD Perspective. *Pharm. Dev. Technol.* **2013**, *18* (3), 719–729.

Gill, S.; Lobenberg, R.; Ku, T.; Azarmi, S.; Roa, W.; Prenner, E. J. Nanoparticles: Characteristics, Mechanisms of Action and Toxicity in Pulmonary Drug Delivery—A Review. *J. Biomed. Nanotechnol.* **2007**, *3*, 107–119.

Girotra, P.; Singh, S. K.; Nagpal, K. Supercritical Fluid Technology: A Promising Approach in Pharmaceutical Research. *Pharm. Dev. Technol.* **2013**, *18*, 22–38.

Hecq, J.; Deleers, M.; Fanara, D.; Vranckx, H.; Boulanger, P.; Le Lamer, S.; Amighi, K. Preparation and *In vitro/In vivo* Evaluation of Nano-sized Crystals for Dissolution Rate Enhancement of UCB-35440-3, A Highly Dosed Poorly Water-soluble Weak Base. *Eur. J. Pharm. Biopharm.* **2006**, *64*, 360–368.

Heyder, J.; Gebhart, J.; Rudolf, G.; Stahlhofen, W. Deposition of Particles in the Human Respiratory Tract in the Size Range 0.005–15 μm. *J. Aerosol Sci.* **1986**, *17* (5), 811–825.

Hiemenz, P. C.; Rajagopalan, R. *Principles of Colloid and Surface Chemistry*; Marcel Dekker Inc.: New York, 1997.

Hui, H. W.; Robinson, J. R. Effect of Particle Dissolution Rate on Ocular Drug Bioavailability. *J. Pharm. Sci.* **1986**, *75*, 280–287.

Jackson, A. G.; Else, L. J.; Mesquita, P. M.; Egan, D.; Back, D. J.; Karolia, Z.; Ringner-Nackter, L.; Higgs, C. J.; Herold, B. C.; Gazzard, B. G.; Boffito, M. A. compartmental Pharmacokinetic Evaluation of Long-acting Rilpivirine in HIV-negative Volunteers for Pre-exposure Prophylaxis. *Clin. Pharmacol. Ther.* **2014**, *96*, 314–323.

Jacobs, C.; Kayser, O.; Muller, R. Nanosuspensions as a New Approach for the Formulation for the Poorly Soluble Drug Tarazepide. *Int. J. Pharm.* **2000**, *196*, 161–164.

Jacobs, C.; Muller, R. H. Production and Characterization of a Budesonide Nanosuspension for Pulmonary Administration. *Pharm. Res.* **2002**, *19* (2), 189–194.

Junghanns, J. U.; Müller, R. H. Nanocrystal Technology, Drug Delivery and Clinical Applications. *Int. J. Nanomedicine* **2008**, *3*, 295–309.

Kassem, M. A.; Abdel Rahman, A. A.; Khalil, R. M. Nanosuspension as an Ophthalmic Delivery System for Certain Glucocorticoid Drugs. *Int. J. Pharm.* **2007**, *340*, 126–133.

Kayaert, P.; Anné, M. Bead Layering as a Process to Stabilize Nanosuspensions: Influence of Drug Hydrophobicity on Nanocrystal Reagglomeration Following In-vitro Release from Sugar Beads. *J. Pharm. Pharmacol.* **2011**, *63* (11), 1446–1453.

Kayser, O. Nanosuspensions for the Formulation of Aphidicolin to Improve Drug Targeting Effects Against Leishmania Infected Macrophages. *Int. J. Pharm.* **2000**, *196* (2), 253–256.

Kayser, O.; Olbrich, C.; Yardley, V.; Kiderlen, A. J.; Croft, S. L. Formulation of Amphotericin B as Nanosuspension for Oral Administration. *Int. J. Pharm.* **2003**, *254*, 73–75.

Kayser, O.; Lemke, A.; Hernández-Trejo N. The Impact of Nanobiotechnology on the Development of New Drug Delivery Systems. *Curr. Pharm. Biotech.* **2005**, *6*, 3–5.

Keck, C.; Müller, R. Characterisation of Nanosuspensions by Laser Diffractometry. AAPS: Nashville, 2005.

Kraft, W. K.; Steiger, B.; Beussink, D.; Quiring, J. N.; Fitzgerald, N.; Greenberg, H. E.; Waldman, S. A. The Pharmacokinetics of Nebulized Nanocrystal Budesonide Suspension in Healthy Volunteers. *J. Clin. Pharmacol.* **2004**, *44*, 67–72.

Kocbek, P.; Baumgartner, S.; Kristl, J. Preparation and Evaluation of Nanosuspensions for Enhancing the Dissolution of Poorly Soluble Drugs. *Int. J. Pharm.* **2006**, *312*, 179–186.

Kovarova, M.; Council, O. D.; Date, A. A.; Long, J. M.; Nochi, T.; Belshan, M.; Shibata, A.; Vincent, H.; Baker, C. E.; Thayer, W. O.; Kraus, G.; Lachaud-Durand, S.; Williams, P.; Destache, C. J.; Garcia, J. V. Nanoformulations of Rilpivirine for Topical Pericoital and Systemic Coitus-independent Administration Efficiently Prevent HIV Transmission. *PLoS Pathog.* **2015,** *11*, e1005170.

Kumar, M. P.; Rao, Y. M.; Apte, S. Formulation of Nanosuspensions of Albendazole for Oral Administration. *Curr. Nanosci.* **2008,** *4*, 53–58.

Kumar, S.; Burgess, D. J. Wet Milling Induced Physical and Chemical Instabilities of Naproxen Nano-crystalline Suspensions. *Int. J. Pharm.* **2014,** *466* (1–2), 223–232.

Labiris, N. R.; Dolovich, M. B. Pulmonary Drug Delivery. Part I: Physiological Factors Affecting Therapeutic Effectiveness of Aerosolized Medications. *Br. J. Clin. Pharmacol.* **2003,** *56*, 588–599.

Landowitz, R. J.; Kofron, R.; McCauley, M. The Promise and Pitfalls of Long-acting Injectable Agents for HIV Prevention. *Curr. Opin. HIV AIDS* **2016,** *11*, 122–128.

Lee, J. Drug Nano and Microparticles Processed into Solid Dosage Forms: Physical Properties. *J. Pharm. Sci.* **2003,** *92* (10), 2057–2068.

Lentz, K. A.; Quitko, M.; Morgan, D. G.; Grace, J. E.; Gleason, C.; Marathe, P. H. Development and Validation of a Preclinical Food Effect Model. *J. Pharm. Sci.* **2007,** *96*, 459–472.

Li, Y.; Dong, L.; Jia, A.; Chang, X.; Xue, H. Preparation and Characterization of Solid Lipid Nanoparticles Loaded Traditional Chinese Medicine. *Int. J. Biol Macromol.* **2006,** *38*, 296–299.

Liu, Y.; Huang, L.; Liu, F. Paclitaxel Nanocrystals for Overcoming Multidrug Resistance in Cancer. *Mol. Pharm.* **2010,** *7*, 863–869.

Liversidge, G. G.; Conzentino, P. Drug Particle Size Reduction for Decreasing Gastric Irritancy and Enhancing Absorption of Naproxen in Rats. *Int. J. Pharm.* **1995,** *125*, 309–313.

Liversidge, G. G.; Cundy, K. C.; Bishop, J. F.; Czekai, D. A. Surface Modified Drug Nanoparticles. U.S. Patent 5,145,684, Assigned to Sterling Drug Inc., Sept. 8, 1992.

Luck, M.; Schroder, W.; Harnisch, S.; Thode, K.; Blunk, T.; Paulke, B. R.; Kresse, M.; Muller, R. H. Identification of Plasma Proteins Facilitated by Enrichment on Particulate Surfaces: Analysis by Two-dimensional Electrophoresis and N-terminal Microsequencing. *Electrophoresis* **1997,** *18*, 2961–2967.

Mansour, H. M.; Rhee, Y. S.; Wu, X. Nanomedicine in Pulmonary Delivery. *Int. J. Nanomed.* **2009,** *4*, 299–319.

Mantzaris, N. Liquid-phase Synthesis of Nanoparticles: Particle Size Distribution Dynamics and Control. *Chem. Eng. Sci.* **2005,** *60*, 4749–4770.

McDonald, T. O.; Giardiello, M.; Martin, P.; Siccardi, M.; Liptrott, N. J.; Smith, D.; Roberts, P.; Curley, P.; Schipani, A.; Khoo, S. H.; Long, J.; Foster, A. J.; Rannard, S. P.; Owen, A. Antiretroviral Solid Drug Nanoparticles with Enhanced Oral Bioavailability: Production, Characterization, and *In vitro-In vivo* Correlation. *Adv. Healthc. Mater.* **2014,** *3*, 400–411.

McKee, J.; Rabinow, B.; Cook, C.; Gass, J. Nanosuspension Formulation of Itraconazole Eliminates the Negative Inotropic Effect of SPORANOX in Dogs. *J. Med. Toxicol.* **2010,** *6*, 331–336.

Merisko-Liversidge, E.; Liversidge, G. G. Nanosizing for Oral and Parenteral Drug Delivery: A Perspective on Formulating Poorly-water Soluble Compounds Using Wet Media Milling Technology. *Adv. Drug. Deliv. Rev.* **2011,** *63*, 427–440.

Mirani, A. G.; Patravale, V. B. Designs of Experiments: Basic Concepts and its Application in Pharmaceutical Product Development. In *Pharmaceutical Product Development: Insights*

Into Pharmaceutical Processes, Management and Regulatory Affairs; Patravale, V. B., Disouza, J. I., Rustomjee M. T., Eds.; CRC Press: Boca Raton, Florida, 2016; pp 117–162.

Mosharraf, M.; Nystrom, C. The Effect of Particle Size and Shape on the Surface Specific Dissolution Rate of Micronized Practically Insoluble Drugs. *Int. J. Pharm.* **1995**, *122*, 35–47.

Möschwitzer, J. P. Drug Nanocrystals in the Commercial Pharmaceutical Development Process. *Int. J. Pharm.* **2013**, *453*, 142–156.

Mouton, J. W.; Van, P. A.; de Beule, K.; Van, V. A.; Donnelly, J. P.; Soons, P. A. Pharmacokinetics of Itraconazole and Hydroxyl Itraconazole in Healthy Subjects After Single and Multiple Doses of a Novel Formulation. *Antimicrob. Agents Chemother.* **2006**, *50*, 4096–4102.

Muller, R. H.; Peters, K.; Becker, R.; Kruss, B. Nanosuspensions for the I.V. Administration of Poorly Soluble Drugs: Stability During Sterilization and Long-term Storage. *Proc. Control. Rel. Soc.* **1995**, *22*, 574–575.

Muller, R. H.; Becker, B.; Kruss, P. K. Pharmaceutical Nanosuspensions for Medicament Administration as System of Increased Saturation Solubility and Rate of Solution. U.S. Patent 5,858,410, January 12, 1999.

Noyes, A.; Whitney, W. The Rate of Solution of Solid Substances in Their Own Solutions. *J. Am. Chem. Soc.* **1897**, *19*, 930–934.

Ostrander, K. D.; Bosch, H. W.; Bondanza, D. M. An *In vitro* Assessment of a Nano Crystal Beclomethasone Dipropionate Colloidal Dispersion via Ultrasonic Nebulization. *Eur. J. Pharm. Biopharm.* **1999**, *48* (3), 207–215.

Pace, S. N.; Pace, G. W.; Parikh, I.; Mishra, A. K. Novel Injectable Formulations of Insoluble Drugs. *Pharm. Tech.* **1999**, *23*, 116–133.

Patel, A.; Cholkar, K.; Agrahari, V.; Ashim K. M. Ocular Drug Delivery Systems: An Overview. *World J. Pharmacol.* **2013**, *9, 2* (2), 47–64.

Patravale, V. B.; Date, A. A.; Kulkarni, R. M. Nanosuspensions: A Promising Drug Delivery Strategy. *J. Pharm. Pharmacol.* **2004**, *56*, 827–840.

Patton, J. S.; Byron, P. R. Inhaling Medicines: Delivering Drugs to the Body Through the Lungs. *Nat. Rev. Drug Discov.* **2007**, *6*, 67–74.

Plakkot, S.; de Matas, M.; Sulaiman, B. Comminution of Ibuprofen to Produce Nano-particles for Rapid Dissolution. *Int. J. Pharm.* **2011**, *415* (1–2), 307–314.

Peltonen, L.; Hirvonen, J. Pharmaceutical Nanocrystals by Nanomilling: Critical Process Parameters, Particle Fracturing and Stabilization Methods. *J. Pharm. Pharmacol.* **2010**, *62*, 1569–1579.

Peltonen, L.; Annika, T.; Jouni, H. Polymeric Stabilizers for Drug Nanocrystals. In *Handbook of Polymers for Pharmaceutical Technologies*; Thakur, V. K., Thakur, M. K., Eds.; Scrivener Publishing LLC: Beverly, Massachusetts, 2016; Vol. 4, pp 67–88.

Peters, K.; Muller, R. H. An Investigation into the Distribution of Lecithins in Nanosuspension Systems Using Low Frequency Dielectric Spectroscopy. *Int. J. Pharm.* **1999**, *184* (1), 53–61.

Petersen, R. (2008). Nanocrystals for Use in Topical Cosmetic Formulations and Method of Production Thereof. Patent: WO2008058755.

Ponchel, G.; Montisci, M.; Dembri, A. Mucoadhesion of Colloidal Particulate Systems in the Gastro-intestinal Tract. *Eur. J. Pharm. Biopharm.* **1997**, *44*, 25–31.

Rabinow, B. E. Nanosuspensions in Drug Delivery. *Nat. Rev. Drug. Discov.* **2004**, *3*, 785–796.

Rabinow, B.; Kipp, J.; Papadopoulos, P.; Wong, J.; Glosson, J.; Gass, J.; Sun, C. S.; Wielgos, T.; White, R.; Cook, C.; Barker, K.; Wood, K. Itraconazole IV Nanosuspension Enhances Efficacy Through Altered Pharmacokinetics in the Rat. *Int. J. Pharm.* **2007,** *339,* 251–260.

Radzio, J.; Spreen, W.; Yueh, Y. L.; Mitchell, J.; Jenkins, L.; García-Lerma, J. G.; Heneine, W. The Long-acting Integrase Inhibitor GSK744 Protects Macaques from Repeated Intravaginal SHIV Challenge. *Sci. Transl. Med.* **2015,** *7,* 2705.

Rodriguez-Aller, M.; Guillarme, D.; Veuthey, J. L.; Gurny, R. Strategies for Formulating and Delivering Poorly Water-soluble Drugs. *J. Drug Deliv. Sci. Tech.* **2015,** *30,* 342–351.

Romero, G. B.; Chen, R.; Keck, C. M.; Müller, R. H. Industrial Concentrates of Dermal Hesperidin smartCrystals®-production, Characterization & Long-term Stability. *Int. J. Pharm.* **2015,** *482,* 54–60.

Rundfeldt, C.; Steckel, H.; Scherliess, H.; Wyska, E.; Wla, P. Inhalable Highly Concentrated Itraconazole Nanosuspension for the Treatment of Broncho Pulmonary Aspergillosis. *Eur. J. Pharm. Biopharm.* **2013,** *83,* 44–53.

Sanganwar, G. P.; Sathigari, S.; Babu, R. J.; Gupta, R. B. Simultaneous Production and Co-mixing of Microparticles of Nevirapine with Excipients by Supercritical Antisolvent Method for Dissolution Enhancement. *Eur. J. Pharm. Biopharm.* **2010,** *39* (1–3), 164–174.

Schoenwald, R. D.; Stewart, P. Effect of Particle Size on Ophthalmic Bioavailability of Dexamethasone Suspensions in Rabbits. *J. Pharm. Sci.* **1980,** *69,* 391–394.

Scholer, N.; Krause, K.; Kayser, O. Atovaquone Nanosuspensions Show Excellent Therapeutic Effect in a New Murine Model of Reactivated Toxoplasmosis. *Antimicrob. Agents Chemother.* **2001,** *45,* 1771–1779.

Scholz, P.; Keck, C. M. Nanocrystals: From Raw Material to the Final Formulated Oral Dosage Form—A Review. *Curr. Pharm. Des.* **2015,** *21,* 4217–4228.

Shegokar, R.; Müller, R. Nanocrystals: Industrially Feasible Multifunctional Formulation Technology for Poorly Soluble Actives. *Int. J. Pharm.* **2010,** *399,* 129–139.

Shegokar, R.; Singh, K. K. Surface Modified Nevirapine Nanosuspensions for Viral Reservoir Targeting: *In vitro* and *In vivo* Evaluation. *Int. J. Pharm.* **2011,** *421,* 341–352.

Sinha, V. R. Emerging Potential of Nanosuspension-enabled Drug Delivery: An Overview. *Crit. Rev. Ther. Drug. Carrier Syst.* **2015,** *32,* 535–557.

Sinha, B.; Müller, R. H.; Möschwitzer, J. P. Bottom-up Approaches for Preparing Drug Nanocrystals: Formulations and Factors Affecting Particle Size. *Int. J. Pharm.* **2013,** *453* (1), 126–141.

Takagi, T.; Ramachandran, C.; Bermejo, M.; Yamashita, S.; Yu, L. X.; Amidon, G. L. A Provisional Biopharmaceutical Classification of the Top 200 Oral Drug Products in the United States, Great Britain, Spain, and Japan. *Mol. Pharm.* **2006,** *3,* 631–643.

Talekar, M.; Gantab, S.; Amiji, M.; Jamieson, S.; Kendall, J.; Dennyd, W.; Garg, S. Development of PIK-75 Nanosuspension Formulation with Enhanced Delivery Efficiency and Cytotoxicity for Targeted Anti-cancer Therapy. *Int. J. Pharma.* **2013,** *450,* 278–228.

Trotta, M.; Gallarate, M. Emulsions Containing Partially Water-miscible Solvents for the Preparation of Drug Nanosuspensions. *J. Control Release* **2001,** *76* (1–2), 119–128.

Trotta, M.; Gallarate, M. Preparation of Griseofulvin Nanoparticles from Water Dilutable Microemulsions. *Int. J. Pharm.* **2003,** *254* (2), 235–242.

Van Eerdenbrugh, B.; Froyen, L; Van, H. J.; Martens, J. A.; Augustijns, P.; Van den, M. G. Drying of Crystalline Drug Nanosuspensions the Importance of Surface Hydrophobicity on Dissolution Behavior Upon Redispersion. *Eur. J. Pharm. Sci.* **2008a,** *35* (1–2), 127–135.

Van Eerdenbrugh, B.; Van den, M. G.; Augustijns, P. Top Down Production of Drug Nano-crystals: Nanosuspension Stabilization, Miniaturization and Transformation into Solid Products. *Int. J. Pharm.* **2008b,** *364* (1), 64–75.

Van Eerdenbrugh, B.; Vermant, J.; Martens, J. A.; Froyen, L.; Van Humbeeck, J.; Augustijns, P.; Van den Mooter, G. A Screening Study of Surface Stabilization During the Production of Drug Nanocrystals. *J. Pharm. Sci.* **2009,** *98* (6), 2091–2103.

van 't Klooster, G.; Hoeben, E.; Borghys, H.; Looszova, A.; Bouche, M. P.; van Velsen, F.; Baert, L. Pharmacokinetics and Disposition of Rilpivirine (TMC278) Nanosuspension as a Long-acting Injectable Antiretroviral Formulation. *Antimicrob. Agents Chemother.* **2010,** *54,* 2042–2050.

Wais, U.; Jackson, A. W.; He, T.; Zhang, H. Nanoformulation and Encapsulation Approaches for Poorly Water-soluble Drug Nanoparticles. *Nanoscale* **2016,** *8,* 1746–1769.

Wallis, K. H.; Muller, R. H. Determination of the Surface Hydrophobicity of Colloidal Dispersions by Mini-hydrophobic Interaction Chromatography. *Pharm. Ind.* **1993,** *55,* 1124–1128.

Wang, Y.; Liu, Z.; Zhang, D.; Gao, X.; Zhang, X.; Duan, C.; Jia, L.; Feng, F.; Huang, Y.; Shen, Y.; Zhang, Q. Development and *In vitro* Evaluation of Deacetymycoepoxydiene Nanosus-pension. *Colloids Surf. B Bioint.* **2011,** *83,* 189–197.

Wiedmann, T. S.; DeCastro, L.; Wood, R. W. Nebulization of NanoCrystals: Production of a Respirable Solid-in-liquid-in-air Colloidal Dispersion. *Pharm. Res.* **1997,** *14* (1), 112–116.

Wu, L.; Zhang, J.; Watanabe, W. Physical and Chemical Stability of Drug Nanoparticles. *Adv. Drug Delivery Rev.* **2011,** *63,* 456–469.

Xu, R. Extracted Polarization Intensity Differential Scattering for Particle Characterization. U. S. Patent 6,859,276, February 22, 2005.

Yang, J.; Young, A.; Chiang, P. C.; Thurston, A.; Pretzer, D. Fluticasone and Budesonide Nanosuspensions for Pulmonary Delivery: Preparation, Characterization, and Pharmacoki-netic Studies. *J. Pharm. Sci.* **2008,** *97* (11), 4869–4878.

Yang, W.; Johnston, K. P.; Williams R. O. Comparison of Bioavailability of Amorphous versus Crystalline Itraconazole Nanoparticles via Pulmonary Administration in Rats. *Eur. J. Pharm. Biopharm.* **2010,** *75,* 33–41.

Yin, T.; Cai, H.; Liu, J.; Cui, B.; Wang, L.; Yin, L.; Zhou, J.; Huo, M. Biological Evaluation of PEG Modified Nanosuspensions Based on Human Serum Albumin for Tumor Targeted Delivery of Paclitaxel. *Eur. J. Pharm. Sci.* **2016,** *83,* 79–87.

Zhang, D.; Tan, T.; Gao, L.; Zhao, W.; Wang, P. Preparation of Azithromycin Nanosuspen-sions by High Pressure Homogenization and its Physicochemical Characteristics Studies. *Drug Dev. Ind. Pharm.* **2007,** *33* (5), 569–575.

CHAPTER 6

NANOSUSPENSIONS FOR DRUG DELIVERY

KIRTI RANI*

Amity Institute of Biotechnology, Amity University Uttar Pradesh, Sector 125, Noida 201303, Uttar Pradesh, India

E-mail: krsharma@amity.edu; kirtisharma2k@rediffmail.com

ABSTRACT

Some of the poor water-soluble drug molecules are major problem for drug formulation practices. So, drug nanosuspensions are considered as an effective emerging solution which can be used for safe delivery of such kind of hydrophobic drugs. Scaling down of biocompatible nanoparticles or nano-biomaterials led to enhance drug aqueous solubility and bioavailability by increasing drug surface area when that comes into contact with biological media. Formulated nanosuspensions are stabilized by various polymers to get effective targeted delivery in cancer tissues and infarct zones with minimal damage to healthy tissues. So, the concept of preparation and application of nanosuspensions in this chapter is depicted in brief for highlighting their effective drug delivery designing and their administration routes, for example, parenteral, pulmonary, oral, and ocular which can be considered further for clinical and pharmaceutical purposes.

6.1 INTRODUCTION

Poor water solubility of some of the drug molecules is found to be a major problem during their formulation for targeted drug delivery (Amidon et al., 1995; Kawabata et al., 2011; Martinez et al., 2002). The Biopharmaceutical

Classification Systems (BCS) has allocated these poorly water-soluble drugs further into four classes: high solubility, high permeability (Class I); low solubility and high permeability (Class II); high solubility and low permeability (Class III); and low solubility and low permeability (Class IV) (Reddy and Karunakar, 2011) (Table 6.1).

TABLE 6.1 Examples of Drugs in Different Biopharmaceutical Classification Systems (BCS) Classes.

Biopharmaceutical classification systems (BCS classes)	Drug examples
I	Metoprolol, diltiazem, verapamil, propranolol
II	Glibenclamide, phenytoin, danazol, mefenamic acid, nifedipine, ketoprofen, naproxen, carbamazepine, ketoconazole
III	Cimetidine, ranitidine, acyclovir, neomycin B, atenolol, captopril
IV	Hydrochlorothiazide, taxol, furosemide

Nowadays, formulated nanosuspensions are prepared to stabilize them by various polymers to get effective targeted delivery in cancer tissues and infarct zones with minimal damage to healthy tissues to achieve effective drug delivery design and their administration routes, for example, parenteral, pulmonary, oral, and ocular drug administration. Approximately 40% of lipophilic compounds have poor aqueous solubility which causes their low bioavailability of loaded drug whose administration or delivery is done via systemic circulation (Mercadante et al., 2010).

Most exploited characteristics of nanosuspensions are their exhibited particle size, polydispersity index, drug saturation solubility, physical stability, dissolution rate (Fig. 6.1), and bioavailability to make them more effective drug delivery system as compared to other conventional clinical and pharmaceutical designs (Gao et al., 2008; Shid et al., 2104).

Nanoformulation of azithromycin has been demonstrated as a significant tool to increase its dissolution percentage more than 65% to get increased drug bioavailability and to achieve the maximum possibility of lipophilic drugs to incorporate them in various nanosuspensions dosage formats such as tablets, pellets, and capsules following standard manufacturing techniques (Kesisoglou et al., 2002; Muller et al., 2001). Previously, ketoprofen nanosuspension has reported whose dissolution percentage was increased by performing its fabrication and it was successfully transformed into pellets with its lower fed/fasted variability (Chaubal, 2004).

Emulsion–diffusion method has been also reported to prepare these kinds of poorly water-soluble drug-loaded nanosuspensions by using partially water-miscible and volatile organic solvents such as butyl lactate, benzyl alcohol, triacetin, and ethyl acetate as the dispersed phase systems (Trotta et al., 2001; Kocbek et al., 2006). These kinds of nanoemulsion formations are prepared by dispersing the drug in a mixture of solvents or an organic solvent by using high-pressure homogenization that leads to the diffusion of the internal phase into the external phase when droplets convert into semi-solid to solid nanoparticles. But, only major drawback of this technology was accounted for their potential environmental hazards and human safety issues due to use of organic solvents such as ethyl acetate, ethanol, methanol, and chloroform (Dandagi et al., 2009; Kocbek et al., 2006).

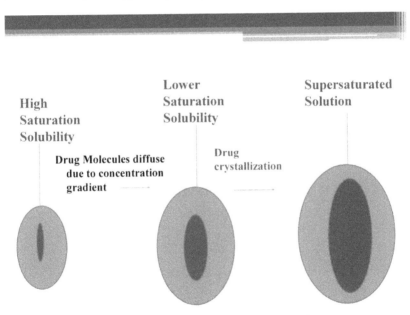

FIGURE 6.1   (See color insert.) Drug diffusivity in nanosuspensions.

6.2   BASIC METHODS FOR PREPARING NANOSUSPENSION USED FOR DRUG DELIVERY

Major reported methods of preparation of nanosuspension are Disso cubes, Nanopure, Nanoedge and Nanojet (Athul, 2013; Nash, 2002; Sharma et al., 2009) and their accounted steps of preparation and various evaluation parameters are given in Figures 6.2 and 6.3.

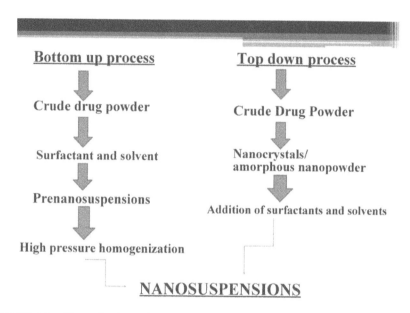

FIGURE 6.2   (See color insert.) Nanosuspensions preparation.

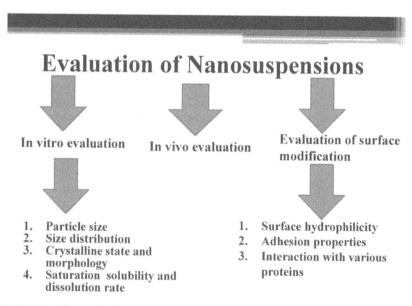

FIGURE 6.3   (See color insert.) Evaluation parameter of nanosuspensions.

6.2.1 DISSO CUBES TECHNOLOGY

Disso cubes (homogenization in aqueous media), a technology which was developed by R. H. Muller using a piston-gap type high-pressure homogenizer. In this method, the suspension containing a drug and surfactant is forced under pressure through a nanosized aperture valve of a high-pressure homogenizer (Muller et al., 2000).

6.2.2 NANOEDGE TECHNOLOGY

Nanoedge technology is the combination of both precipitation and homogenization. In this method, water starts boiling at room temperature and forms gas bubbles, which implode when the suspension leaves the gap, called cavitation, till normal air pressure is reached. These particles cavitation forces are sufficiently high to convert the drug microparticles into nanoparticles. (Dearns, 2000)

6.2.3 NANOPURE TECHNOLOGY

Nanopure (homogenization in nonaqueous media) is one of the reported technologies for preparation of nanosuspensions in which suspension is homogenized in water-free media or water mixtures such as PEG 400, PEG 1000, etc. at room temperature and is further subjected to below freezing point ($-200°C$), called, "deep freeze" homogenization (Keck et al., 2006).

6.2.4 NANOJET TECHNOLOGY

Nanojet is also known as opposite stream technology where a stream of suspension is divided into two or more parts in different chambers which collide with each other at high pressure, due to the high shear forces produced till the particle size of the suspension is reduced to achieve its desired maximum nanosized fabricated nanosuspensions (Prassanna and Giddam, 2010).

6.3 APPLICATION OF NANOSUSPENSIONS IN DRUG DELIVERY

Nanosuspensions are commonly used as oral, parenteral, ocular, and pulmonary drug delivery systems where drugs have poor solubility and dissolution in water.

6.3.1 ORAL ADMINISTRATION

Oral administration is the first clinical choice because of its painless and noninvasive administration given to patients. Oleanolic acid, which has many applications such as hepatoprotective, antitumor, antibacterial, antiinflammatory, and antiulcer effects, has low aqueous solubility which results in erratic pharmacokinetics oral administration. So, applying oleanolic acid in the form of nanosuspension increases dissolution rate of nanopowder of drug up to 90% (Chen et al., 2005). This reduction of drug particle size to the nanoscale leads to an increased dissolution rate which can improve adhesion of the drug particles to the mucosa and intestinal cells, called bioadhesive phase, that has greater concentration gradient between blood and gastrointestinal tract (GIT) to increase drug intestinal absorption (Arunkumar et al., 2009; Venkatesh et al., 2011).

6.3.2 PARENTERAL ADMINISTRATION

Parenteral administration includes administration of dosage forms by subcutaneous administration routes using intramuscular and intra-arterial methods which include the avoidance of first-pass metabolism, reliable doses and higher bioavailability. This controlled rate of drug delivery allows more predictable pharmacodynamic and pharmacokinetic profiles as compared to oral administration (Bhalla, 2007).

In addition, a study on mice investigated tumor growth inhibition rate that showed the oridonin in the form of nanosuspension decreased considerably the volume and weight of the tumor. Oridonin in the form of nanosuspension was observed to raise the rate of tumor inhibition to 60.23% as compared to 42.49% for the conventional form (Lou et al., 2009). So, these kinds of fabricated nanosuspensions can improve therapeutic efficiency and reduce the cost of therapy through improved dosing efficiency and smaller injection volumes.

6.3.3 PULMONARY DRUG DELIVERY

Pulmonary drug delivery has been reported to be more advantageous over oral and parenteral drug administration which includes direct delivery to the site of action which leads to decreased dosage and side effects (Liao and Wiedmann, 2003). Other reported conventional pulmonary delivery systems were noted to provide only rapid drug release, poor residence time, and lack of selectivity (Beck-Broichsitter, 2011).

Nanosuspensions can solve problems of poor drug solubility in pulmonary secretions and lack of selectivity through direct delivery to target pulmonary cells by improving drug diffusion and dissolution rate which can consequently increase bioavailability and prevent undesirable drug deposition in the mouth and pharynx.

Surface engineered nanosuspensions provide quick onset followed by controlled drug release for achieving optimal drug delivery pattern for most pulmonary diseases. Moreover, nanosuspensions for treating lung infections have demonstrated well for their observed good proportion between actual and delivered drug concentrations in each actuation (Dhiman et al., 2011).

6.3.4 OCULAR ADMINISTRATION

Nanosuspensions as ocular drug delivery systems also offer several advantages over other conventional methods used for administration of drugs to combat several diseases. For this, appropriate bioerodible polymers were used for modifications of nanoparticle surface which causes prolonged residual time for getting effective treatment of targeted diseased cell. Commonly reported polymers in ocular nanosuspensions are poly(alkyl cyanoacrylates), polycaprolactone, and poly(lactic acid)/poly(lactic-co-glycolic acid) (Gupta et al., 2010). Employing these polymers in ocular drug delivery significantly prolongs ocular residence time of drug and improves bioavailability. Among them, chitosan was reported to be a more potent mucoadhesive cationic polymer used in ocular drug delivery which bind well with negatively charged mucin and enhance drug residence time (Nagarwal et al., 2009). Flurbiprofen nanosuspensions covered by Eudragit polymers RS 100 and RL 100 were found to exhibit prolonged drug release during administration (Pignatello et al., 2002).

6.4 CONCLUSION

So, the information provided in this chapter can be very helpful for depicting the methods using chemical entities which lead to form more potent nano-suspensions which might be used for future clinical therapies and pharmaceutical trials including their high surface area, controllable nanosize dimensions, and tailored surface chemistry. In addition, nanosuspensions may have good ease for advanced fabrication of nanoparticles to increase their surface properties for getting optimized bioavailability response for treating a number of diseases, for example, cardiovascular disease, cancer, diabetes, Parkinson's, and Alzheimer's disease. Hence, this approach might be useful to understand the behavior of fabricated nanosuspensions *in vivo* systems, including their interactions with cells and different biological barriers such as the blood–brain barrier to achieve effective active or passive targeting to combat the various clinical challenges that are associated with poorly soluble drugs.

ACKNOWLEDGMENT

The author would like to express her cordially appreciation to Amity University Uttar Pradesh, Noida, India.

KEYWORDS

- dispersions
- nanotechnology
- drug delivery
- routes of drug administration
- therapeutics
- biopharmaceutical classification system

REFERENCES

Amidon, G. L.; Lennernas, H.; Shah, V. P.; Crison, J. R. A Theoretical Basis for a Biopharmaceutic Drug Classification: The Correlation of *In vitro* Drug Product Dissolution and *In vivo* Bioavailability. *Pharm. Res.* **1995,** *12* (3), 413–420.

Arunkumar, N., Deecaraman, M.; Rani, C. Nanosuspension Technology and Its Applications in Drug Delivery. *A. J. Pharm.* **2009,** *3* (3), 168–173.

Athul, P. V. Preparation and Characterization of Simvastatin Nanosuspension by Homogenization Method. *Int. J. Pharm. Tech. Res.* **2013,** *5* (1), 189–197.

Beck-Broichsitter, M. Pulmonary Drug Delivery with Nanoparticles. In *Nanomedicine in Health and Disease*; Hunter, R. J., Preedy, V. R., Eds.; CRC Press: NY, USA, 2011; pp 229–248

Bhalla, S. Parenteral Drug Delivery. In *Gibaldi's Drug Delivery Systems in Pharmaceutical Care*; Lee, M.; Desai, A., Eds.; ASHP: Bethesda, MD, USA, 2007; p 107

Chaubal, M. V. Application of Formulation Technologies in Lead Candidate Selection and Optimization. *Drug Discov. Today* **2004,** *9* (14), 603–609.

Chen, Y.; Liu, J.; Yang, X.; Zhao, X.; Xu, H. Oleanolic Acid Nanosuspensions: Preparation, *In vitro* Characterization and Enhanced Hepatoprotective Effect. *J. Pharm. Pharmacol.* **2005,** *57* (2), 259–264.

Keck, C. M; Muller, R. H. Drug Nanocrystals of Poorly Soluble Drugs Produced by High Pressure Homogenization. *Eur. J. Pharm. Biopharm.* **2006,** *62,* 3–16.

Dandagi, P.; Kerur, S.; Mastiholimath, V.; Gadad, A.; Kulkarni, A. Polymeric Ocular Nanosuspension for Controlled Release of Acyclovir: *In vitro* Release and Ocular Distribution. *Iranian J. Pharm. Res.* **2009,** *8* (2), 79–86.

Dearns R. Atovaquone Pharmaceutical Compositions. US Patent US 6,018,080, January 25, 2000.

Dhiman, S.; Singh, T. G.; Dharmila. Nanosuspension: A Recent Approach for Nano Drug Delivery System. *Int. J. Curr. Pharm. Res.* **2011,** *3* (4), 96–101.

Gao, L.; Zhang, D.; Chen, M. Drug Nanocrystals for the Formulation of Poorly Soluble Drugs and Its Application as a Potential Drug Delivery System. *J. Nanopart. Res.* **2008,** *10* (5), 845–862.

Gupta, H.; Aqil, M.; Khar, R. K.; Ali, A.; Bhatnagar, A.; Mittal, G. Sparfloxacin-loaded PLGA Nanoparticles for Sustained Ocular Drug Delivery. *Nanomed: Nanotechnol. Biol. Med.* **2010,** *6* (2), 324–333.

Kawabata, Y.; Wada, K.; Nakatani, M.; Yamada, S.; Onoue, S. Formulation Design for Poorly Water-soluble Drugs Based on Biopharmaceutics Classification System: Basic Approaches and Practical Applications. *Int. J. Pharm.* **2011,** *420* (1), 1–10.

Kesisoglou, F.; Panmai, S.; Wu, Y. Nanosizing—Oral Formulation Development and Biopharmaceutical Evaluation. *Adv. Drug Deliv. Rev.* **2002,** *59* (7), 631–644.

Kocbek, P.; Baumgartner, S.; Kristl, J. Preparation and Evaluation of Nanosuspensions for Enhancing the Dissolution of Poorly Soluble Drugs. *Int. J. Pharm.* **2006,** *312* (1–2), 179–186.

Liao, X.; Wiedmann, T. S. Solubilization of Cationic Drugs in Lung Surfactant. *Pharm. Res.* **2003,** *20* (11), 1858–1863.

Lou, H.; Zhang, X.; Gao, L. *In vitro* and *In vivo* Antitumor Activity of Oridonin Nanosuspension. *Int. J. Pharm.* **2009,** *379* (1–2), 181–186.

Martinez, M.; Augsburger, L.; Johnston, T.; Jones, W. W. Applying the Biopharmaceutics Classification System to Veterinary Pharmaceutical Products. Part I: Biopharmaceutics and Formulation Considerations. *Adv. Drug Deliv. Rev.* **2002,** *54* (6), 805–824.

Mercadante, S.; Vitrano, V. Pain in Patients with Lung Cancer: Pathophysiology and Treatment. *Lung Cancer* **2010,** *68,* 10–15.

Muller, R. H.; Jacobs, C.; Kayer, O. Nanosuspensions for the Formulation of Poorly Soluble Drugs. In *Pharmaceutical Emulsion and Suspension*; Nielloud, F., Marti-Mestres, G., Eds., Marcel Dekker: New York, 2000; pp 383–407.

Muller, R. H.; Jacobs, C.; Kayser, O. Nanosuspensions as Particulate Drug Formulations in Therapy: Rationale for Development and What We Can Expect for the Future. *Adv. Drug Deliv. Rev.* **2001,** *47* (1), 3–19.

Nagarwal, R. C.; Kant, S.; Singh, P. N.; Maiti, P.; Pandit, J. K. Polymeric Nanoparticulate System: A Potential Approach for Ocular Drug Delivery. *J. Control. Release* **2009,** *136* (1), 2–13.

Nash, R. A. Suspensions. In *Encyclopedia of Pharmaceutical Technology*, 2nd ed.; Swarbrick J, Boylan J. C., Eds.; Marcel Dekker: New York, 2002; Vol. 3, pp 2045–3032.

Pignatello, R.; Bucolo, C.; Spedalieri, G.; Maltese, A.; Puglisi, G. Flurbiprofen-loaded Acrylate Polymer Nanosuspensions for Ophthalmic Application. *Biomaterials* **2002,** *23* (15), 3247–3255.

Prassanna, L; Giddam, A. K. Nanosuspensions Technology: A Review. *Int. J. Pharm.* **2010,** *2* (4), 35–40.

Reddy, B. B.; Karunakar, A. Biopharmaceutics Classification System: A Regulatory Approach. *Dissolut. Technol.* **2011,** 31–37 dx.doi.org/10.14227/DT180111P31.

Sharma, D.; Soni, M.; Kumar, S.; Gupta, G. D. Solubility Enhancement-eminent Role in Poorly Soluble Drugs. *Res. J. Pharm. Tech.* **2009,** *2* (2), 220–224.

Shid, R. L.; Dhole, S. N.; Kulkarni, N. ; Shid, S. L. Formulation and Evaluation of Nanosuspensions Formulation for Drug Delivery of Simvastin. *Int. J. Pharm. Sci. Nanotech.* **2014,** *7* (4), 2650–2665.

Trotta, M.; Gallarate, M.; Pattarino, F.; Morel, S. Emulsions Containing Partially Water-miscible Solvents for the Preparation of Drug Nanosuspensions. *J. Control. Release* **2001,** *76* (1–2), 119–128.

Venkatesh, T., Reddy, A. K.; Uma Maheswari, J.; Deena Dalith, M.; Ashok Kumar, C. K. Nanosuspensions: Ideal Approach for the Drug Delivery of Poorly Water Soluble Drugs. *Der Pharmacia Lettre* **2011,** *3* (2), 203–213.

FIGURE 1.1   (A) Nanoemulsion system and (B) microemulsion system.

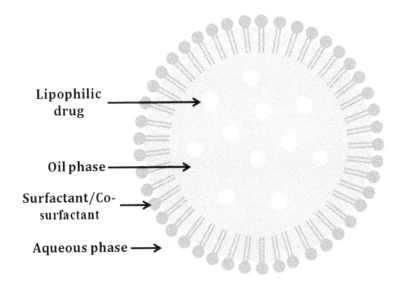

FIGURE 2.1   Lipophilic drug encapsulated inside nanoemulsion system.

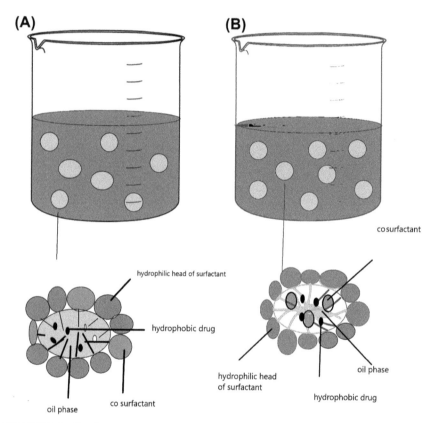

FIGURE 3.2　Depicting (A) nanoemulsion droplet; round and showing polydispersity in size and (B) microemulsion droplet; shape which may be or may be not sphere-shaped and entire droplets of equal diameter and shape.

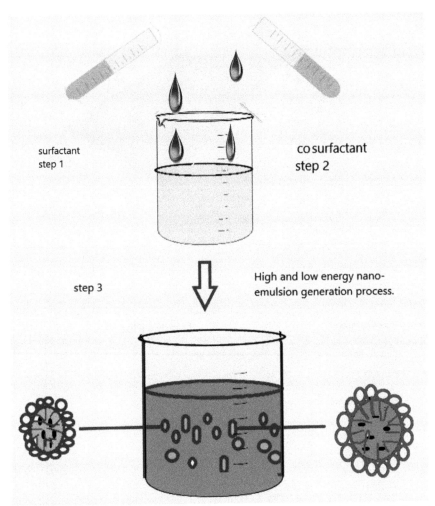

FIGURE 3.3 Nanoemulsion preparation requires a specific mixing order where oil phase is first mixed with surfactant; however, microemulsions do not require any particular mixing order.

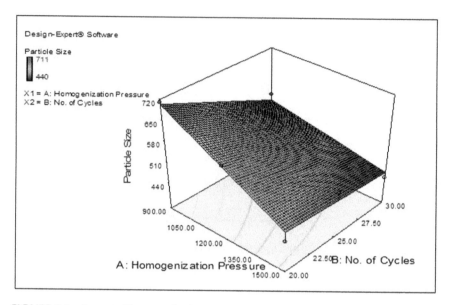

FIGURE 5.6   Impact of homogenization pressure and number of cycles on particle size.

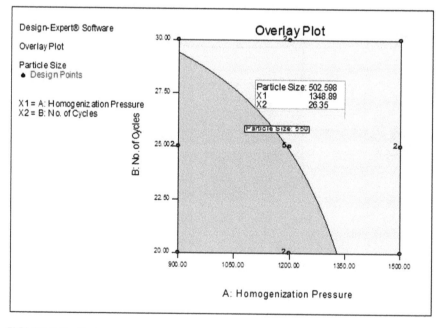

FIGURE 5.7   Design space.

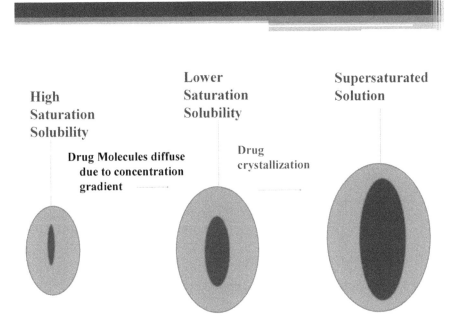

FIGURE 6.1   Drug diffusivity in nanosuspensions.

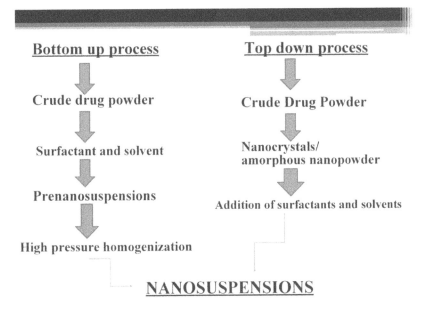

FIGURE 6.2   Nanosuspensions preparation.

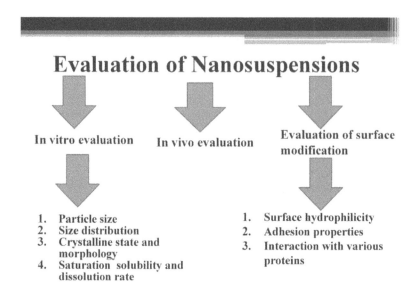

Evaluation of Nanosuspensions

In vitro evaluation   **In vivo evaluation**   **Evaluation of surface modification**

1. Particle size
2. Size distribution
3. Crystalline state and morphology
4. Saturation solubility and dissolution rate

1. Surface hydrophilicity
2. Adhesion properties
3. Interaction with various proteins

FIGURE 6.3   Evaluation parameter of nanosuspensions.

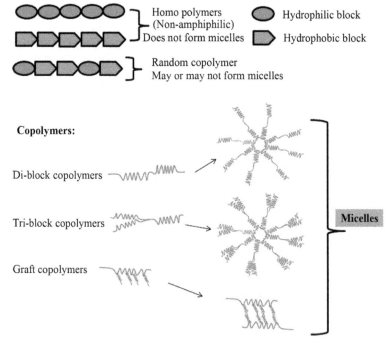

Homo polymers (Non-amphiphilic) Does not form micelles

Hydrophilic block

Hydrophobic block

Random copolymer May or may not form micelles

Copolymers:

Di-block copolymers

Tri-block copolymers

Graft copolymers

Micelles

FIGURE 7.2   Main structural types of copolymers and micelles formed from amphiphilic copolymers.

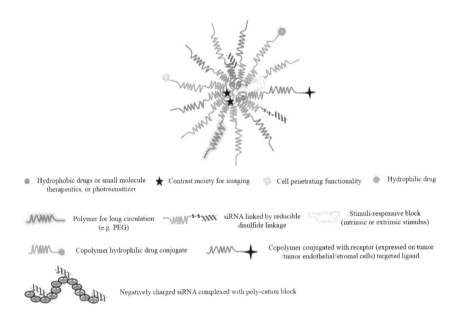

FIGURE 7.7 A hypothetical multifunctional micelles. Micelles can be designed to incorporate two or more of the above functions.

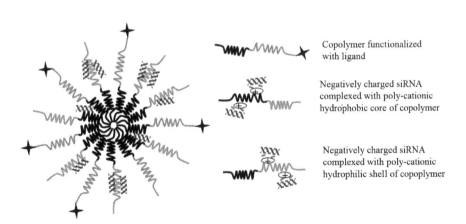

FIGURE 7.8 Schematic illustration of functionalized micelles designed for siRNA delivery.

SECTION III
DIVERSE DISPERSED SYSTEMS

CHAPTER 7

NANOMICELLAR APPROACHES FOR DRUG DELIVERY

A. S. MANJAPPA[1], P. S. KAKADE[2], J. I. DISOUZA[1], and
VANDANA B. PATRAVALE[2*]

[1]*Department of Pharmaceutics, Tatyasaheb Kore College of Pharmacy,
Warananagar, Panhala, Kolhapur, Maharashtra, India*

[2]*Department of Pharmaceutical Sciences and Technology, Institute
of Chemical Technology, Nathalal Parekh Marg, Matunga, Mumbai
400019, Maharashtra, India*

Corresponding author. E-mail: vbp_muict@yahoo.co.in

ABSTRACT

Micellar system, a self-assembling nanoconstruct of amphiphilic copolymers with a core–shell structure, has received growing scientific attention
as an efficient carrier of therapeutics in recent years owing to its solubilization, biocompatibility, high stability, selective targeting, P-glycoprotein
inhibition, and altered drug internalization route and subcellular localization properties. This chapter summarizes the micellar systems tested orally
for improving drug absorption either by increasing drug membrane permeability and/or by inhibiting drug efflux transporters, bioadhesive micelles
which lengthen the residence time in the gastrointestinal tract and promote
drug permeation, and pH-sensitive micelles which ensure therapeutic
release at its absorption site. This chapter focuses also on the recent developments related to targeted delivery of anticancer therapeutics to tumor sites
using micellar approaches via passive and active mechanisms. Furthermore,
it elaborates on "smart" multifunctional micelles designed for enhanced
biological performance by modification of copolymers to contain chemistries that can sense a specific biological environment. This chapter also

enumerates the strategies developed to enhance *in vivo* stability of micelles and performances of micelles tested via pulmonary, nasal, and rectal route. Finally, the clinical toxicity of micelles and strategies to improve clinical translation of micelles or other nanomedicines are pinpointed.

7.1   INTRODUCTION

7.1.1   MICELLES AND MICELLE FORMATION

Micelles are spherical, supramolecular, self-assembled, nanosized colloidal particles with an internal hydrophobic core and an external hydrophilic corona (shell) formed by the aggregation of amphiphilic molecules or surfactants in aqueous medium. Amphiphilic molecules or surfactant monomers that have a polar head and a lipophilic tail could show changes in their physicochemical properties in solutions. These changes are associated with the orientation and association of amphiphilic molecules in solution and result in the formation of structures called micelles (Fig. 7.1). The radius of a spherical micelle is 1–3 nm, and thus they lie in the colloidal range (Zana, 2005). The main driving force behind self-association of amphiphilic molecules is the decrease of system free energy. The decrease in free energy is a result of removal of hydrophobic fragments from the aqueous surroundings with the formation of a micelle core stabilized with hydrophilic fragments exposed into water. The factors that affect the process of micelle formation are the size of the hydrophobic domain in the amphiphilic molecule, concentration of amphiphiles, temperature, and solvent. The minimum concentration of amphiphilic molecules required to form assembly is called critical micelle concentration (CMC). These amphiphilic molecules, at low concentrations in the medium, exist separately and are so small that they appear to be subcolloidal. Below CMC, the amphiphiles undergo adsorption at the air–water interface. As the total amphiphile concentration is increased up to CMC, the interface as well as bulk phase is saturated with monomers. Any further amphiphile added in excess of CMC results in the aggregation of monomers in the bulk phase, such that the system free energy is reduced. The temperature above which amphiphilic molecules exist as aggregates is the critical micellization temperature (Adams et al., 2003; Moroi, 1992; Zana, 2005). The size and shape of micellar aggregates vary as spheres, vesicles, lamellae, rods, and tubules depending on the balance between the hydrophilicity and lipophilicity of a given amphiphilic molecule.

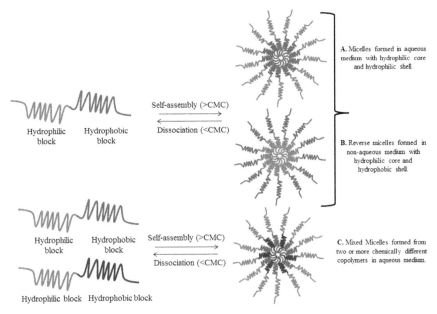

FIGURE 7.1   Schematic diagram for the formation and dissociation of (A) micelles, (B) reverse micelles, and (C) mixed micelles prepared from amphiphilic unimers.

7.1.2   SELF-ASSEMBLY OF AMPHIPHILIC MOLECULES INTO MICELLES

7.1.2.1   SURFACTANTS

The surfactants considered for drug delivery are characterized by at least two segments or blocks of a distinct chemical nature, where one of the segments is relatively hydrophobic or nonpolar (called *tail*; usually a long alkyl substituent) and the other segment is relatively hydrophilic or polar (*head*). These multiple segments are linked together chemically to form an individual block copolymer molecule termed a "unimer." The simultaneous existence of chemical entities having an affinity and antipathy for water within the same molecule is the key phenomenon of amphipathicity. The presence of a hydrophilic group makes the surfactants slightly soluble in water. Aqueous exposure induces the hydrophobic and hydrophilic segments to phase-separate and form nanoscopic, supramolecular core/shell structures, at a minimum particular surfactant concentration, termed micelles (Biswas et al., 2013).

Surfactants are classified on the basis of the charge carried by the polar head group as anionic, cationic, nonionic, or amphoteric. Table 7.1 shows the typical structures of the few surfactants of these individual classes. Micellization is easier for nonionic surfactants (show very low CMC) because the aggregation takes place mainly due to the hydrophobic attraction among nonpolar chains whereas hydrophilic chains are easily separated in an aqueous environment. Ionic surfactants are more difficult with regard to micellization in aqueous solutions than the nonionic surfactants of identical alkyl chain length, because higher concentrations are necessary to overcome the electrostatic repulsion between the ionic head groups of ionic surfactants during aggregation.

TABLE 7.1   Chemical Structure of Hydrophilic Groups of Some Common Surfactants.

Type of surfactant	Chemical structure
Nonionic	
Polyoxyethylene alcohol	$R\text{-}(OCH_2CH_2)_n\text{-}OH$
Polyoxyethylene ester	$R\text{-}COO\text{-}(CH_2CH_2O)n\text{-}H$
Polyoxyethylene thioether	$R\text{-}S\text{-}(CH_2CH_2O)n\text{-}H$
Anionic	
Carboxylate	$R\text{-}(COO^-)_n M^{n+}$
Sulfonate	$R\text{-}SO_3^- M^+$
Polyoxyethylene sulfate	$R\text{-}(OCH_2CH_2)_n\text{-}OSO_3^- M^+$
Sulfocarboxylate	$^-O_3S\text{-}R\text{-}COO^- M^{2+}$
Cationic	
Ammonium	$R\text{—}\overset{\displaystyle R_1}{\underset{\displaystyle R_3}{N^+}}\text{—} R_2 X^-$

TABLE 7.1 *(Continued)*

Type of surfactant	Chemical structure
Sulfonium	$R-\overset{\underset{\displaystyle R_2}{\mid}}{S^+}-R_1X^-$
Phosphonium	$R-\overset{\underset{\displaystyle R_3}{\mid}}{\overset{\overset{\displaystyle R_1}{\mid}}{P^+}}-R_2X^-$
Pyridinium	$R\diagdown_{N^+}$ pyridine ring X^-
Amphoteric:	
Triglycerine	$R-\overset{\underset{\displaystyle CH_2COO^-}{\mid}}{\overset{\overset{\displaystyle CH_2COOH}{\mid}}{N^+}}-CH_2COOH$
Phosphatidyl choline	$R_2\text{-}COO-\overset{\underset{\displaystyle OH_2C}{\mid}}{\overset{\overset{\displaystyle CH_2O\text{-}OCR_1}{\mid}}{C}}-HO$ $OH_2C-\overset{\underset{\displaystyle OH}{\mid}}{\overset{\cdot\cdot}{P}}-O-CH_2CH_2N^+(CH_3)_3$

7.1.2.2 COPOLYMERS (POLYMERIC MICELLES)

Synthetic amphiphilic block copolymers (di, tri, or tetra) or graft copolymers self-assemble into macromolecular polymeric micelles (PMs) in aqueous solution. There are several types of block copolymers, such as homo and hetero, depending on the arrangements of hydrophilic and hydrophobic units in a single unimer (Fig. 7.2). Homopolymers are made up of identical monomeric units that do not have the ability to form micelles. The hydrophobic and hydrophilic monomeric units can be oriented in many different ways to provide random, block, and graft copolymers. In random copolymers, the hydrophobic and hydrophilic units are distributed throughout the polymeric chain in a random manner. It may or may not form micellar aggregates depending on the thermodynamics of micellization. In the case of block copolymers, the hydrophobic and hydrophilic segments are linked together in an organized fashion. The orientation of hydrophobic and hydrophilic blocks in diblock, triblock and graft copolymers are shown in (Fig. 7.2) Micelles made up of di- and triblock copolymers are of particular interest for the delivery of poorly water soluble drugs (Kwon and Kataoka, 1995; Kwon, 1998). In grafted copolymers, multiple hydrophobic chains are anchored along the main chain composed of hydrophilic units (Schild et al., 1991). Amphiphilic diblock AB-type or triblock ABA-type copolymers, with the hydrophilic block length exceeding to some extent than that of a hydrophobic one, would self-assemble to form spherical micelles in aqueous solutions. If the hydrophilic block length is too large, the copolymers exist in water as individual molecules (unimers) and the molecules with lengthy hydrophobic blocks develop various structures.

In aqueous medium the amphiphilic block copolymers can principally self-assemble into spherical, worm-like, or cylindrical micelles and polymer vesicles or polymersomes. The main factor that governs the morphology of micelles is the hydrophilic–hydrophobic balance of block copolymer defined by the hydrophilic volume fraction (f). For amphiphilic block copolymers with the f value of nearly 35%, the polymer vesicles are formed, whereas, for f value more than 45%, spherical micelles are formed from self-assembly (Erhardt et al., 2001; Riess et al., 1985). By using amphiphiles of more complicated molecular design (like miktoarm star copolymers) or by varying experimental conditions for self-assembly, the micelles with more complex morphologies (such as crew-cuts, multicompartment, toroids, etc.) may be obtained (Xu et al., 2006). Factors such as molecular weight of the amphiphilic block copolymer, aggregation number of the amphiphiles, relative proportion of hydrophilic and hydrophobic chains, and the preparation process will control the size of the PMs (Jones and Leroux, 1999). The properties of the core-forming block of the PMs have the most profound impact on the CMC and aggregation/association number of PMs. Increasing hydrophobicity and size of the hydrophobic block serve to lower the CMC (Alexandridis et al., 1994; Kabanov and Alakhov, 2002). When the hydrophobic to hydrophilic block ratio is kept constant, an increase in the molecular weight leads to a decrease in the CMC (Yokoyama et al., 1998). Also, the additives in the solution may affect the CMC of PMs. Some highly hydrophobic drugs incorporated in the micelle core increase the hydrophobicity of the core and thereby lower the CMC and increase the association number (Rakshit and Palepu, 2003; Yokoyama et al., 1998). The most commonly used core and corona-forming blocks of copolymer are shown in Table 7.2.

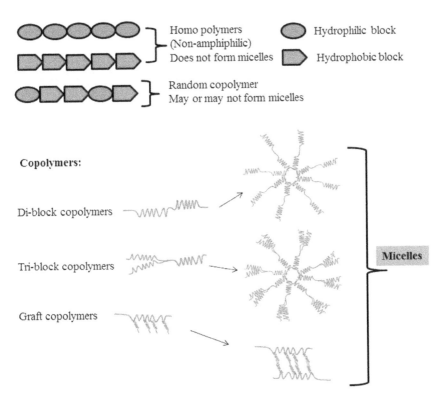

FIGURE 7.2 (**See color insert.**) Main structural types of copolymers and micelles formed from amphiphilic copolymers.

TABLE 7.2 Commonly Used Copolymer Blocks Used to Make Polymeric Micelles.

Copolymers and their abbreviation	Chemical structure of repeating units
Hydrophobic core	
PPO	

TABLE 7.2 *(Continued)*

Copolymers and their abbreviation	Chemical structure of repeating units
Poly(esters)	
PLA	
PDLLA	
PLGA	
PCL	
PAE	
Polyamides	
pHis	
pAsp	

TABLE 7.2   *(Continued)*

Copolymers and their abbreviation	Chemical structure of repeating units
pGlu	
Hydrophilic shell	
PEG, PEO	
pNIPAM, NIPAM	
PVP	

PAE, poly(beta-amino ester); pAsp, poly(L-aspartic acid); PCL, poly(epsilon-caprolactone); PDLLA, poly(D,L-lactide); PEG, PEO, poly(ethylene glycol/oxide); pGlu, poly(L-glutamic acid); pHis, poly(L-histidine); PLA, poly(L-lactide); PLGA, poly(lactide-co-glycolide); pNIPAM, NIPAM, poly(N-isopropyl acrylamide); PPO, poly(propylene oxide); PVP, poly(N-vinyl pyrrolidone).

7.1.3   NATURE OF INTERACTIONS CAUSING MICELLE FORMATION

Micelles self-assembled from copolymers interacting through weak hydrophobic interactions and interactions via molecular recognition such as stereocomplexation, H-bonding, ionic interactions, as well as chemical crosslinking in the core are expected to behave differently (Fig. 7.3).

Hydrophobic interaction: It is the most frequently studied type of noncovalent interaction in micellar systems. In this case, the copolymers self-assemble to form micelles in aqueous solution through hydrophobic interactions between the hydrophobic blocks of the copolymers.

Stereocomplexation: Here, the micelle formation is because of stereo-complexation between individual enantiomers having different stereochemistry. Typical examples of iso- and syndiotactic stereo complexes include polylactides (PLAs), poly(methyl methacrylate), polylactones, polythiiranes, and polyoxiranes (Tsuji, 2005). Among these examples, the PLAs have attracted a great attention because the PLA homo- and block copolymers are biodegradable, are produced from renewable resources, and are nontoxic (Fukushima and Kimura, 2006; Kim et al., 1997). In addition, since both polylactide and polyethylene oxide are FDA-approved and block copolymers can be readily dispersed in water, the significant effort has been devoted toward the sequestration and delivery of therapeutics using polylactide-b-polyethylene oxide micelles. Cyclic lactide monomers have two stereocenters (denoted as L- and D-enantiomers) that generate a number of polylactides such as optically active poly(D-(+)-R-lactide) (D-PLA) and poly(L-(−)-S-lactide) (L-PLA), racemic poly(D, L-lactide) (rac-PLA) and meso-poly(D, L-lactide) (mes-PLA). The polymer mixtures described above are of polylactide having identical chemical composition but opposite enantiomeric configurations and form homostereocomplexes. Slager and coworkers demonstrated the stereoselective hetero-complexes between two dissimilar polymers, specifically L-peptides and poly(D-PLA) (Slager, 2003; Slager and Domb, 2004).

Hydrogen bonding: Like stereocomplexation, the hydrogen bonds are short-range interactions that only form if the two interacting groups (i.e., hydrogen bonding receptor and donor) are in the close proximity to each other. It is hypothesized that the amphiphilic copolymers self-assemble into micelles first via hydrophobic interaction and then hydrogen bonds form when the interacting groups in the hydrophobic blocks of the copolymers come together and stabilize the hydrophobic core. Although hydrogen bonding has the potential to drive the formation of micelles, the research on hydrogen bonding micelles is relatively less pervasive than other types of micelles.

Ionic interactions: In this case, the driving force for micellization is the strong electrostatic attraction between a pair of oppositely charged copolymers. This class of micelles is referred to as polyion complex (PIC) micelles. When copolymers are dissolved in water, the oppositely charged blocks of the copolymers come together to from aggregates through long-range electrostatic interaction, driving the micelles formation. The PIC micelles, compared to conventional micelles formed through hydrophobic or hydrogen bonding interactions, have added advantage of encapsulating ionic drugs/compounds

such as folic acid (Luo et al., 2010), coenzyme A (Luo et al., 2009), oligonucleotides and plasmid DNA (Katayose and Kataoka 1997), and siRNA
(Lee et al., 2007) in the core of the micelles. The ionic interaction within PIC
micelles is also reversible and can be influenced by pH, salt concentration,
and chain length of the charged block. The pH affects the degree of ioniza-
tion of the interacting blocks. As pH increases, the basic (positively charged)
block polymer will become less ionized and eventually becomes neutral.
Similarly, the acidic (negatively charged) block polymer will become less
ionized with protonation as pH decreases. Upon pH changes , the increasing
neutralization of one block copolymer of PIC micelles results in a decrease
in the electrostatic attraction between the block copolymers while hydrophobic interactions become dominant in the micellization.

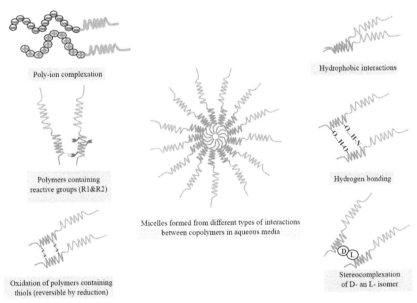

FIGURE 7.3 Self-assembly of copolymers into micelles through different types of
interactions.

7.1.4 REVERSE MICELLES

PMs are core–shell structures formed through the self-assembly of amphiphilic copolymers in a solvent that is considered hostile toward either moiety.
In water, these micelles are characterized by a hydrophobic core shielded
from the external medium by a hydrophilic shell. This type of micelles has
the ability to improve the aqueous solubility of hydrophobic therapeutic

agents (Liggins and Burt, 2002; Gaucher et al., 2005). In contrast, the self-association of amphiphilic copolymers in nonaqueous solvents yields nanostructures with a polar core and a hydrophobic shell. Such assemblies are commonly referred to as reverse micelles (RMs). More recently, the formation of unimolecular RMs from arborescent polymeric structures has attracted increasing interest. The RMs have also been prepared from dendrimers (Cooper et al., 1997; Sayed et al., 1997) or hyperbranched polymers (Chen et al., 2005; Sunder et al., 1999). In all instances, the resulting RMs were soluble in organic solvents, where they were shown to improve the solubility of hydrophilic compounds.

7.2 CHARACTERIZATION OF MICELLES

Successful and safe drug delivery using PMs can be achieved by their characterization for several important parameters as mentioned in Table 7.3. The parameters mentioned in the table are important as they depict the performance level of the PMs. The rate of dissociation of PMs into single chains and their interaction with plasma components is as important as the CMC and their size. Once injected, the PMs should maintain their integrity for adequate period of time in order to deliver the drug at its site of action. Physical stability of PMs could be assessed by gel permeation chromatography (GPC) (Riess, 2003; Thipparaboina et al., 2015).

TABLE 7.3   Characterization Techniques for Micelles.

Parameter	Evaluation method/s
Molecular weight	Light scattering, GPC, SANS, SAXS, and ultracentrifugation
CMC	Light scattering, GPC, and fluorescent probe
Intrinsic viscosity of the micellar core	Fluorescent probes and 1-H NMR
Particle size distribution	DLS
Zeta potential	DLS
Morphology	AFM, TEM, and SEM
Aggregation number	SANS, light scattering, and GPC

1-H NMR, proton-nuclear magnetic resonance; AFM, Atomic force microscopy; CMC, critical micelle concentration; DLS, dynamic light scattering; GPC, gel permeation chromatography; SANS, small angle neutron scattering; SAXS, small-angle X-ray scattering; SEM, scanning electron microscopy; TEM, transmission electron microscopy.

7.3 APPLICATIONS OF MICELLAR DELIVERY SYSTEMS

7.3.1 ORAL DELIVERY

7.3.1.1 IMPROVED ORAL BIOAVAILABILITY OF POORLY SOLUBLE DRUGS

Oral administration, especially in the case of repeated dosing for chronic therapy, is the most preferred route for drug delivery. The main mechanisms involved in the enhancement of oral drug absorption by PMs are: (1) increased aqueous solubility of poorly water soluble drugs and the high drug loading facilitates passive diffusion of drug into the gastrointestinal (GI) membrane, (2) protect the loaded drug from the harsh GI environment, (3) Pluronic micelles inhibit cytochrome P450 which may reduce the metabolic elimination of drug in the gut and the liver (Guan et al., 2011; Huang et al., 2008), (4) the PMs (less than 100 nm) could be absorbed via the adsorptive endocytosis pathway (Nam et al., 2003) and fluid phase pinocytosis (Dahmani et al., 2012), (5) PMs release the loaded drug in a controlled manner at target sites, (6) PMs cause prolongation of the residence time in the gut by mucoadhesion, and (7) inhibit efflux pumps to improve drug accumulation (Rieux et al., 2006). Several physicochemical parameters including surface hydrophobicity, polymer nature, and particle size seem to influence the translocation of PMs across the epithelium (Rieux et al., 2006). Furthermore, to achieve a good bioavailability, drugs may be delivered at a specific area in the GI tract (the so-called absorption window). To reach the absorption window, PMs can be manipulated by coupling different types of polymers or by grafting various functional groups at the hydrophilic end of the copolymer such as pH-sensitive (Lee et al., 2003) and receptor-sensitive groups (Zhao et al., 2008).

7.3.1.2 RECEPTOR-MEDIATED UPTAKE

In order to improve the oral absorption of an encapsulated drug, receptor-mediated endocytosis may prove to be beneficial (Russel, 2001; 2004). The numbers of ligands including bacterial toxins, immunoglobulins, and vitamin B12 have been investigated with respect to their role in enhancing the receptor-mediated endocytosis of various therapeutic entities (Daugherty and Mrsny, 1999; Russel, 2004). The role of receptor-enhanced endocytosis in the permeability of cyclosporine A (CyA) through Caco-2 cell membranes was investigated by Francis et al. (2005). Here, they decorated dextran-g-PEO-C16 micelles with VB12 moieties. When VB12 is absorbed, it conjugates to the intrinsic factor (IF), a protein produced in the stomach. This VB12-IF complex then binds to IF receptors on the surface of the enterocytes in the small intestine and is transported through the mucosa via receptor-mediated endocytosis. The permeation coefficient of the drug transported by VB12-tagged micelles was 3.3 versus 1.4 cm/s for bare micelles.

7.3.1.3 INHIBITION OF EFFLUX TRANSPORTERS

The presence of intestinal efflux transporters (P-glycoprotein, Pgp; most widely studied) is associated with a decrease in oral bioavailability. Therefore, modulation of its activity is seen as a potential means to improve oral drug bioavailability. The inhibition of efflux transport appears to be related to a modification of the fluidity of the cellular membrane (Rege et al., 2002). This inhibitory effect has been established with both low-molecular weight and polymeric surfactants among which D-α-tocopheryl polyethylene-glycol 1000 succinate (TPGS) and poloxamers have been the most extensively studied.

The TPGS effect on the bioavailability of a Pgp substrate was first reported with CyA. Chang et al. (1996) confirmed an increased CyA absorption at TPGS concentrations below the CMC. The data suggest that TPGS does not act as Pgp substrate or competitive inhibitor of Pgp substrate efflux nor does it trigger intracellular ATP depletion. Instead, TPGS is thought to function as an allosteric modulator not involving the cis(Z)-flupentixol binding site (Collnot et al., 2010). In case of poloxamers, a combination of decreased ATPase activity and ATP depletion due to increased membrane fluidity is thought to be responsible for intestinal Pgp inhibition (Batrakova et al., 2004). This effect is maximal at concentrations just below the CMC where unimer concentration is the most important. Usually, drug permeability increases consistently with polymer concentration until a maximum is reached near the CMC. Zhu et al. (2009) demonstrated that the poly(ethylene glycol (PEG) stearate could inhibit the Pgp in Caco-2 cells with greater efficacy at concentrations just below the CMC. PEG stearate unimers are thought to interact with activated Pgp, thereby reducing its substrate-stimulated ATPase activity and finally inhibiting the efflux of Pgp substrates.

7.3.1.4   MUCOADHESIVE POLYMERIC MICELLES

Mucoadhesive polymers have applications in a variety of pharmaceutical formulations as a means to enhance drug absorption and/or provide sustained drug levels. By achieving an intimate contact with the mucosa, they reduce drug presystemic metabolism, extend the residence time at the site of drug absorption or action, and give a steep concentration gradient at the absorption membrane. Synthetic mucoadhesive polymers currently investigated in pharmaceutical formulations include PEG, cellulose derivatives such as methylcellulose and hydroxypropyl cellulose (HPC) and polyelectrolytes such as poly(acrylic acid) (PAA) and chitosan (Grabovac and Guggi, 2005). These polymers bind to mucus via noncovalent bonds such as electrostatic interactions, van der Waals forces, and hydrogen bonding. Chayed and Winnik (2007) reported the mucoadhesive behavior of PMs prepared from hydrophobically-modified dextran and HPC. They found that grafting of alkyl chains (C16) to HPC and dextran enhanced the polymer adsorption on the mucin layer. Bromberg and coworkers proved the mucoadhesive properties of poloxamer-PAA graft polymers (Barreiro-Iglesias et al., 2005). Much stronger bioadhesion can be achieved by decorating polymers with targeting ligands (e.g., lectins) or reactive groups such as thiols. Through strong adherence to glycoproteins and glycolipids in the enterocytes membrane, lectins may prove useful in both extending the small intestine transit time of a host cargo as well as promoting its uptake via receptor-mediated endocytosis (Russel, 2001). The surface-exposed thiols form disulfide bonds with cysteine-rich subdomains of mucus glycoproteins. Thiolated polymers also show increased permeation-enhancing effect as well as enzyme inhibitory properties (Bernkop et al., 2005). Thiol-decorated polyion complex (PIC) micelles (45 nm) prepared via complexation between PEG-b-poly(2-(N, N-dimethylamino) ethyl methacrylate) and a 20-mer oligonucleotide have been shown to interact with mucin by means of the formation of disulfide bonds (Dufresne et al., 2005). A similar strategy was applied through the use of thiol-functionalized PEG-b-PLA or PEG-b-PCL PMs (Kalarickal et al., 2007).

7.3.1.5   pH-SENSITIVE POLYMERIC MICELLES

Several PMs designed to increase the oral bioavailability of hydrophobic compounds show release times which largely exceed the transit time in the small intestine (Pierri et al., 2005; Ould et al., 2005). This is also true for surfactant micelles that they have been found to impede, in some cases, the absorption of hydrophobic drugs due to excessive retention in the micellar phase (Jonge et al., 2004). Therefore, it is very important to adequately control the release rate in order to avoid either precipitation upon dilution or sequestration within the micellar phase which may lead to incomplete adsorption.

One approach to ensure progressive and complete drug release in the GI tract is using PMs that exhibit a pH-dependent ionization/dissociation profile. Such pH-responsive PMs minimize the initial burst release and possible precipitation in the stomach by releasing small amounts of their cargo at acidic stomach pH. At the basic intestinal pH (pH > 5), they partially or completely ionize and thereby release the residual entrapped drug in a molecularly dispersed form in the small intestine where absorption is maximum. These pH-responsive PMs can be either multimolecular or unimolecular (Jones et al., 2003; Sant et al., 2004). To release their payload in a pH-dependent fashion, the inner core of unimolecular PMs contains weakly acidic groups which ionize at the intestinal pH. The polymer deprotonation increases the core polarity and thereby lowering its affinity toward the encapsulated drug (Jones et al., 2003). The pH-responsive multimolecular PEG-b-poly(vinylbenzyloxy)-N, N-diethylnicotinamide) [PEG-b-P(VBODENA)] micelles showed significantly faster paclitaxel (PTX) release through micelle destabilization upon incubation in simulated intestinal fluid (SIF) (pH 6.5) compared to their plain hydrotropic counterparts (Kim et al., 2008). Alternatively, PEGb-P(alkyl(meth)acrylate-co-methacrylic acid) s (PEG-b-P(Al(M)Aco-MAA)s) diblock copolymers were found to self-assemble at pH ranging from 4.5 to 5.5 and completely dissociated at near-neutral pH owing to the deprotonation of MAA (Sant et al., 2004; Sant et al., 2005; Satturwar et al., 2007).

7.3.2 PARENTERAL DELIVERY

7.3.2.1 CANCER THERAPY AND PASSIVE TARGETING

Chemotherapy is the most commonly used cancer treatment. Because of their low molecular weight and/or high hydrophobicity, the majority of routinely used anticancer agents are characterized by short circulation times and a large volume of distribution, leading to very low concentrations at the target site (usually less than 0.1% of the injected dose accumulates in tumors; resulting in poor efficacy) and also to prominent localization in healthy non-target tissues (resulting in significant toxicity).

To overcome the limitations of conventional chemotherapy, PMs (nanosized colloidal particles), amongst many different nanosized drug delivery systems, have been under intense investigation during the past decade for drug delivery purposes (Aliabadi and Lavasanifar, 2006; Oerlemans et al., 2010). The mechanism of action of these systems is based on the enhanced permeability and retention (EPR) effect (Fig. 7.4), which relates to the physiological fact that solid tumors have leaky blood vessels that lead to extravasation and accumulation (i.e., retention because of lacking lymphatic drainage) of ~1–200 nm-sized PMs. For the EPR effect to happen, a prolonged circulation time is an absolute necessity (Maeda, 2012). In this regard, the hydrophilic corona of PMs shields their core after intravenous administration and protects it from interactions with blood components resulting in the prolonged residence time of PMs in blood and reduced recognition by the mononuclear phagocytic system (MPS) (Owens III and Peppas, 2006). Consequently, by accommodating/solubilizing highly hydrophobic anticancer agents within their hydrophobic core and thereby increasing their circulation time and reducing their volume of distribution (i.e., their off-target localization), the PMs can increase the therapeutic index of anticancer agents (including small molecule therapeutics).

The above mentioned advantages (improved drug pharmacokinetic and biodistribution characteristics) could also be obtained for hydrophobic drugs of other category (such as antiviral, antimicrobial, etc.) by loading them in PMs.

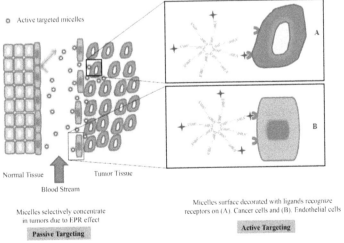

FIGURE 7.4   Passive and active targeting of micelles.

7.3.2.2 APPROACHES TO IMPROVE IN VIVO EFFICACY OF PARENTERAL POLYMERIC MICELLES

7.3.2.2.1 Active Targeting of Tumor Endothelial and Tumor Cell

The therapeutic efficacy of drug-loaded passively-targeted PMs can be further enhanced by conjugating a rationally selected cancer cell specific targeting moiety to the PMs surface. Tumor cells and/or tumor vasculature often show increased expression of antigens or receptors which are generally not expressed or present at low levels on the surface of normal cells and surrounding normal tissues (Kamaly et al., 2012). Active targeting exploits this feature of cancer cells to allow selective accumulation of anticancer therapeutics in the tumor tissue, tumor cells, or intracellular organelles of the tumor cell (Nie et al., 2007). PMs can be functionalized for active targeting by chemically modifying their surface with targeting ligands that show a strong specificity for antigens or receptors overexpressed on cancer cells and/or tumor endothelial cells (Fig. 7.4). Usually, the targeting ligands can be attached to the water-exposed free termini of hydrophilic blocks of the micelles, so that they extend above the hydrophilic block and avoid steric hindrance when binding to their target receptors (Torchilin, 2007). Actively targeted PMs decrease the drugs side-effects by allowing their preferential accumulation in diseased cells and also facilitate cellular uptake by receptor-mediated endocytosis (Danhier et al., 2010). Active targeting especially benefits the intracellular delivery of macromolecules such as siRNA, DNA, and proteins.

Antibodies have been the most popular targeting ligands for PMs to date because of the diversity of their targets and the specificity of their interaction (Torchilin, 2004). Over the years, antibody engineering technologies have enabled the development of murine, chimeric, humanized, and human monoclonal antibodies (mAbs) as well as antibody fragments (e.g., Fab, scFv, diabodies, triabodies, and single-domain antibodies) (Weiner, 2007). Although very popular as targeting ligands, antibodies do face some challenges. These challenges include their large size (~150 kDa) which result in limited ligand densities on micelles, potential immunogenicity which may lead to rapid clearance, stability considerations, and engineering challenges during scale-up manufacturing (Goldenberg and Sharkey, 2012; Kamaly et al., 2012).

Proteins and peptides have also been used extensively as targeting ligands for PMs. The transferrin receptor (TfR) is over-expressed in many cancers and offers an attractive option for the development of transferrin (Tf)-targeted nanocarriers (Singh, 1999). PMs may be modified either with endogenous ligand (Tf) or antibodies against TfR (Torchilin, 2006). Other protein ligands such as tumor necrosis factor-related apoptosis-inducing ligand (TRAIL), that binds to death receptors overexpressed in cancer cells to induce apoptosis (Riehle et al., 2013) and epidermal growth factor (EGF) which targets the EGF receptors overexpressed in many cancer cells (Zeng et al., 2006), are also used for modification of PMs for active targeting.

Peptides are used as targeting ligands due to their small size, lower immunogenicity compared to proteins, better *in vivo* stability, lower cost, and relative ease of conjugation to PM surface (Kamaly et al., 2012). The arginine–glycine–aspartic acid (RGD) tripeptide which targets integrin receptors ($\alpha v\beta 3$ and $\alpha v\beta 5$) has been widely investigated. Other peptides such as vasoactive intestinal peptide (VIP) to target VIP receptors over-expressed in breast cancer cells (Dagar et al., 2012; Gülçür et al., 2013), Lyp-1 (Cys–Gly–Asn–Lys–Arg–Thr–Arg–Gly–Cys) which targets the p32 receptors (p32/gC1qR) overexpressed on some tumor cells (Wang et al., 2012), cell-penetrating peptides such as transactivating transcriptional transactivator (TAT) from HIV-1 (Kanazawa et al., 2012; Taki et al., 2012), and octreotide that targets somatostatin receptors have also been used to modify PMs (Xu et al., 2013).

Folate receptors (FR) are overexpressed in a number of cancer cells, and hence folate is widely used as a targeting ligand for cell-specific delivery in these cancers (Sudimack and Lee, 2000). Other small molecules such as biotin (Lin et al., 2013), galactose (Zhong et al., 2013), and mannose (Freichels et al., 2012) have also been reported for PM surface modification.

Aptamers—oligonucleotides, which have the ability to fold into defined 3D structures and bind with high affinity and specificity to their target molecules (proteins, peptides, or small molecules), are also gaining momentum as targeting ligands (Zhang et al., 2011). A10 aptamer (Apt), that recognizes the extracellular domain of the prostate-specific membrane antigen (PSMA), decorated micelles showed a significantly higher therapeutic efficacy against prostate cancer.

7.3.2.2.2 *Stimuli-responsive Polymeric Micelles*

An added sophistication to selective delivery of PMs can be brought by utilizing certain cues, inherently characteristic of the tumor microenvironment (intrinsic), or by applying certain stimuli to this region from outside the body (extrinsic). The commonly encountered intrinsic stimuli in tumors include low pH, redox status of the cell, and the presence of certain up-regulated enzymes. Similarly, the extrinsic stimuli include magnetic fields, light (UV, infrared, or visible), and ultrasound. Hyperthermia is a stimulus that could be either intrinsic (from inflammation) or extrinsic (upon application of ultrasound or alternating magnetic fields in conjunction with magnetic nanoparticles which release heat) (Torchilin, 2009). PMs can be designed so as to respond to various stimuli (intrinsic or extrinsic) of physical, chemical, or biochemical origins to achieve spatial and temporal control over the therapeutic payload release (Cheng et al., 2013). These "smart" PMs can release their therapeutic payloads upon structural changes in response to the stimulus. The response may result in disintegration/destabilization, isomerization, polymerization, or supramolecular aggregation of PMs (Fleige et al., 2012).

The acidic pH in tumors result from the extensive hypoxia and cell death which leads to production and accumulation of lactic acid (Rotin and Tannock, 1989). The pH in tumors is ~6.5 when compared to ~7.4 in the normal tissues and decreases even further in the intracellular organelles such as endosomes (~5–6) and lysosomes (~4–5) (Casey et al., 2010). These pH-gradients have been utilized successfully to design pH-responsive PMs which can release their cargos when they encounter a change in the pH of their microenvironment.

The redox potential in cancer cells is elevated (100–1000 fold higher) due to the high intracellular glutathione tripeptide (γ-glutamyl-cysteinyl-glycine) (GSH) concentration (2–10 mM) when compared to its concentration (2–10 μM) outside the cells (Saito et al., 2003). PMs with disulfide bonds have been designed to hold the cargos (drugs, siRNA, DNA, or proteins) under normal conditions and release their payloads upon destabilization in the reducing conditions found inside cancer cells that can convert disulfide linkages to thiols (Wei et al., 2012). The disulfide linkages can be incorporated in the hydrophobic backbone (Li et al., 2013b), at the junction of hydrophobic and hydrophilic blocks (Li et al., 2012), or by incorporating reduction sensitive crosslinks in the micelle core (Li et al., 2011), shell (Koo et al., 2012), or the core–shell interface (Hossain et al., 2010).

Gradients of oxygen tension within the tumors can be utilized to design hypoxia responsive PMs. Perche et al. used azobenzene as a hypoxia-responsive, bio-reductive linker for hypoxia-targeted delivery of siRNA from the PEGylated nanopreparations upon PEG cleavage (Perche et al., 2014). Enzyme-responsive micelles take advantage of the altered expression profile of certain enzymes in tumors or other diseases to deliver therapeutic payloads to the desired targets (Mura et al., 2013). Enzyme responsive moieties can be used to modify the polymers (main chain or side groups), which upon recognition by the enzyme cause structural modifications in the micelles. Another option is to modify the micelle surface with peptides or oligonucleotides that can cause physical changes in the micelles upon enzymatic transformation (De et al., 2012; Hu et al., 2012). The enzymes most frequently dysregulated in cancer are hydrolases (proteases, lipases, and glycosidases), metabolic enzymes including those involved in glycolysis and fatty acid synthesis, and oxidoreductases (De et al., 2012). The matrix metalloproteinase (MMP) family of enzymes (MMP-2 and MMP-9 in particular) is primarily linked to cancer progression and metastasis. PMs containing MMP responsive linkers have been reported for tumor-specific delivery of drugs and siRNA in response to the up-regulated MMPs (Li et al., 2013a; Zhu et al., 2014). Phosphatase (Wang et al., 2010a) and acetylcholinesterase (Xing et al., 2012) responsive polyion complex micelles (PICs) have also been reported.

Among the extrinsic stimuli, ultrasound has been investigated widely as a trigger for drug release from PMs. Ultrasound refers to the application of pressure waves above a frequency of 20 kHz to spatially and temporally control cargo release. While low-frequency ultrasound (20–100 kHz) can penetrate deeper into the body tissues than high-frequency ultrasound (1–3 MHz), it cannot be focused as well (Rapoport, 2012). *In vitro*, ultrasound can perturb the micelle structure and cause the release of therapeutic cargos due to cavitation. *In vivo*, this mechanical effect of ultrasound may also be accompanied by local hyperthermia, which could lead to increased micelle extravasation and accumulation in the tumor tissues (Rapoport, 2012).

Magnetic field has also been discovered as an extrinsic stimulus for PMs. Micelles can be loaded with drugs as well as superparamagnetic iron oxide nanoparticles such as magnetite (Fe_3O_4) or maghemite (Fe_2O_3), which allow them to be manipulated under the guidance of an externally applied permanent magnet or an alternating magnetic field to control either the drug release, result in a temperature increase, or even both when used alternately (Mura et al., 2013; Torchilin, 2009). Temperature is one of the most widely studied stimuli for drug delivery and has been extensively discovered for cancer treatment. Thermoresponsive micelles are constructed from thermoresponsive blocks which can undergo a sharp change in phase that destabilizes the micelles and triggers the release of the cargo (Mura et al., 2013; Torchilin, 2009). Light-responsive micelles can utilize ultraviolet (UV), visible, or near infrared (NIR) light to trigger cargo release with excellent remote spatiotemporal control (Mura et al., 2013). Some recent reviews have discussed the design principles of such photoresponsive micelles and mechanisms of photo-induced drug release from delivery carriers (Fomina et al., 2012; Gohy and Zhao, 2013; Schumers et al., 2010). Photoresponsive groups (Azobenzenes and their derivatives) can be incorporated within the micelle core, corona, or at the core–shell interface in the design of light-responsive micelles, (Gohy and Zhao, 2013). They undergo a reversible trans–cis photoisomerization upon UV light irradiation which converts the apolar trans-isomer to a polar cis-isomer, whereas the visible light reverses this isomerization (Zhao, 2007). Recently, spiropyran-initiated hyper-branched polyglycerol (SP-hb-PG) micelles were reported which responded to UV/visible light and could dissociate due to the conversion of the hydrophobic chromophore SP to zwitterionic and hydrophilic merocyanine (ME) (Son et al., 2014). Chromophores such as coumarin, o-nitrobenzyl, stilbene, dithienylethene, and 2-diazo-1,2-napthoquinone (DNQ) have also been employed in light-responsive micelles, which can respond either to UV/visible or NIR irradiation to undergo structural or phase changes and trigger the cargo release from micelles (Cao et al., 2013; Chen et al., 2011; Jin et al., 2011; Liu et al., 2012; Menon et al., 2011).

7.3.2.2.3 By Circumventing Stability Problems

7.3.2.2.3.1 Crosslinked Micelles

Physically self-assembled PMs rapidly disintegrate in systemic circulation either due to dilution to concentrations below their CMC or due to interactions of the hydrophobic blocks with plasma proteins (thereby disturbing the micelle equilibrium), resulting in rapid plasma clearance of the individual polymers (which have a MW allowing for renal filtration).

To avoid premature destabilization, the PMs have been developed with increased *in vivo* stability via core- or shell-crosslinking (Fig. 7.5) (Aliabadi and Lavasanifar, 2006; Armes and Read, 2007; O'Reilly et al., 2006; Van Nostrum, 2011; Yang et al., 2011). Core-crosslinking has been performed through bifunctional reagents (Zhang et al., 2007), radical polymerization of the side chains of the hydrophobic block when polymerizable groups are present (Shuai et al., 2004), dimerization of side groups upon UV exposure (photo-crosslinking) (Saito et al., 2007; Shi et al., 2009), or through disulfide bridges (Meng et al., 2009). Shell-crosslinking methods include bifunctional agents, radical polymerization, and disulfide bridges (all methods are extensively reviewed in Read and Armes (2007) and the crosslinking is performed either in the outer shell when AB-type block copolymers are used or in the interfacial block in the case of ABC-type block copolymer micelles (Van Nostrum, 2011).

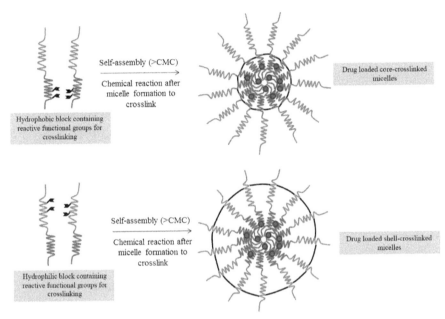

FIGURE 7.5 Core- and shell-crosslinked micelles.

7.3.2.2.3.2 Polymer–Drug Conjugate Micelles

Most of the drug molecules have reactive or derivable groups. The coupling group on the polymer chains and the comonomer functionality were chosen to satisfy the reactive group on the drug molecule. Because the drugs used are usually hydrophobic and are linked to the hydrophobic segments the polymer–drug conjugates remain amphiphilic and can self-assemble into nanomicelles in aqueous medium with the drugs in the micellar core and hydrophilic segments in the micellar shell. Usually, amphiphilic block copolymers of the A-B or B-A-B type, where "A" represents a hydrophilic block and "B" represents a hydrophobic block, are used to construct PMs (Fig. 7.6). Other examples, including star-shaped copolymers (Danhauser-Riedl et al., 1993), graft copolymers (Chen et al., 2008; Peng et al., 2008), and dendrimers (Bougard et al., 2008; Kono et al., 2008; Park 2009) have also been reported.

Polymer–drug conjugates have demonstrated several advantages over their parent drugs such as fewer side effects, ease of drug administration, and enhanced therapeutic efficacy. Polymer–drug conjugate micelles show improved release kinetics, especially elimination of burst release at the early stage. Conjugation of targeting moieties (ligands) to the carrier polymer endows polymer–drug conjugate micelles' targeting capability (Fig. 7.6). Therefore, polymer–drug conjugate micelles are a new generation of drug delivery systems. Interestingly, the polymer–drug conjugate micelles can be potentially used for the loading of another therapeutic molecule. Drug loading using a biologically active carrier is a smart strategy as it represents a unique form of combination. Combination therapy with multiple agents working at several signaling pathways at the same time could not only maximize the anticancer effect but also help to overcome the drug resistance (Broxterman and Georgopapadakou, 2005). For instance, the combination of PGA–PTX conjugate with cisplatin (Verschraegen et al., 2009) or carboplatin (Langer et al., 2008) has shown improved therapeutic benefits with reduced toxicity in phase I clinical study. In addition, the anticancer effect of N-(2-hydroxypropyl) methacrylamide (HPMA) copolymer–doxorubicin conjugate in combination with HPMA copolymer-mesochlorin e6 was found to be more efficacious when compared to either conjugate alone (Shiah et al., 2001).

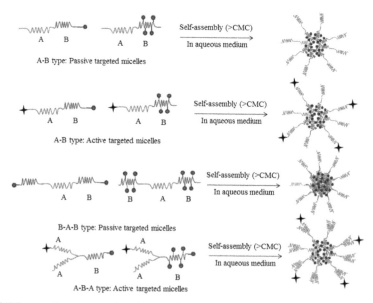

FIGURE 7.6 Structural and architectural types of polymer–drug conjugate micelles: (A) hydrophilic block and (B) hydrophobic block.

7.3.2.2.4 Mixed Micelle Approach

PMs formed from single copolymers are often lacking in one or more departments primarily due to limitations in the number of building blocks. The combination of two or more dissimilar block copolymers (e.g., AB + CB type (Lo et al., 2009) or AB + AC type (Wu et al., 2009) or AB + CD type (Xie et al., 2007) to form mixed micelles (MMs) (Fig. 7.1C) is a practical and efficient approach to address many of the issues without the need for complicated synthetic schemes. Shim and coworkers (Shim et al., 1991) have proposed a model for the comicellization of diblock copolymers in solution, which was described to be dependent on the relative concentrations of the two species of diblock copolymers. In addition, the CMC of a mixed micelle system can be mathematically derived from the CMC values and molar fractions of its constituents (Torchilin, 2001). Some of the merits of MMs over their individual constituents include significant improvements in thermodynamic (by lowering CMC) (Kim et al., 2009) and kinetic (through enhancing the hydrophobic interactions, stereocomplexation, H-bonding, ionic interaction, or chemical crosslinking between the core-forming blocks) stability (Lo et al., 2009), improved drug loading (Yang et al., 2009), exquisite size control to avoid precipitation (Alakhov et al., 1999), and ease of incorporating multiple functionalities (for instance, both temperature- and pH-responsive drug release, combined delivery of drug and/or contrast agent/siRNA, cellular targeting, etc.) (Yoo and Park, 2004). MMs could be fabricated via simply mixing monofunctional copolymers bearing desirable attributes in aqueous medium. The nature of interactions between different block copolymers would remain same as discussed for PMs prepared from single copolymer (Fig. 7.3).

7.3.2.2.5 Multifunctional Micelles

PMs, while allowing for the different modifications individually, also offer a platform that allows for incorporation of multiple components within a single micelle. We can thus design micelles to have two or more different modifications that enable them to simultaneously or sequentially perform important therapeutic and diagnostic functions (theranostic micelles). Such micelles which can combine a number of distinct functions or properties within a single carrier, with each individual component functioning seamlessly and in perfect coordination with the other components, give rise to the so-called "multifunctional" micelles. In addition to the various modifications just mentioned, contrast or reporter moieties can be incorporated within micelles which allow real-time imaging of these micelles and their accumulation within cells (Torchilin, 2006). Therefore, an ideal multifunctional micelle may simultaneously deliver drugs/biologics, circulate in the body for prolonged periods, allow for passive or active targeting-mediated accumulation, respond to various stimuli to release encapsulated cargoes, and may also give diagnostic and therapeutic monitoring abilities (Fig. 7.7). The key to developing such multifunctional micelles is to ensure that each of the components function in a coordinated manner so that the combined contribution of each adds up to something better than the individual components themselves.

Cancer is a complex disease characterized by molecular and phenotypic heterogeneity within and between tumor types which makes chemotherapy very challenging. Although, molecularly targeted therapies have been developed, a selection of tumor cells may still escape the targeted pathway and lead to adaptive resistance, causing the failure of therapy (Blanco et al., 2009). It follows from the discussion above that using a multifaceted approach for targeting cancer seems imperative. While it may be difficult to incorporate all the features of the "ideal" multifunctional PMs, a combination of two or more of the desirable features is possible and also necessary for the success of cancer therapy using PMs. A huge focus of the current research has been in the direction of development of such multifunctional micelles for their obvious advantages in enhancing the efficacy, maximizing the safety and specificity of existing as well as novel chemotherapy regimens.

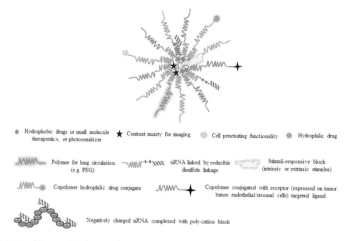

FIGURE 7.7   (See color insert.) A hypothetical multifunctional micelles. Micelles can be designed to incorporate two or more of the above functions.

7.3.2.3 MICELLEPLEXES FOR siRNA DELIVERY

Since the detection of ribonucleic acid (RNA) interference (RNAi), there has been an increased interest in developing siRNA-based therapies to achieve sequence-specific posttranscriptional silencing of aberrant genes in diseases such as cancer (Elbashir et al., 2001; Fire et al., 1998). RNAi is an endogenous pathway which is utilized by all eukaryotic cells to silence genes posttranscriptionally and can be triggered by double-stranded RNAs (dsRNA) such as endogenous microRNA (miRNA), short hairpin RNA (shRNA), and synthetic siRNA (Wang et al., 2010b). The detailed mechanism of the RNAi pathway has been reviewed elsewhere (Hannon, 2002; Rana, 2007). To access and activate the RNAi pathway, the siRNA must be delivered to the cytoplasm of the cell. A number of barriers prevent the successful systemic delivery of siRNA. After IV injection, the naked siRNA exhibits low *in vivo* stability due to quick degradation by nucleases. It has a short circulation half-life due to rapid clearance by the kidneys and uptake by the phagocytes. The hydrophilic nature and negative charge of siRNAs prevent them from crossing the plasma membrane easily, despite their relatively small size (about 13 kDa) (Liu et al., 2013; Navarro et al., 2013). Other challenges with siRNA include the potential to generate off-target effects due to silencing of genes that have partial homology with the siRNA and cause immune stimulation (Bumcrot et al., 2006). Moreover, because siRNAs also share the RNAi pathway with endogenous microRNAs (miRNAs), they may compete for the RNAi pathway, saturate it and inhibit normal gene regulation by miRNAs (Kanasty et al., 2012).

PMs, which overcome some of the aforementioned siRNA delivery challenges, have prompted their use as vehicles for siRNA delivery. Moreover, by designing micelles with suitable changes they may likely meet most criteria for an "ideal" nanocarrier for siRNA delivery (Fig. 7.8). So far, two main strategies have been used to design PMs for siRNA delivery. The first one includes the direct conjugation of hydrophilic (PEG) or hydrophobic (lipid) moieties to siRNA via degradable (e.g., disulfide) or nondegradable linkages, followed by their condensation with polycationic ions to form micellar structures known as polyion complex (PIC) micelles or polyelectrolyte complex micelles. In PIC micelles, the polyion segments are usually made of poly(amino acids) such as poly(L-lysine) or poly(aspartic acid) or polyethyleneimine (PEI) (Kim et al., 2008; Oishi et al., 2005; Suma et al., 2012).

In the second strategy, the siRNA is complexed with an amphiphilic block copolymer containing polycation (or lipid) segment followed by micellization of the block copolymer–siRNA complex (Falamarzian et al., 2012; Navarro et al., 2013). PMs and other nanoparticles enter cells by the process of endocytosis (Decuzzi and Ferrari, 2008). One of the major intracellular barriers for siRNA delivery is the endosomal escape following its delivery by various carriers. Following endocytosis, the siRNA-loaded carriers in membrane-bound endocytic vesicles fuse with early endosomes to become increasingly acidic as they mature into late endosomes (pH 5–6). Finally, the endosomal contents are delivered to the lysosome where the pH further decreases to ~4.5, at this pH the hydrolysis of proteins and nucleic acids take place (Dominska and Dykxhoorn, 2010; Singh et al., 2011). To avoid lysosomal degradation, it is necessary for the siRNA to escape the endosome and be released into the cytosol and interact with the RNAi machinery. To overcome this "endosomal escape barrier," PMs can be engineered to incorporate cationic polymers such as PEI which act as "proton sponges" to disrupt the endosomes and release siRNA in the cytosol. Alternatively, pH-responsive polymers can be used to construct PMs so that they disrupt and release the siRNA at the endosomal pH. Finally, the fusogenic lipids, cell penetrating peptides, other polymers with high buffering capacity, and photosensitizers (PSs) (upon light activation they induce endosomal disruption via singlet oxygen production) can be used to design PMs to overcome the endosomal escape barrier (Dominska and Dykxhoorn, 2010).

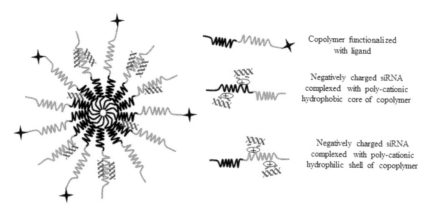

Copolymer functionalized with ligand

Negatively charged siRNA complexed with poly-cationic hydrophobic core of copolymer

Negatively charged siRNA complexed with poly-cationic hydrophilic shell of copolymer

FIGURE 7.8 (**See color insert.**) Schematic illustration of functionalized micelles designed for siRNA delivery.

7.3.2.4 PHOTODYNAMIC THERAPY

Photodynamic therapy (PDT) is a cancer treatment option in which activation of PS drugs with specific light wavelengths leads to energy transfer to oxygen molecules or other substrates in the surrounding areas, generating cytotoxic reactive oxygen species (ROS) which can trigger apoptotic and necrotic cell death (Dolmans et al., 2003; Hopper, 2000). In the absence of external photo-activating light the toxicity of PS drugs is negligible. Therefore, PDT gives a safe and effective way to selectively eradicate cancerous cells while avoiding systemic toxicity and side effects on healthy tissues (Detty et al., 2004). Compared to conventional chemotherapy and radiotherapy, the PDT-based cancer treatment significantly decreases side effects and improves target specificity because only the lesion under light irradiation is treated (Dolmans et al., 2003; Dougherty et al., 1998). Moreover, PDT-based cancer therapy is more beneficial to patients in which location or size of lesions limits the acceptability of conventional therapy (Choudhary et al., 2009). Despite the many positive features of PDT, the PDT is still not fully adapted in the clinical settings because of some inherent characteristics of PS drugs. Most existing PS drugs are hydrophobic with poor water solubility (Bechet et al., 2008). Therefore, they are easily aggregated in physiological conditions, drastically lowering the quantum yields of ROS production (Keene et al., 1986). Even in the case of some modified PS drugs for increased water solubility, their selective accumulation at target tissues or cells remains insufficient for successful clinical use. In this regard, the development of effective delivery systems that incorporate PS drugs and carry them into target tissues or cells is very much needed.

Recently PMs, amongst other type of nanoparticles, carrying PS drugs find considerable attention in PDT because they can overcome critical limitations of conventional PS drugs discussed above (Konan et al., 2002; Lim et al., 2013). PMs significantly enhance the solubility of hydrophobic PS drugs by dissolving them in hydrophobic core of their structure and thus increase their cellular uptake via endocytosis. PMs carrying PS drugs can achieve passive targeting to tumor by the EPR effect (Maeda et al., 2000). Moreover, cell specificity of PS drugs can be significantly improved by the surface modification of PMs with different targeting ligands (Allison et al., 2008; Paszko et al., 2011). This also increases the bioavailability of PS drugs and reduces the undesirable side effects of PS drugs to healthy tissues. All other approaches discussed under parenteral delivery of PMs can be applied for the effective delivery of PS drugs to cancer tissues.

7.3.3   MICELLES FOR OPHTHALMIC DRUG DELIVERY

PMs that can deliver drug to intended sites of the eye have attracted much scientific attention recently. PMs have the potential to target ocular tissues at high therapeutic value presenting several favorable biological properties such as biodegradability, biocompatibility, and mucoadhesiveness, which fulfill the ophthalmic application requirements. Physicochemical characteristics such as size, surface charge, morphology, physical state of the encapsulated drug, drug release characteristics, and stability of the PMs are of particular importance for topical ophthalmic application. The PMs can provide the biopharmaceutical advantages of higher permeation and enhancement of residence time at ocular surface for better absorption of drugs through ocular barriers. Adequate PM surface charge can be achieved to produce a stable colloidal dispersion. Furthermore, surface charge of topically applied PMs determines their performance at the absorbing surfaces, for example, their interactions with the cell membranes as well as the glycoproteins of the eye tissue and/or fluids (e.g., cornea, conjunctiva, tears, vitreous humor) forming a depot with extended release of the loaded cargo. Once the drug is released from the PMs, the drug molecule size, charge, lipophilicity, solubility in the eye fluid, and metabolic stability determine its *in vivo* fate in the eye (e.g., transcorneal, transconjunctival/scleral absorption, metabolic degradation, loss through the tear drainage, etc.) (Pepiae and Lovriae, 2012).

Mucoadhesive characteristics of biopolymers forming micelle enhance their contact time and minimize their elimination from the absorbing surface, consequently increasing the drug bioavailability. Particle size of topically-applied colloidal carriers influences absorption or permeation through the ocular barriers. For instance, the PMs of 100 nm are able to permeate across the corneal barrier. Surface modification or coating by biocompatible hydrophilic polymers improves PMs uptake and enhances their stability. The formulation of PMs allow easy application in the form of eye drops without blurring of vision and discomfort, thus achieving the requirements of patient compliance (Pepiae and Lovriae, 2012).

In addition, Duan and coworkers (Duan et al., 2015) developed a novel ion-sensitive in situ mixed micelle gel system composed of Pluronic P123/TPGS/gellan gum to prolong ocular retention time and improve corneal permeability of curcumin. The gellan gum, anionic polysaccharide, undergoes phase transition (sol–gel) in the presence of Na^+, Ca^{2+}, and Mg^{2+} ions present in human tear fluid and exhibits excellent ion-sensitive property (Gong et al., 2013). The gel formulations containing 20–30% poloxamers had displayed no ocular irritation and toxicity (Jianbin and Jingxue, 2008), and the presence of TPGS could increase the hydrophobicity of micellar core when mixed with P123. Also, TPGS is an antioxidant and could enhance the stability of formulations (Constantinides et al., 2006). Furthermore, P123 and TPGS have served as the inhibitors of P-glycoprotein in ocular tissues (Katragadda et al., 2006), and both of them can influence the fluidity of cell membrane and thereby promote drug penetration (Jiao, 2008; Krasavage and Terhaar, 1977).

7.3.4 MICELLES FOR PULMONARY AND OTHER TRANSMUCOSAL DELIVERY

The reported PMs for pulmonary, nasal, and rectal delivery of therapeutics are summarized in Table 7.4.

TABLE 7.4   Polymeric Micelles Reported for Pulmonary and Other Transmucosal Delivery.

Pulmonary delivery					
Therapeutic agent	Copolymer	PS (nm)	EE (%)	Performance	References
Insulin	Soluplus®, Pluronic® F68, Pluronic F108 and Pluronic F127	≤200	>60	Enhanced the permeation of insulin through pulmonary epithelial models. Minimal internalization by macrophages *in vitro*. Increased bioavailability	Andrade et al., 2015
Beclo-methasone dipropionate	PHEA-PEG$_{2000}$-EDA-LA	200	5 wt%	Increased cell uptake of BDP, higher biocompatibility, increased drug solubility and sustained drug release	Triolo et al., 2017
Curcumin	L-lactide grafted xyloglucan	102	96	Sustained release, improved bioavailability and pharmacokinetics	Mahajan et al., 2016
Curcumin acetate	PEG-PLGA	28	99	Increased stability, sustained drug release, and improved pharmacokinetics. Micelles efficiently translocated across the air-blood barrier into the bloodstream and distributed to extra-pulmonary organs including the brain as intact micellar vesicles	Hu et al., 2014

TABLE 7.4   *(Continued)*

Beclo-methasone dipropionate (BDP)	PHEA-PEG2000-DSPE	69	3 wt%	Increased drug solubility, excellent stability and sustained drug release	Craparo et al., 2011
Paclitaxel	PEG5000–DSPE	5	95	Sustained release, improved pharmaco-kinetics and targeting efficiency, and reduced systemic toxicity	Gill et al., 2011
Fasudil	Peptide-PEG-DSPE	14	58	Controlled release, improved cellular uptake, and pharmacokinetics	Gupta et al.,2014
Nasal delivery					
Clonazepam	P123/Pluronic L121	124	86	Kinetically and ther-modynamically stable, improved pharmaco-kinetics, and effective brain uptake	Nour et al., 2016
(siRaf-1) and camptoth-ecin (CPT)	MPEG-PCL-Tat	129	89	Improved intracellular uptake and acceler-ated the nose-to-brain delivery of siRNA/drug	Kanazawa et al., 2014
Olanzapine	Pluronic L121 and P123	59	75	Sustained drug release, improved kinetic and thermodynamic stability and pharmacokinetics	Abdelbary et al., 2013
Camptoth-ecin	MPEG-PCL-TAT	88	63	Higher cytotoxicity, improved cellular uptake efficiency, promoted the delivery of CPT into the brain and the mean survival time clearly increased	Taki et al., 2012
Coumarin	MPEG-PCL-Tat	73	86	Improved cellular uptake and greater brain distribution of coumarin in rats	Kanazawa et al., 2011
siRNA or dextran	MPEG-PCL-Tat	50–100	–	Enhanced delivery to the brain, increased the transfer of nucleic acid to the brain via the olfac-tory nerve pathway	Kanazawa et al., 2013

TABLE 7.4 *(Continued)*

Rectal delivery

Docetaxel	P 188/P 407	13	–	Improved bioavailability, pharmacokinetics and anti-tumor potential	Seo et al., 2013
Docetaxel	P188, P 407, Tween 80 and 0.1% NaTC	~15	–	Improved antitumor efficacy, bioavailability and pharmacokinetics	Kim et al., 2014
Diazepam	P407, Poloxamer 188 and TPGS	8–12	85	Increased physical stability and maintained chemical stability	Suksiri-worapong et al., 2014

%EE, % entrapment efficiency; BDP, Beclomethasone dipropionate; EDA, ethylenediamine; LA, lipoic acid; MPEG-PCL, methoxy poly(ethylene glycol)- poly(ε-caprolactone); NaTC, sodium taurocholate; PEG_{2000}-DSPE, 1,2-distearoyl-sn-glycero-3-phosphoethanolamine (polyethylene glycol)$_{2000}$; PHEA, α,β-poly(N-2-hydroxyethyl)-D,L-aspartamide; PS, average particle size; siRNA, Small interfering RNA; Tat, cell-penetrating peptide; TPGS, D-α-tocopheryl polyethylene-glycol 1000 succinate.

7.4 CLINICAL TOXICITY OF POLYMERIC MICELLES

PTX-loaded PMs formed from PEG-b-poly(D, L-lactide) (Genexol-PM) (Samyang Biopharm Co., Seoul, Korea) have been approved for the treatment of breast, nonsmall cell lung and ovarian cancer (Kim et al., 2007; Lee et al., 2008). Expansion of clinical applications of Genexol-PM to other cancers is under clinical evaluation (Kim et al., 2015; Lee et al., 2012). Amongst nanoparticulate systems, five micellar formulations incorporating PTX (NK105), cisplatin (NC-6004), SN-38 (NK102), dachplatin (active complex of oxaliplatin) (NC-4016), and epirubicin (NC-6300/K-912) are currently under clinical evaluation (Cabral and Kataoka, 2014; Matsumura and Kataoka, 2009).

In a phase I study (patients with colon or gastric cancers), NK105 was given by intravenous infusion for 1 h without antiallergic premedication (Hamaguchi et al., 2007). Of 19 patients, only one experienced allergic reactions. Dose-limiting toxicity (DLT) of NK105 was grade 4 neutropenia. In a phase II study against advanced stomach cancer, as a second-line therapy, patients received NK105 by intravenous infusion for 0.5 h once every 3 weeks without antiallergic premedication (Kato et al., 2012). The most observed grade 3/4 hematological toxicity was neutropenia; there was no grade 3/4 nonhematological toxicity and no hypersensitive reactions without antiallergic premedication. Importantly, grade 3 neurotoxicity was observed for only one in 57 patients (1.8%) which is in contrast with other PTX delivery systems such as Xyotax (CTI BioPharm Co., Seattle, WA, USA) and Abraxane, showing grade 3 neurotoxicity in 10–15% of patients. Currently, a multi-national Phase III clinical study comparing NK105 versus paclitaxel in patients with metastatic or recurrent breast cancer has been completed. The primary endpoint of the study is statistical noninferiority of progression-free survival. Detailed efficacy and safety analyses from this study are expected to be submitting soon.

Phase I study of NC-6004 was carried out in the United Kingdom using 1 h intravenous infusion once every 3 weeks and hydration with 1000 mL saline on the day of drug treatment (Plummer et al., 2011). In all the patients, NC-6004 was given without hospitalization. Although nausea and vomiting caused by NC-6004 were milder than conventional cisplatin treatment, the hypersensitive reactions were more frequently observed with NC-6004 treatment. In a phase I/II study, NC-6004 in combination with gemcitabine was tested in advanced pancreatic cancer patients (Su et al., 2012). In these studies, the dose of NC-6004 reached 90 mg/m^2 without hydration, and hypersensitivity observed in phase I study was prevented by dexamethasone premedication. Currently, NC-6004 has progressed to phase III clinical study for comparison between the combination of NC-6004 with gemcitabine and gemcitabine alone in advanced or metastatic pancreatic cancer in Asian countries. A phase I/II clinical study of NC-6004 against nonsmall cell lung cancer patients is ongoing in the United States.

NC-6300/K-912 is a pH-responsive PM that can selectively release epirubicin under acidic conditions of intratumoral microenvironment or the endo-/lysosomal compartment in cancer cells (Bae et al., 2005). Currently, a phase I clinical study of NC-6300/K-912 is ongoing at the National Cancer Center, Japan.

SP1049C is a Pluronic-based PM formulation containing doxorubicin. Phase I clinical study was conducted in Canada in 1999 (Danson et al., 2004). SP1049C showed similar spectrum of toxicities as conventional Doxil treatment at doses of 35 mg/m^2 and above. Neutropenia was the primary toxicity observed. Unlike Doxil treatment, hand-foot syndrome was not observed for SP1049C. Phase II clinical study was initiated in 2002 in patients with advanced adenocarcinoma of the esophagus and gastroesophageal junction (Valle et al., 2011). The dosing regimen consisted of 30 min infusion at dose of 75 mg/m^2 given once every 3 weeks for up to 6 cycles. 61.9% of patients experienced grade 3/4 neutropenia, with one patient requiring granulocyte colony-stimulating factor treatment for grade 4 neutropenia and fever. One patient experienced grade 3/4 mucositis. Gradual absolute decrements in left ventricular ejection fraction (LVEF, a measure of how much blood is pumped out of the left ventricle with each contraction) were also observed with cumulative treatment. The drug is currently in phase III clinical studies and was designated an orphan drug by the FDA in 2008.

NK012 contains 7-ethyl-10-hydroxy-camptothecin (SN38) (aqueous solubility <5 g/mL) (Meyer-Losic et al., 2008), which is an active metabolite of irinotecan hydrochloride (CPT-11) with powerful cytotoxic effects against various cancerous cell lines *in vitro*. NK012 is formed by covalently conjugating SN38 with poly(glutamic acid) P(Glu) segment of PEG-P(Glu) copolymer followed by self-assembling of the amphiphilic copolymer, PEG-P(Glu)(SN38), in aqueous media (Koizumi et al., 2006). In phase I clinical study, the NK012 was administered intravenously for 30 min every 3 weeks with an SN38 equivalent dose range of 2–28 mg/m^2 and 9–28 mg/m^2 in Japan and the United States, respectively. Prior to infusion, NK012 was diluted to total volume of 250 mL with 5% glucose solution. No DLT was observed for both until 28 mg/m^2 with the exception of one elevated glutamyl transpeptidase at 20 mg/m^2 in the Japanese trial. Nonhematologic toxicities were found minimal. In comparison to CPT-11 trials, cholinergic reactions appeared less frequently. Phase II clinical studies are currently ongoing in Japan and the United States (Nagano et al., 2010).

7.5 STRATEGIES TO IMPROVE CLINICAL TRANSLATION OF NANOMEDICINES

Progressing nanomedicines therapeutics to market is often slow. This may be because their clinical efficacy is not sufficient to warrant accelerated development or that technical or cost challenges in scale-up and manufacturing can delay (or necessitate further) investment. However, the greatest drivers of failure may be our poor understanding of the disease heterogeneity in the patient population, inability to fine-tune the system based on the disease biology or stage of the target patients, and failure to build a platform of evidence supporting a specific end clinical application. The nanomedicine research projects have been structured to adapt the physicochemical parameters of a delivery system such as loading, chemistry, size, charge, and surface modifications to control its *in vivo* behavior. What has been largely lacking is insight into the features of patient tumors that present unique challenges for nanomedicines to display optimal performance. However, the successful clinical translation of nanomedicines would be improved by greater focus in four key areas as mentioned in Figure 7.9.

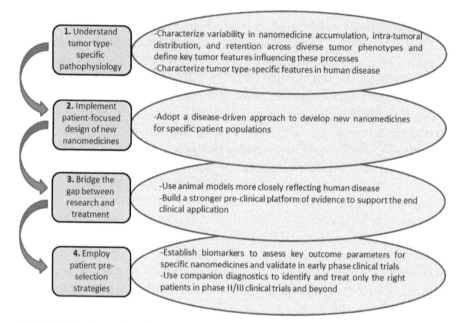

FIGURE 7.9   Strategies to improve clinical translation of nanomedicines.

7.6   PHARMACEUTICAL INDUSTRY'S PERSPECTIVES ABOUT NANOMEDICINES

To make a project attractive for pharmaceutical industries to take into development there must be an opportunity to test a focused hypothesis and the ability to make a decision on whether the agent is likely to succeed with a minimal initial investment. The projects are progressed only when there is confidence in the five "R"s such as right target/efficacy, right tissue/exposure, right safety, right patients, and right commercial potential (Cook, 2014). This general approach has been shown to improve success for projects to develop classical small molecules. The probability of success of a small molecule drug progressing from preclinical proof-of-principle to commercial launch is in the region of 6% (Cook, 2014). There is a continuing demand for innovative, patient personalized therapeutics to improve clinical outcomes. Nanomedicines have a tremendous potential to achieve this goal. However, pharmaceutical industries, correctly or incorrectly, look for additional challenges when considering investment in nanomedicines and other classes of therapeutics such as antibodies, peptides, and DNA/RNA-based agents. For these agents, the formulations are often more complex, there is poor understanding of the correlation between critical quality attributes and efficacy, the regulatory perceptions and environment are constantly evolving, manufacturing and scale-up are technically challenging, important analytical methods require further maturation, and the cost of goods is high (up to 15% higher than for tablets and standard parenterals).

As a result, the development costs for novel classes of therapeutics can be disproportionately high. Moreover, with a lack of *in vitro* and preclinical tests to predict performance in humans adequately, it requires a substantial investment to frontload formulation and process optimization to avoid repeating long and complex clinical studies. Looking to the future, rather than using nanomedicines to develop a solution to formulate the drug, it may be more rational to develop nanomedicine-friendly active pharmaceutical ingredients. For instance, it is possible for synthetic chemists to engineer specific features into the design of novel small molecules to develop drugs that are more compatible for conjugation, active-loading, encapsulation, or complexation with nanomedicine technologies.

For any company, in particular smaller biotech companies or academic labs, the expense of taking a prototype nanomedicine into the clinic and beyond can be prohibitive. For large pharmaceutical companies, this cost can be a major blocker for investment due to a lower perceived probability of success with added technical complexity potentially compounding the risk already associated with a novel drug. Therefore, the preclinical data sets supporting a nanomedicine therapeutic need to be stronger compared to those associated with classic drug therapeutics, which have more conventional formulations, defined patient populations, and fewer treatment obstacles.

7.7 CONCLUSIONS

The major factors that must be taken into account while formulating aqueous-based PMs include required drug therapeutic concentration and toxicity of PMs, compatibility of PMs with other formulation ingredients as well as packaging components, the effect of pH and ionic strength on stability and/or solubility of drug-loaded micelles, and choice of formulation preservative. The aqueous-based PMs can be easily manufactured by the direct dissolution method. After the simple mixing of the drug and PMs in aqueous media and sterilization via sterile filtration, this sterile PMs may further be mixed with previously sterilized solutions of additional components such as bioadhesive, viscosity enhancing, lubricating, buffering, tonicity, antioxidant, preservative, and chelating agents. Then the batch is brought to final volume with additional sterile water. For comparison, the manufacture of sterile nanoparticles is more complicated as compared to PMs including more and much complicated manufacturing steps (e.g., milling, homogenization, emulsion techniques, etc.). The major benefit associated with PMs is related to their sterilization processes in pharmaceutical productions. PMs are simply and inexpensively sterilized by filtration using typical sterilization filters with 0.45 or 0.22 μm pores owing to a fact that PMs are essentially free of microsized particles contamination.

Although the micellar formulations have been clinically investigated in the past decade, the major challenges still inhibit their widespread clinical application. Comprehensive structure–function relationships between the micelle structure and anticancer efficacy still need to be fully established. The size, shape, composition, ligands, responsiveness, and surface properties of micelles need to be precisely regulated and their effects on drug pharmacokinetics, tumor accumulation, and intracellular delivery need to be clearly elucidated. There are many parameters of PMs required to be screened for the optimization with respect to their efficiency and safety concerns.

Even though polymer chemistry provides a flexible strategy to construct tailor-made PMs with clinical potential, the complex structure of the micelles (e.g., stimulus-responsive or actively targeted micelles) may result in some new challenges. For instance, the large-scale production technique, product quality, and pharmacological evaluations of absorption, distribution, metabolism, and excretion (ADME) are emerging as the new concerns, even though many existing guidelines for preclinical and clinical studies for free drugs will also be applicable to micelles. Other technological challenges include the lack of precise characterization methods and evaluation of micelle metabolite. Accurate understanding of the risk and benefits of micelles with complex structure will be essential to achieve their clinical potential.

KEYWORDS

- self-assembly
- nanomicelles
- drug delivery
- multifunctional
- polymeric micelles

REFERENCES

Abdelbary, G. A.; Tadros, M. I. Brain Targeting of Olanzapine via Intranasal Delivery of Core-shell Difunctional Block Copolymer Mixed Nanomicellar Carriers: *In vitro* Characterization, Ex Vivo Estimation of Nasal Toxicity and *In vivo* Biodistribution Studies. *Int. J. Pharm.* **2013,** *452,* 300–310.

Adams, M. L.; Lavasanifar, A; Kwon, G. S. Amphiphilic Block Copolymers for Drug Delivery. *J. Pharm. Sci.* **2003,** *92,* 1343–1355.

Alakhov, V.; Klinski, E.; Li, S.; Pietrzynski, G.; Venne, A.; Batrakova, E; Bronitch, T.; Kabanov, A. Block Copolymer-based Formulation of Doxorubicin. From Cell Screen to Clinical Trials. *Colloids Surf. B* **1999,** *16,* 113–134.

Alexandridis, P.; Holzwarth, J. F.; Hatton, T. A. Micellization of Poly(Ethylene Oxide)-poly(Propylene Oxide)-poly(Ethylene Oxide) Triblock Copolymers in Aqueous Solution: Thermodynamics of Copolymer Association. *Macromolecules* **1994,** *27,* 2414–2425.

Aliabadi, H. M.; Lavasanifar, A. Polymeric Micelles for Drug Delivery. *Expert Opin. Drug Deliv.* **2006,** *3,* 139–162.

Allison, R. R.; Mota, H. C.; Bagnato, V. S.; Sibata, C. H. Bio-nanotechnology and Photodynamic Therapy-state of the Art Review. *Photodiagn. Photodyn. Ther.* **2008,** *5,* 19–28.

Andrade, F.; Neves, J. D.; Gener, P.; Schwartz, J. S.; Ferreira, D.; Oliva, M.; Sarmento, B. Biological Assessment of Self-assembled Polymeric Micelles for Pulmonary Administration of Insulin. *Nanomed. Nanotech. Biol. Med.* **2015,** *11,* 1621–1631.

Bae, Y.; Nishiyama, N.; Fukushima, S.; Koyama, H.; Matsumura, Y.; Kataoka, K. Preparation and Biological Characterization of Polymeric Micelle Drug Carriers with Intracellular pH-triggered Drug Release Property: Tumor Permeability, Controlled Subcellular Drug Distribution, and Enhanced *In vivo* Antitumor Efficacy. *Bioconjugate Chem.* **2005,** *16,* 122–130.

Barreiro-Iglesias, R.; Bromberg, L.; Temchenko, M.; Hatton, T. A.; Alvarez, C. L.; Concheiro, A. Pluronic-g-poly(Acrylic Acid) Copolymers as Novel Excipients for site Specific, Sustained Release Tablets. *Eur. J. Pharm. Sci.* **2005,** *26,* 374–385.

Batrakova, E. V.; Li, S.; Li, Y.; Alakhov, V. Y.; Kabanov, A. V. Effect of Pluronic P85 on ATPase Activity of Drug Efflux Transporters. *Pharm. Res.* **2004,** *21,* 2226–2233.

Bechet, D.; Couleaud, P.; Frochot, C.; Viriot, M. L.; Guillemin, F.; Barberi-Heyob, M. Nanoparticles as Vehicles for Delivery of Photodynamic Therapy Agents. *Trends Biotechnol.* **2008**, *26*, 612–621.

Bernkop, A. S.; Pinter, Y.; Guggi, D.; Kahlbacher, H.; Schöffmann, G.; Schuh, M.; Schmerold, I.; Dorly, M.; D'antonio, M.; Esposito, P.; Huck, C. The Use of Thiolated Polymers as Carrier Matrix in Oral Peptide Delivery-proof of Concept. *J. Control. Release* **2005**, *106*, 26–33.

Biswas, S; Onkar, S; Sara, M; Torchilin, V. P. Polymeric Micelles for the Delivery of Poorly Soluble Drugs. In *Drug Delivery Strategies for Poorly Water Soluble Drugs*; Douroumis D, Fahr A., Eds.; Wiley: UK, 2013; pp 411–476.

Blanco, E.; Kessinger, C. W.; Sumer, B. D.; Gao, J. Multifunctional Micellar Nanomedicine for Cancer Therapy. *Exp. Biol. Med.* **2009**, *234*, 123–131.

Bougard, F.; Giacomelli, C.; Mespouille, L. Influence of the Macromolecular Architecture on the Self-assembly of Amphiphilic Copolymers Based on Poly(N, N-dimethylamino-2-ethyl methacrylate) and Poly(Epsilon-caprolactone). *Langmuir* **2008**, *24*, 8272–8279.

Broxterman, H. J.; Georgopapadakou, N. H. Anticancer Therapeutics: "Addictive" Targets, Multi-targeted Drugs, New Drug Combinations. *Drug Resist. Update* **2005**, *8*, 183–197.

Bumcrot, D.; Manoharan, M.; Koteliansky, V.; Sah, D. W. RNAi Therapeutics: A Potential New Class of Pharmaceutical Drugs. *Nat. Chem. Biol.* **2006**, *2*, 711–719.

Cabral, H.; Kataoka, K. Progress of Drug-loaded Polymeric Micelles into Clinical Studies. *J. Control. Release* **2014**, *190*, 465–476.

Cao, J.; Huang, S.; Chen, Y.; Li, S; Li, X.; Deng, D.; et al. Near-infrared Light-triggered Micelles for Fast Controlled Drug Release in Deep Tissue. *Biomaterials* **2013**, *34*, 6272–6283.

Casey, J. R.; Grinstein, S.; Orlowski, J. Sensors and Regulators of Intracellular pH. *Nat. Rev. Mol. Cell Biol.* **2010**, *11*, 50–61.

Chang, T.; Benet, L. Z.; Hebert, M. F. The Effect of Water-soluble Vitamin E on Cyclosporine Pharmacokinetics in Healthy Volunteers. *Clin. Pharmacol. Ther.* **1996**, *59*, 297–303.

Chayed, S.; Winnik, F. M. *In vitro* Evaluation of the Mucoadhesive Properties of Polysaccharide-based Nanoparticulate Oral Drug Delivery Systems. *Eur. J. Pharm. Biopharm.* **2007**, *65*, 363–370.

Chen, Y.; Shen, Z.; Frey, H.; Perez-Prieto, J.; Stiriba, S. E. Synergistic Assembly of Hyperbranched Polyethylenimine and Fatty Acids Leading to Unusual Supramolecular Nanocapsules. *Chem. Commun.* **2005**, *6*, 755–757.

Chen, Z.; He, Y.; Wang, Y.; Wang, X. Amphiphilic Diblock Copolymer with Dithienylethene Pendants: Synthesis and Photo-modulated Self-assembly. *Macromol. Rapid Commun.* **2011**, *32*, 977–982.

Chen, A. L.; Ni, H. C.; Wang, L. F.; Chen, J. S. Biodegradable Amphiphilic Copolymers Based on Poly(Epsilon-caprolactone)-graft Chondroitin Sulfate as Drug Carriers. *Biomacromolecules* **2008**, *9*, 2447–2457

Cheng, R.; Meng, F.; Deng, C.; Klok, H. A.; Zhong, Z. Dual and Multi-stimuli Responsive Polymeric Nanoparticles for Programmed Site-specific Drug Delivery. *Biomaterials* **2013**, *34*, 3647–3657.

Choudhary, S.; Nouri, K.; Elsaie, M. L. Photodynamic Therapy in Dermatology: A Review. *Lasers Med. Sci.* **2009**, *24*, 971–980.

Collnot, E. M.; Baldes, C.; Schaefer, U. F.; Edgar, K. J.; Wempe, M. F.; Lehr, C. M. Vitamin E TPGS P-glycoprotein Inhibition Mechanism: Influence on Conformational Flexibility,

Intracellular ATP levels, and Role of Time and Site of Access. *Mol. Pharm.* **2010,** *7,* 642–651.

Constantinides, P. P.; Han, J.; Davis, S. S. Advances in the Use of Tocols as Drug Delivery Vehicles. *Pharm. Res.* **2006,** *23,* 243–255.

Cook, D. Lessons Learned from the Fate of Astra Zeneca's Drug Pipeline: A Five Dimensional Framework. *Nat. Rev. Drug Discov.* **2014,** *13,* 419–431.

Cooper, A. I.; Londono, J. D.; Wignall, G.; McClain, J. B.; Samulski, E. T.; Lin, J. S.; Dobrynin, A.; Rubinstein, M.; Burke, A. L. C.; Frechet, J. M. J. Desimone, Extraction of a Hydrophilic Compound from Water into Liquid CO_2 Using Dendritic Surfactants. Nature **1997,** *389,* 368–371.

Craparo, E. F.; Teresi, G.; Bondi, M. L.; Licciardi, M.; Cavallaro, G. Phospholipid-polyaspartamide Micelles for Pulmonary Delivery of Corticosteroids. *Int. J. Pharm.* **2011,** *406,* 135–144.

Dagar, A.; Kuzmis, A.; Rubinstein, I.; Sekosan, M.; Onyuksel, H. VIP-targeted Cytotoxic Nanomedicine for Breast Cancer. *Drug Deliv. Transl. Res.* **2012,** *2,* 454–462.

Dahmani, F. Z.; Yang, H.; Zhou, J.; Yao, J.; Zhang, T.; Zhang, Q. Enhanced Oral Bioavailability of Paclitaxel in Pluronic/LHR Mixed Polymeric Micelles: Preparation, *In vitro* and *In vivo* Evaluation. *Eur. J. Pharm. Sci.* **2012,** *47,* 179–189.

Danhauser-Riedl, S.; Hausmann, E.; Schick, H. D. Phase-I Clinical and Pharmacokinetic Trial of Dextran Conjugated Doxorubicin (Ad-70, Dox-Oxd). *Invest. New Drugs* **1993,** *11,* 187–195.

Danhier, F.; Feron, O.; Preat, V. To Exploit the Tumor Microenvironment: Passive and Active Tumor Targeting of Nano Carriers for Anticancer Drug Delivery. *J. Control Release* **2010,** *148,* 135–146.

Danson, S.; Ferry, D.; Alakhov, V.; Margison, J.; Kerr, D.; Jowle, D.; Brampton, M.; Halbert, G.; Ranson, M. Phase I Dose Escalation and Pharmacokinetic Study of Pluronic Polymerbound Doxorubicin (SP 1049C) in Patients with Advanced Cancer. *Br. J. Cancer* **2004,** *90,* 2085–2091.

Daugherty, A. L.; Mrsny, R. J. Transcellular Uptake Mechanisms of the Intestinal Epithelial Barrier Part One. *Pharm. Sci. Technol. Today* **1999,** *2,* 144–151.

De, L. R.; Aili, R. D.; Stevens, M. M. Enzyme-responsive Nanoparticles for Drug Release and Diagnostics. *Adv. Drug. Deliv Rev.* **2012,** *64,* 967–978.

Decuzzi, P.; Ferrari, M. The Receptor-mediated Endocytosis of Non-spherical Particles. *Biophys. J.* **2008,** *94,* 3790–3797.

Detty, M. R.; Gibson, S. L.; Wagner, S. J. Current Clinical and Preclinical Photosensitizers for Use in Photodynamic Therapy. *J. Med. Chem.* **2004,** *47,* 3897–915.

Dolmans, D. E.; Fukumura, D.; Jain, R. K. Photodynamic Therapy for Cancer. *Nat. Rev. Cancer* **2003,** *3,* 380–387.

Dominska, M.; Dykxhoorn, D. M. Breaking Down the Barriers: siRNA Delivery and Endosome Escape. *J. Cell Sci.* **2010,** *123,* 1183–1189.

Dougherty, T. J.; Gomer, C. J.; Henderson, B. W.; Jori, G.; Kessel, D.; Korbelik, M. Photodynamictherapy. *J. Nat. Cancer Inst.* **1998,** *90,* 889–905.

Duan, Y.; Cai, X.; Du, H.; Zhai, G. Novel *In Situ* Gel Systems Based on P123/TPGS Mixed Micelles and Gellan Gum for Ophthalmic Delivery of Curcumin. *Colloids Surf B* **2015,** *128,* 322–330.

Dufresne, M. H.; Gauthier, M. A.; Leroux. J. C. Thiol-functionalized Polymeric Micelles: From Molecular Recognition to Improved Mucoadhesion. *Bioconjugate Chem.* **2005,** *16,* 1027–1033.

Elbashir, S. M.; Harborth, J.; Lendeckel, W.; Yalcin, A.; Weber, K.; Tuschl, T. Duplexes of 21-nucleotide RNA Mediated RNA Interference in Cultured Mammalian Cells. *Nature* **2001,** *411,* 494–498.

Erhardt, R.; Böker, A.; Zettl, H.; Kaya, H.; Pyckhout-Hintzen, W.; Krausch, G.; Abetz, V.; Müller, A. H. Janus Micelles. *Macromolecules* **2001,** *34,* 1069–1075.

Falamarzian, A.; Xiong, X. B.; Uludag, H.; Lavasanifar, A. Polymeric Micelles for siRNA Delivery. *J. Drug Deliv. Sci. Technol.* **2012,** *22,* 43–54.

Fire, A.; Xu, S. Q.; Montgomery, M. K.; Kostas, S. A.; Driver, S. E.; Mello, C. C. Potent and Specific Genetic Interference by Double-stranded RNA in *Caenorhabditis elegans. Nature* **1998,** *391,* 806–811.

Fleige, E.; Quadir, M. A.; Haag, R. Stimuli-responsive Polymeric Nanocarriers for the Controlled Transport of Active Compounds: Concepts and Applications. *Adv. Drug Deliv. Rev.* **2012,** *64,* 866–884.

Fomina, N.; Sankaranarayanan, J.; Almutairi, A. Photochemical Mechanisms of Light-triggered Release from Nanocarriers. *Adv. Drug Deliv. Rev.* **2012,** *64,* 1005–1020.

Francis, M. F.; Cristea, M.; Winnik, F. M. Exploiting the Vitamin B-12 Pathway to Enhance Oral Drug Delivery via Polymeric Micelles. *Biomacromolecules* **2005,** *6,* 2462–2467.

Freichels, H.; Alaimo, D; Auzely-Velty, R.; Jerome, C. Alpha-acetal, Omega-alkyne Poly(Ethyleneoxide) as a Versatile Building Block for the Synthesis of Glycoconjugated Graft-copolymers Suited for Targeted Drug Delivery. *Bioconjugate Chem.* **2012,** *23,* 1740–1752.

Fukushima, K.; Kimura, Y. Stereocomplexed Polylactides (Neo-PLA) as High Performance Bio-based Polymers: Their Formation, Properties, and Application. *Polym. Int.* **2006,** *55,* 626–642.

Gaucher, G.; Dufresne, M. H.; Sant, V.; Maysinger, D.; Leroux, J. C. Block Copolymer Micelles: Preparation, Characterization and Application in Drug Delivery. *J. Control Release* **2005,** *109,* 169–188.

Gill, K. K.; Nazzal, S.; Kaddoumi, A. Paclitaxel Loaded PEG5000-DSPE Micelles as Pulmonary Delivery Platform: Formulation Characterization, Tissue Distribution, Plasma Pharmacokinetics, and Toxicological Evaluation. Eur. J. Pharm. Biopharm. **2011,** *79,* 276–284.

Gohy, J. F.; Zhao, Y. Photo-responsive Block Copolymer Micelles: Design and Behavior. *Chem. Soc. Rev.* **2013,** *42,* 7117–7129.

Goldenberg, D. M.; Sharkey, R. M. Using Antibodies to Target Cancer Therapeutics. *Expert Opin. Biol. Ther.* **2012,** *12,* 1173–1190.

Gong, C.; Wu, Q.; Wang, Y.; Zhang, D.; Luo, F.; Zhao, X.; Wei, Y.; Qian, Z. A Biodegradable Hydrogel System Containing Curcumin Encapsulated in Micelles for Cutaneous Wound Healing. *Biomaterials* **2013,** *34,* 6377–6387.

Grabovac, V.; Guggi, D. A. Comparison of the Mucoadhesive Properties of Various Polymers. *Adv. Drug. Deliv. Rev.* **2005,** *57,* 1713–1723.

Guan, Y.; Huang, J.; Zuo, L.; Xu, J.; Si, L.; Qiu, J.; Li, G. Effect of Pluronic P123 and F127 Block Copolymer on P-glycoprotein Transport and CYP3A Metabolism. *Arch. Pharm. Res.* **2011,** *34,* 1719–1728.

Gülçür, E.; Thaqi, M.; Khaja, F.; Kuzmis, A.; Önyüksel, H. Curcumin VIP-targeted Sterically Stabilized Phospholipid Nanomicelles: A Novel Therapeutic Approach for Breast Cancer and Breast Cancer Stem Cells. *Drug Deliv. Transl. Res.* **2013**, *3*, 562–574.

Gupta, N.; Ibrahim, H. M.; Ahsan, F. Peptide-micelle Hybrids Containing Fasudil for Targeted Delivery to the Pulmonary Arteries and Arterioles to Treat Pulmonary Arterial Hypertension. *J. Pharm. Sci.* **2014**, *103*, 3743–3753.

Hamaguchi, T.; Kato, K.; Yasui, H. A Phase I and Pharmacokinetic Study of NK105, a Paclitaxel-incorporating Micellar Nanoparticle Formulation. *Br. J. Cancer* **2007**, *97*, 170–176.

Hannon, G. J. RNA Interference. *Nature* **2002**, *418*, 244–251.

Hopper, C. Photodynamic Therapy: A Clinical Reality in the Treatment of Cancer. *Lancet Oncol.* **2000**, *1*, 212–219.

Hossain, M. D.; Tran, L. T. B.; Park, J. M.; Lim, K. T. Facile Synthesis of Core-Surface Cross Linked Nanoparticles by Inters Block RAFT Polymerization. *J. Polym. Sci. A Polym. Chem.* **2010**, *48*, 4958–4964.

Hu, J.; Zhang, G.; Liu, S. Enzyme-responsive Polymeric Assemblies, Nanoparticles and Hydrogels. *Chem. Soc. Rev.* **2012**, *41*, 5933–5949.

Hu, X.; Yang, F. F.; Quan, L. H.; Liu, C. Y.; Liu, X. M.; Ehrhardt, C.; Liao, Y. H. Pulmonary Delivered Polymeric Micelles-pharmacokinetic Evaluation and Biodistribution Studies. *Eur. J. Pharm. Biopharm.* **2014**, *88*, 1064–1075.

Huang, J.; Si, L.; Jiang, L.; Fan, Z.; Qiu, J.; Li, G. Effect of Pluronic F68 Blocks Copolymer on P-glycoprotein Transport and CYP3A4 Metabolism. *Inter. J. Pharm.* **2008**, *356*, 351–353.

Jianbin, A. N.; Jingxue, M. A. Research Progress of Pharmacological Functions and Ophthalmic Applications of Curcumin. *China J. Chi. Ophthal.* **2008** 18, 360–362.

Jiao, J. Polyoxyethylated Nonionic Surfactants and Their Applications in Topical Ocular Drug Delivery. *Adv. Drug Deliv. Rev.* **2008**, *60*, 1663–1673.

Jin, Q.; Mitschang, F.; Agarwal, S. Biocompatible Drug Delivery System for Photo-triggered Controlled Release of 5-Fluorouracil. *Biomacromolecules* **2011**, *12*, 3684–3691.

Jones, M.; Leroux, J. Polymeric Micelles: A New Generation of Colloidal Drug Carriers. *Eur. J. Pharm. Biopharm.* **1999**, *48*, 101–111.

Jones, M. C.; Ranger, M.; Leroux, J. C. pH-sensitive Unimolecular Polymeric Micelles: Synthesis of a Novel Drug Carrier. *Bioconjugate Chem.* **2003**, *14*, 774–781.

Jonge, M. E.; Huiteman, A. D.; Schellens, J. H. M.; Rodenhuis, S.; Beijnen, J. H. Population Pharmacokinetics of Orally Administered Paclitaxel Formulated in Cremophor EL. *Br. J. Clin. Pharmacol.* **2004**, *59*, 325–334.

Kabanov, A.; Alakhov, V. Pluronic Block Copolymers in Drug Delivery: From Micellar Nanocontainers to Biological Response Modifiers. *Crit. Rev. Ther. Drug* **2002**, *19*, 173.

Kalarickal, N. C.; Rimmer, S.; Sarker, P.; Leroux, J. C. Thiol-functionalized Poly(Ethylene Glycol)-b-polyesters: Synthesis and Characterization. *Macromolecules* **2007**, *40*, 1874–1880.

Kamaly, N.; Xiao, Z.; Radovic, A. F.; Farokhzad, O. C. Targeted Polymeric Therapeutic Nano Particles: Design, Development and Clinical Translation. *Chem. Soc. Rev.* **2012**, *41*, 2971–3010.

Kanasty, R. L.; Whitehead, K. A.; Vegas, A. J.; Anderson, D. G. Action and Reaction: The Biological Response to siRNA and Its Delivery Vehicles. *Mol. Ther.* **2012**, *20*, 513–524.

Kanazawa, T.; Taki, H.; Tanaka, K.; Takashima, Y.; Okada, H. Cell-penetrating Peptide-modified Block Copolymer Micelles Promote Direct Brain Delivery via Intranasal Administration. *Pharm. Res.* **2011**, *28*, 2130–2139.

Kanazawa, T.; Sugawara, K.; Tanaka, K.; Horiuchi, S.; Takashima, Y.; Okada, H. Suppression of Tumor Growth by Systemic Delivery of anti-VEGF siRNA with Cell-penetrating Peptide-modified MPEG-PCL Nanomicelles. *Eur. J. Pharm. Biopharm.* **2012**, *81*, 470–477.

Kanazawa, T.; Morisaki, K.; Suzuki, S.; Takashima, Y. Prolongation of Life in Rats with Malignant Glioma by Intranasal siRNA/Drug Codelivery to the Brain with Cell-penetrating Peptide-modified Micelles. *Mol. Pharmaceutics* **2014**, *11*, 1471–1478.

Kanazawa, T.; Akiyama, F.; Kakizaki, S.; Takashima, Y.; Seta, Y. Delivery of siRNA to the Brain Using a Combination of Nose-to-brain Delivery and Cell-penetrating Peptide-modified Nano-micelles. *Biomaterials* **2013**, *34*, 9220–9226.

Katayose, S.; Kataoka, K. Water-soluble Polyion Complex Associates of DNA and Poly(Ethylene Glycol)-poly(l-lysine) Block Copolymer. *Bioconjugate Chem.* **1997**, *8*, 702–707.

Kato, K.; Chin, K.; Yoshikawa, K.; et al. Phase II Study of NK105, a Paclitaxel-incorporating Micellar Nanoparticle, for Previously Treated Advanced or Recurrent Gastric Cancer. *Invest. New Drugs* **2012**, *30*, 1621–1627.

Katragadda, S.; Talluri, R. S.; Mitra, A. K. Modulation of P-glycoprotein-mediated Efflux by Prodrug Derivatization: An Approach Involving Peptide Transporter-mediated Influx Across Rabbit Cornea. *J. Ocular Pharmacol. Ther.* **2006**, *22*, 110–120.

Keene, J. P.; Kessel, D.; Land, E. J.; Redmond, R. W.; Truscott, T. G. Direct Detection of Single to Oxygen Sensitized by Haematoporphyrin and Related Compounds. *Photochem. Photobiol.* **1986**, *43*, 117–120.

Kim, S. W.; Jeong, B.; Bae, Y. H.; Lee, D. S. Biodegradable Block Copolymers as Injectable Drug-delivery Systems. *Nature* **1997**, *388*, 860–862.

Kim, D. W.; Kim, S. Y.; Kim, H. K. Multicenter Phase II Trial of Genexol-PM, a Novel Cremophor-free, Polymeric Micelle Formulation of Paclitaxel, with Cisplatin in Patients with Advanced Non-small-cell Lung Cancer. *Ann. Oncol.* **2007**, *18*, 2009–2014.

Kim, S.; Kim, J. Y.; Moo, H. K.; Acharya, G.; Park, K. Hydrotropic Polymer Micelles Containing Acrylic Acid Moieties for Oral Delivery of Paclitaxel. *J. Control. Release* **2008**, *132*, 222–229.

Kim, S. H.; Jeong, J. H.; Lee, S. H.; Kim, S. W.; Park, T. G. Local and Systemic Delivery of VEGF siRNA Using Polyelectrolyte Complex Micelles for Effective Treatment of Cancer. *J. Control. Release* **2008**, *129*, 107–116.

Kim, S. H.; Tan, J. P. K.; Nederberg, F.; Fukushima, K.; Yang, Y. Y.; Waymouth, R. M.; Hedrick, J. L. Mixed Micelle Formation Through Stereocomplexation Between Enantiomeric Poly(Lactide) Block Copolymers. *Macromolecules* **2009**, *42*, 25–29.

Kim, D. W.; Ramasamy, T.; Choi, J. Y.; Kim, J. H.; Yong, C. S.; Kim, J. O.; Choi, H. G. The Influence of Bile Salt on the Chemotherapeutic Response of Docetaxel-loaded Thermosensitive Nanomicelles. *Int. J. Nanomed.* **2014**, *9*, 3815–3824.

Kim, H. S.; Lee, J. Y.; Lim, S. H. A Prospective Phase II Study of Cisplatin and Cremophor EL-free Paclitaxel (Genexol-PM) in Patients with Unresectablethymic Epithelial Tumors. *J. Thorac. Oncol.* **2015**, *10*, 1800–1806.

Koizumi, F.; Kitagawa, M.; Negishi, T.; Onda, T.; Matsumoto, S.; Hamaguchi, T.; Matsumura, Y. Novel SN-38-incorporating Polymeric Micelles, NK012, Eradicate Vascular Endothelial Growth Factor-secreting Bulky Tumors. *Cancer Res.* **2006**, *66*, 10048–10056.

Konan, Y. N.; Gurny, R.; Allémann, E. State of the Art in the Delivery of Photosensitizers for Photodynamic Therapy. *J. Photochem. Photobiol. B* **2002**, *66*, 89–106.

Kono, K.; Kojima, C.; Hayashi, N. Preparation and Cytotoxic Activity of Poly(Ethylene Glycol)-modified Poly(Amidoamine) Dendrimers Bearing Adriamycin. *Biomaterials* **2008**, *29*, 1664–1675.

Koo, A. N.; Min, K. H.; Lee, H. J.; Lee, S. U.; Kim, K.; Kwon, I. C. Tumor Accumulation and Antitumor Efficacy of Docetaxel-loaded Core-shell-corona Micelles with Shell-specific Redox-responsive Cross-links. *Biomaterials* **2012**, *33*, 1489–1499.

Krasavage, W. J.; Terhaar, C. J. D-alpha-tocopheryl Poly(Ethylene Glycol) 1000 Succinate: Acute Toxicity, Subchronic Feeding, Reproduction, and Teratologic Studies in the Rat. *J. Agric. Food Chem.* **1977**, *25*, 273–278.

Kwon, G. S. Diblock Copolymer Nanoparticles for Drug Delivery. *Crit. Rev. Ther. Drug Carrier Syst.* **1998**, *15*, 481–512.

Kwon, G. S; Kataoka, K. Block Copolymer Micelles as Long Circulating Drug Vehicles. *Adv. Drug Deliv. Rev.* **1995**, *16*, 295–309.

Langer, C. J.; O'Byrne, K. J.; Socinski, M. A.; Mikhailov, S. M.; Lesniewski-Kmak, K.; Smakal, M. Phase III Trial Comparing Paclitaxel Poliglumex (CT-2103, PPX) in Combination with Carboplatin Versus Standard Paclitaxel and Carboplatin in the Treatment of PS 2 Patients with Chemotherapy-naive Advanced Non-small Cell Lung Cancer. *J. Thorac. Oncol.* **2008**, *3*, 623–30.

Lee, E. S.; Shin, H. J.; Na, K.; Bae, Y. H. Poly(l-histidine)-PEG Blocks Copolymer Micelles and pH-induced Destabilization. *J. Control. Release* **2003**, *90*, 363–374.

Lee, S. H.; Kim, S. H.; Park, T. G. Intracellular siRNA Delivery System Using Polyelectrolyte Complex Micelles Prepared from VEGF siRNA-PEG Conjugate and Cationic Fusogenic Peptide. *Bioche. Biophy. Res. Commun.* **2007**, *357*, 511–516.

Lee, K. S.; Chung, H. C.; Im, S. A. Multicenter Phase II Trial of Genexol-PM, a Cremophor-free, Polymeric Micelle Formulation of Paclitaxel, in Patients with Metastatic Breast Cancer. *Breast Cancer Res. Treat.* **2008**, *108*, 241–250.

Lee, J. L.; Ahn, J. H.; Park, S. H. Phase II Study of a Cremophor-free, Polymeric Micelle Formulation of Paclitaxel for Patients with Advanced Urothelial Cancer Previously Treated with Gemcitabine and Platinum. *Invest. New Drugs* **2012**, *30*, 1984–1990.

Li, Y.; Xiao, K.; Luo, J.; Xiao, W.; Lee, J. S.; Gonik, A. M.; et al. Well-defined, Reversible Disulfide Cross-linked Micelles for On-demand Paclitaxel Delivery. *Biomaterials* **2011**, *32*, 6633–6645.

Li, J.; Huo, M.; Wang, J.; Zhou, J.; Mohammad, J. M.; Zhang, Y.; et al. Redox-sensitive Micelles Self-assembled from Amphiphilic Hyaluronic Acid-Deoxycholic Acid Conjugates for Targeted Intracellular Delivery of Paclitaxel. *Biomaterials* **2012**, *33*, 2310–2320.

Li, J.; Ge, Z.; Liu, S. PEG-sheddable Polyplex Micelles as Smart Gene Carriers Based on MMP-cleavable Peptide-linked Block Copolymers. *Chem. Commun. (Camb).* **2013a**, *49*, 6974–6976.

Li, Y.; Liu, T.; Zhang, G.; Ge, Z.; Liu, S. Tumor-targeted Redox-responsive Nonviral Gene Delivery Nanocarriers Based on Neutral-cationic Brush Block Copolymers. *Macromol. Rapid Commun.* **2013b**, *35*, 466–473.

Liggins, R. T.; Burt, H. M. Polyether-polyester Diblock Copolymers for the Preparation of Paclitaxel Loaded Polymeric Micelle Formulations. *Adv. Drug Deliv. Rev.* **2002**, *54*, 191–202.

Lim, C. K.; Heo, J.; Shin, S.; Jeong, K.; Seo, Y. H.; Jang, W. D.; Park, C. R.; Park, S. Y.; Kim, S.; Kwon, I. C. Nanophotosensitizers toward Advanced Photodynamic Therapy of Cancer. *Cancer Lett.* **2013,** *334,* 176–87.

Lin, G. Y.; Lv, H. F.; Lu, C. T.; Chen, L. J.; Lin, M.; Zhang, M.; et al. Construction and Application of Biotin-poloxamer Conjugate Micelles for Chemotherapeutics. *J. Microencapsul.* **2013,** *30,* 538–545.

Liu, G. Y.; Chen, C. J.; Li, D. D.; Wang, S. S.; Ji, J. Near-infrared Light-sensitive Micelles for Enhanced Intracellular Drug Delivery. *J. Mater. Chem.* **2012,** *22,* 16865–16871.

Liu, X. Q.; Sun, C. Y.; Yang, X. Z.; Wang, J. Polymeric-micelle-based Nanomedicine for siRNA Delivery. *Part. Syst. Charact.* **2013,** *30,* 211–228.

Lo, C. L.; Lin, S. J.; Tsai, H. C.; Chan, W. H.; Tsai, C. H.; Cheng, H. D.; Hsiue, G. H. Mixed Micelle Systems Formed from Critical Micelle Concentration and Temperature Sensitive Diblock Copolymers for Doxorubicin Delivery. *Biomaterials* **2009,** *30,* 3961–3970.

Luo, Y.; Yao, X.; Yuan, J.; Ding, T.; Gao, Q. Preparation and Drug Controlled-release of Polyion Complex Micelles as Drug Delivery Systems. *Colloids Surf. B.* **2009,** *68,* 218–224.

Luo, Y. L.; Yuan, J. F.; Liu, X. J.; Xie, H.; Gao, Q. Y. Self-assembled Polyion Complex Micelles Based on PVP-b-PAMPS and PVP-b-PDMAEMA for Drug Delivery. *J. Bioact. Compatible Poly.* **2010,** *25,* 292–304.

Maeda, H. Macromolecular Therapeutics in Cancer Treatment: The EPR Effect and Beyond. *J. Control Release* **2012,** *164,* 138–44.

Maeda, H.; Wu, J.; Sawa, T.; Matsumura, Y.; Hori, K. Tumor Vascular Permeability and the EPR Effect in Macromolecular Therapeutics: A Review. *J. Control Release* **2000,** *65,* 271–284.

Mahajan, H. S.; Mahajan, P. R. Development of Grafted Xyloglucan Micelles for Pulmonary Delivery of Curcumin: *In vitro* and *In vivo* Studies. *Int. J. Biol. Macromol.* **2016,** *82,* 621–627.

Matsumura, Y.; Kataoka, K. Preclinical and Clinical Studies of Anticancer Agent-incorporating Polymer Micelles. *Cancer Sci.* **2009,** *100,* 572–579.

Meng, F.; Hennink, W. E.; Zhong, Z. Reduction-sensitive Polymers and Bioconjugates for Biomedical Applications. *Biomaterials* **2009,** *30,* 2180–2198.

Menon, S.; Thekkayil, R.; Varghese, S.; Das, S. Photoresponsive Soft Materials: Synthesis and Photophysical Studies of a Stilbene-based Diblock Copolymer. *J. Polym. Sci. A Polym. Che.* **2011,** *49,* 5063–5073.

Meyer-Losic, F.; Nicolazzi, C.; Quinonero, J.; Ribes, F.; Michel, M.; Dubois, V.; de Coupade, C.; Boukaissi, M.; et al. DTS-108, a Novel Peptidic Prodrug of SN38: *In vivo* Efficacy and Toxicokinetic Studies. *Clin. Cancer Res.* **2008,** *14,* 2145–2153.

Moroi, Y. *Micelles Theoretical and Applied Aspects*; Springer: New York, 1992.

Mura, S.; Nicolas, J.; Ouvreur, P. Stimuli-responsive Nanocarriers for Drug Delivery. *Nat. Mater.* **2013,** *12,* 991–1003.

Nagano, T.; Yasunaga, M.; Goto, K.; Kenmotsu, H.; Koga, Y.; Kuroda, J.; Nishimura, Y.; Sugino, T.; Nishiwaki, Y.; Matsumura, Y. Synergistic Antitumor Activity of the SN-38-incorporating Polymeric Micelles NK012 with S-1 in a Mouse Model of Non-small Cell Lung Cancer. *Int. J. Cancer* **2010,** *127,* 2699–2706.

Nam, Y. S.; Kang, H. S.; Park, J. Y.; Park, T. G.; Han, S. H.; Chang, I. S. New Micelle-like Polymer Aggregates Made from PEI-PLGA Diblock Copolymers: Micellar Characteristics and Cellular Uptake. *Biomaterials* **2003,** *24,* 2053–2059.

Navarro, G.; Essex, S.; Torchilin, V. P. "The 'Non-viral' Approach for siRNA Delivery in Cancer Treatment: A Special Focus on Micelles and Liposomes," in DNA and RNA and biotechnologies in Medicine. In *Diagnosis and Treatment of Diseases*; Volker, A., Barcisze-wski, J., Eds.; *Springer:* Berlin, Heidelberg, 2013; pp 241–261.

Nie, S.; Xing, Y.; Kim, G. J.; Simons, J. W. Nanotechnology Applications in Cancer. *Annu. Rev. Biomed. Eng.* **2007**, *9*, 257–288.

Nour, S. A.; Abdelmalak, N. S.; Naguib, M. J.; Rashed, H. M.; Ibrahim, A. B. Intranasal Brain-targeted Clonazepam Polymeric Micelles for Immediate Control of Status Epilep-ticus: *In vitro* Optimization, Ex Vivo Determination of Cytotoxicity, *In vivo* Biodistribution and Pharmacodynamics Studies. *Drug Deliv.* **2016**, *23*, 3681–3695.

O'Reilly, R. K.; Hawker, C. J.; Wooley, K. L. Cross-linked Block Copolymer Micelles: Functional Nanostructures of Great Potential and Versatility. *Chem. Soc. Rev.* **2006**, *35*, 1068–1083.

Oerlemans, C.; Bult, W.; Bos, M.; Storm, G.; Nijsen, J. F. W.; Hennink, W. E. Polymeric Micelles in Anticancer Therapy: Targeting, Imaging and Triggered Release. *Pharm. Res.* **2010**, *12*, 1–21.

Oishi, M.; Nagasaki, Y.; Itaka, K.; Nishiyama, N.; Kataoka, K. Lactosylated Poly(Ethyleneglycol)-siRNA Conjugate Through Acid-labile Beta Thiopropionate Linkage to Construct pH-sensitive Polyion Complex Micelles Achieving Enhanced Gene Silencing in Hepatoma Cells. *J. Am. Chem. Soc.* **2005**, *127*, 1624–1625.

Ould, O. L.; Noppe, M.; Langlois, X.; Willems, B.; Riele, P. T.; Timmerman, P.; Brewster, M. E.; Arien, A.; Preat, V. Self-assembling PEG-p(CL-co-TMC) Copolymers for Oral Delivery of Poorly Water-soluble Drugs: A Case Study with Risperidone. *J. Control. Release* **2005**, *102*, 657–668.

Owens III, D. E.; Peppas, N. A. Opsonization, Biodistribution, and Pharmacokinetics of Poly-meric Nanoparticles. *Int. J. Pharm.* **2006**, *307*, 93–102.

Park, K. Dendrimers Polymeric Micelles for Enhanced Photodynamic Cancer Treatment. *J. Control. Release* **2009**, *133*, 171.

Paszko, E.; Ehrhardt, C.; Senge, M. O.; Kelleher, D. P.; Reynolds, J. V. Nano Drug Applica-tions in Photodynamic Therapy. *Photodiagn. Photodyn. Ther.* **2011**, *8*, 14–29.

Peng, C. L.; Shieh, M. J.; Tsai, M. H. Self-assembled Star-shaped Chlorin-core Poly(C-caprolactone)-poly(Ethylene Glycol) Diblock Copolymer Micelles for Dual Chemo-photo-dynamic Therapies. *Biomaterials* **2008**, *29*, 3599–3608

Pepiae, I.; Lovric, J. Filipovic-Greic. Polymeric Micelles in Ocular Drug Delivery: Rationale, Strategies and Challenges. *Chem. Biochem. Eng.* **2012**, *26*, 365–377.

Perche, F.; Biswas, S.; Wang, T.; Zhu, L.; Torchilin, V. P. Hypoxia-targeted siRNA Delivery. *Angew Chem. Int. Ed. Engl.* **2014**, *53*, 3362–3366.

Pierri, E.; Avgoustakis, K. Poly(Lactide)-poly(Ethylene Glycol) Micelles as a Carrier for Griseofulvin. *J. Biomed. Mater. Res A* **2005**, *75*, 639–647.

Plummer, R.; Wilson, R. H.; Calvert, H. A Phase I Clinical Study of Cisplatin-incorporated Polymeric Micelles (NC-6004) in Patients with Solid Tumors. *Br. J. Cancer* **2011**, *104*, 593–598.

Rakshit, A. K.; Palepu, R. M. Additive Effects on Non-ionic Surfactant Assemblies: A review. *Recent Dev. Coll. Interface Res.* **2003**, *1*, 203–219.

Rana, T. M. Illuminating the Silence: Understanding the Structure and Function of Small RNAs. *Nat. Rev. Mol. Cell Biol.* **2007**, *8*, 23–36.

Rapoport, N. Ultrasound-mediated Micellar Drug Delivery. *Int. J. Hyperthermia* **2012**, *28*, 374385.

Read, E. S.; Armes, S. P. Recent Advances in Shell Cross-linked Micelles. *Chem. Commun.* **2007**, *29*, 3021–3035.

Rege, B. D.; Kao, J. P. Y.; Polli, J. E. Effects of Nonionic Surfactants on Membrane Transporters in Caco-2 cell Monolayers. *Eur. J. Pharm. Sci.* **2002**, *16*, 237–246.

Riehle, R. D.; Cornea, S.; Degterev, A.; Torchilin, V. Micellar Formulations of Pro-apoptotic DM-PIT-1 Analogs and TRAIL *In vitro* and *In vivo*. *Drug Deliv.* **2013**, *20*, 78–85.

Riess, G.; Hurtrez, G.; Bohadur, P. Block Copolymers. In *Encyclopedia of Polymer Science and Engineering*; Wiley-Interscience: New York, 1985; Vol. 2, pp 324–434.

Riess, G. Micellization of Block Copolymers. *Prog. Polym. Sci.* **2003**, *28*, 1107–1170.

Rieux, A. D.; Fievez, V.; Garinot, M.; Schneider, Y. J.; Préat, V. Nanoparticles as Potential Oral Delivery Systems of Proteins and Vaccines: A Mechanistic Approach. *J. Control. Release* **2006**, *116*, 1–27.

Russel, G. J. The Potential Use of Receptor-mediated Endocytosis for Oral Drug Delivery. *Adv. Drug Deliv. Rev.* **2001**, *46*, 59–73.

Russel, G. J. Use of Targeting Agents to Increase Uptake and Localization of Drugs to the Intestinal Epithelium. *J. Drug Target.* **2004**, *12*, 113–123.

Saito, G.; Swanson, J. A.; Lee, K. D. Drug Delivery Strategy Utilizing Conjugation via Reversible Disulfide Linkages: Role and Site of Cellular Reducing Activities. *Adv. Drug Deliv. Rev.* **2003**, *2*, 55, 199–215.

Saito, K.; Ingalls, L. R.; Lee, J.; Warner, J. C. Core-bound Polymeric Micellar System Based on Photocrosslinking of Thymine. *Chem. Commun.* **2007**, *24*, 2503–2505.

Sant, V. P.; Smith, D.; Leroux, J. C. Novel pH-sensitive Supramolecular Assemblies for Oral Delivery of Poorly Water Soluble Drugs: Preparation and Characterization. *J. Control. Release* **2004**, *97*, 301–312.

Sant, V. P.; Smith, D.; Leroux, J. C. Enhancement of Oral Bioavailability of Poorly Water-soluble Drugs by Poly(Ethylene Glycol)-block-poly(Alkyl Acrylate-comethacrylic Acid) Self-assemblies. *J. Control. Release* **2005**, *104*, 289–300.

Satturwar, P.; Eddine, M. N.; Ravenelle, F.; Leroux, J. C. pH-responsive Polymeric Micelles of Poly(Ethylene Glycol)-b-poly(Alkyl (Meth) Acrylate-co-methacrylic Acid): Influence of the Copolymer Composition on Self-assembling Properties and Release of Candesartan Cilexetil. *Eur. J. Pharm. Biopharm.* **2007**, *65*, 379–387.

Sayed, S. Y.; Hedstrand, D. M.; Spinder, R.; Tomalia, D. A. Hydrophobically Modified Poly(Amidoamine) (PAMAM) Dendrimers: Their Properties at the Air Water Interface and Use as Nanoscopic Container Molecules. *J. Mater. Chem.* **1997**, *7*, 1199–1205.

Schild, H. G.; Tirrel, D. A. Microheterogenous Solutions of Amphiphilic Copolymers of N-isopropylacrylamide: An Investigation via Fluorescence Methods. *Langmuir* **1991**, *7*, 1319–1324.

Schumers, J. M.; Fustin, C. A.; Gohy, J. F. Light-responsive Block Copolymers. *Macromol. Rapid Commun.* **2010**, *31*, 1588–1607.

Seo, Y. G.; Kim, D. W.; Yeo, W. H.; Ramasamy, T.; et al. Docetaxel-loaded Thermosensitive and Bioadhesive nanomicelles as a Rectal Drug Delivery System for Enhanced Chemotherapeutic Effect. *Pharm. Res.* **2013**, *30*, 1860–1870.

Shi, D.; Matsusaki, M.; Akashi, M. Photo-cross-linking Induces Size Change and Stealth Properties of Water-dispersible Cinnamic Acid Derivative Nanoparticles. *Bioconjugate Chem.* **2009**, *20*, 1917–1923.

Shiah, J. G.; Sun, Y.; Kopečková, P.; Peterson, C. M.; Straight, R. C.; Kopeček, J. Combination Chemotherapy and Photodynamic Therapy of Targetable N-(2-hydroxypropyl) Methacrylamide Copolymer-doxorubicin/Mesochlorin 6-OV-TL 16 Antibody Immunoconjugates. *J. Control. Release* **2001**, *74*, 249–253.

Shim, D. F. K.; Marques, C.; Cates, M. E. Diblock Copolymers: Comicellization and Coadsorption. *Macromolecules* **1991**, *24*, 5309–5314.

Shuai, X.; Merdan, T.; Schaper, A. K.; Xi, F.; Kissel, T. Core-cross-linked Polymeric Micelles as Paclitaxel Carriers. *Bioconjugate Chem.* **2004**, *15*, 441–448.

Singh, M. Transferrin as a Targeting Ligand for Liposomes and Anticancer Drugs. *Curr. Pharm. Des.* **1999**, *5*, 443–451.

Singh, S.; Narang, A. S.; Mahato, R. I. Subcellular Fate and Off-target Effects of siRNA, shRNA and miRNA. *Pharm. Res.* **2011**, *28*, 2996–3015.

Slager, A. J. Heterostereocomplexes Prepared from D-poly(Lactide) and Leuprolide. I. Characterization. *Biomacromolecules* **2003**, *4*, 1308–1315.

Slager, J.; Domb, A. J. Hetero-stereocomplexes of D-poly(Lactic Acid) and the LHRH Analogue Leuprolide: Application in Controlled Release. *Eur. J. Pharm. Biopharm.* **2004**, *58*, 461–469.

Son, S.; Shin, E.; Kim, B. S. Light-responsive Micelles of Spiropyran Initiated Hyper Branched Polyglycerol for Smart Drug Delivery. *Biomacromolecules* **2014**, *15*, 628–634.

Su, W. C.; Chen, L. T.; Li, C. P. A Novel Micellar Formulation of Cisplatin, in Combination with Gemcitabine in Patients with Pancreatic Cancer in Asia-Results of Phase I. 2012, *ESMO.* (Abstract 863).

Sudimack, J.; Lee, R. J. Targeted Drug Delivery via the Folate Receptor. *Adv. Drug Deliv. Rev.* **2000**, *41*, 147–162.

Suksiriworapong, J.; Rungvimolsin, T.; A-gomol, A. Junyaprasert, V. B.; Chantasart, D. Development and Characterization of Lyophilized Diazepam-loaded Polymeric Micelles. *AAPS Pharm. Sci. Tech.* **2014**, *15*, 52–64.

Suma, T.; Miyata, K.; Ishii, T; Uchida, S.; Uchida, H.; Itaka, K.; et al. Enhanced Stability and Gene Silencing Ability of siRNA-loaded Polyion Complexes Formulated from Polyaspartamide Derivatives with a Repetitive Array of Amino Groups in the Side Chain. *Biomaterials* **2012**, *33*, 2770–2779.

Sunder, A.; Krämer, M.; Hanselmann, R.; Mülhaupt, R.; Frey, H. Molecular Nanocapsules Based on Amphiphilic Hyperbranched Polyglycerol. *Angew. Chem. Int. Ed.* **1999**, *38*, 3552–3555.

Taki, H.; Kanazawa, T.; Akiyama, F.; Takashima, Y.; Okada, H. Intranasal Delivery of Camptothecin-loaded Tat-modified Nanomicells for Treatment of Intracranial Brain Tumors. *Pharmaceuticals* **2012**, *5*, 1092–1102.

Tannock, I. F.; Rotin, D. Acid pH in Tumors and its Potential for Therapeutic Exploitation. *Cancer Res.* **1989**, *49*, 4373–4384.

Thipparaboina, R.; Rahul, B.; Chavan, D.; Kumar, S.; Modugula, N.; Shastri, R. Micellar Carriers for the Delivery of Multiple Therapeutic Agents. *Colloids Surf. B* **2015**, *135*, 291–308.

Torchilin, V. P. Structure and Design of Polymeric Surfactant-based Drug Delivery Systems. *J. Control. Release* **2001**, *73*, 137–172.

Torchilin, V. P. Targeted Polymeric Micelles for Delivery of Poorly Soluble Drugs. *Cell Mol. Life Sci.* **2004**, *61*, 2549–2559.

Torchilin, V. P. Multifunctional Nanocarriers. *Adv. Drug Deliv. Rev.* **2006**, *58*, 1532–1555.

Torchilin, V. P. Micellar Nano Carriers: Pharmaceutical Perspectives. *Pharm. Res.* **2007**, *24*, 1–16.

Torchilin, V. Multifunctional and Stimuli-sensitive Pharmaceutical Nanocarriers. *Eur. J. Pharm. Biopharm.* **2009**, *71*, 431–444.

Triolo, D.; Craparo, E. F.; Porsio, B.; Fiorica, C.; Giammona, G.; Cavallaro, G. Polymeric Drug Delivery Micelle-like Nanocarriers for Pulmonary Administration of Beclomethasone Dipropionate. *Colloids Surf. B.* **2017**, *151*, 206–214.

Tsuji, H. Poly(Lactide) Stereocomplexes: Formation, Structure, Properties, Degradation, and Applications. *Macromol. Biosci.* **2005**, *5*, 569–597.

Valle, J. W.; Armstrong, A.; Newman, C.; Alakhov, V.; Pietrzynski, G.; Brewer, J.; Campbell, S.; Corrie, P.; Rowinsky, E. K.; Ranson, M. A Phase 2 Study of SP1049C, Doxorubicin in P-glycoprotein-targeting Pluronics, in Patients with Advanced Adenocarcinoma of the Esophagus and Gastroesophageal Junction. *Invest. New Drugs* **2011**, *29*, 1029–1037.

Van Nostrum, C. F. Covalently Cross-linked Amphiphilic Block Copolymer Micelles. *Soft Matter.* **2011**, *7*, 3246–3259.

Verschraegen, C. F.; Skubitz, K.; Daud, A.; Kudelka, A. P.; Rabinowitz, I.; Allievi, C. A Phase I and Pharmacokinetic Study of Paclitaxel Poliglumex and Cisplatin in Patients with Advanced Solid Tumors. *Cancer Chemother. Pharmacol.* **2009**, *63*, 903–910.

Wang, C.; Chen, Q.; Wang, Z.; Zhang, X. An Enzyme-responsive Polymeric Superamphiphile. *Angew. Chem. Int. Ed. Eng.* **2010**a, 49, 8612–8615.

Wang, J.; Lu, Z.; Wientjes, M. G.; Au, J. L. Delivery of siRNA Therapeutics: Barriers and Carriers. *AAPS J.* **2010**b, *12*, 492–503.

Wang, Z.; Yu, Y.; Ma, J.; Zhang, H.; Wang, X.; Wang, J. et al. LyP-1 Modification to Enhance Delivery of Artemisinin or Fluorescent Probe Loaded Polymeric Micelles to Highly Metastatic Tumor and Its Lymphatics. *Mol. Pharm.* **2012**, *9*, 2646–2657.

Wei, H.; Zhuo, R. X.; Zhang, X. Z. Design and Development of Polymeric Micelles with Cleavable Links for Intracellular Drug Delivery. *Prog. Polym. Sci.* **2012**, *38*, 503–535.

Weiner, L. M. Building Better Magic Bullets Improving Unconjugated Monoclonal Antibody Therapy for Cancer. *Nat. Rev. Cancer* **2007**, *7*, 701–706.

Wu, C.; Ma, R.; He, H.; Zhao, L.; Gao, H.; An, Y.; Shi, L. Fabrication of Complex Micelles with Tunable Shell for Application in Controlled Drug Release. *Macromol. Biosci.* **2009**, *9*, 1185–1193.

Xie, D.; Xu, K.; Bai, R.; Zhang, G.; Structural Evolution of Mixed Micelles Due to Inter Chain Complexation and Segregation Investigated by Laser Light Scattering. *J. Phys. Chem. B.* **2007**, *111*, 778–781.

Xing, Y.; Wang, C.; Han, P.; Wang, Z.; Zhang, X. Acetylcholinesterase Responsive Polymeric Supra Amphiphiles for Controlled Self-assembly and Disassembly. *Langmuir* **2012**, *28*, 6032–6036.

Xu, J.; Ge, Z.; Zhu, Z.; Luo, S.; Liu, H.; Liu, S. Synthesis and Micellization Properties of Double Hydrophilic A2BA2 and A4BA4 Non-linear Block Copolymers. *Macromolecules* **2006**, *39*, 8178–8185.

Xu, W.; Burke, J. F.; Pilla, S.; Chen, H.; Jaskula, S. R.; Gong, S. Octreotide-functionalized and Resveratrol-loaded Unimolecular Micelles for Targeted Neuroendocrine Cancer Therapy. *Nanoscale* **2013**, *5*, 9924–9933.

Yang, L.; Wu, X.; Liu, F.; Duan, Y.; Li, S. Novel Biodegradable Polylactide/Poly(Ethylene Glycol) Micelles Prepared by Direct Dissolution Method for Controlled Delivery of Anti-cancer Drugs. *Pharm. Res.* **2009**, *26*, 2332–2342.

Yang, R.; Meng, F.; Ma, S.; Huang, F.; Liu, H.; Zhong, Z. Galactose-Decorated Crosslinked Biodegradable Poly(Ethylene Glycol)-b-poly(E-caprolactone) Block Copolymer Micelles for Enhanced Hepatoma-targeting Delivery of Paclitaxel. *Biomacromolecules* **2011**, *12*, 3047–3055.

Yokoyama, M.; Fukushima, S.; Uehara, R.; Okamoto, K.; Kataoka, K.; Sakurai, Y.; Okano, T. Characterization of Physical Entrapment and Chemical Conjugation of Adriamycin in Polymeric Micelles and Their Design For *In vivo* Delivery to a Solid Tumor. *J. Control. Release* **1998**, *50*, 79–92.

Yoo, H. S.; Park, T. G. Folate Receptor Targeted Biodegradable Polymeric Doxorubicin Micelles. *J. Control. Release* **2004**, *96*, 273–283.

Zana, R. *Dynamics of Surfactant Self-assemblies: Micelles, Microemulsions, Vesicles and Lyotropic Phases.* CRC Press: FL, 2005.

Zeng, F.; Lee, H.; Allen, C. Epidermal Growth Factor Conjugated Poly(Ethylene Glycol)-block-poly(Delta-valerolactone) Copolymer Micelles for Targeted Delivery of Chemotherapeutics. *Bioconjugate Chem.* **2006**, *17*, 399–409.

Zhang, J.; Jiang, X.; Zhang, Y.; Li, Y.; Liu, S. Facile Fabrication of Reversible Core Cross-linked Micelles Possessing Thermosensitive Swellability. *Macromolecules* **2007**, *40*, 9125–9132.

Zhang, Y.; Hong, H.; Cai, W. Tumor Targeted Drug Delivery with Aptamers. *Curr. Med. Chem.* **2011**, *18*, 4185–4194.

Zhao, Y. Rational Design of Light-controllable Polymer Micelles. *Chem Rec.* **2007**, *7*, 286–294.

Zhao, X.; Li, H.; Lee, R. J. Targeted Drug Delivery via Folate Receptors. *Expert Opin. Drug Deliv.* **2008**, *5*, 309–319.

Zhong, Y.; Yang, W.; Sun, H; Cheng, R.; Meng, F.; Deng, C.; et al. Ligand-directed Reduction-sensitive Shell-sheddable Biodegradable Micelles Actively Deliver Doxorubicin into the Nuclei of Target Cancer Cells. *Biomacromolecules* **2013**, *14*, 3723–3730.

Zhu, S.; Huang, R.; Hong, M.; Jiang, Y.; Hu, Z.; Liu, C.; Pei, Y. Effects of Polyoxyethylene (40) Stearate on the Activity of P-glycoprotein and Cytochrome P450. *Eur. J. Pharm. Sci.* **2009**, *37*, 573–580.

Zhu, L.; Perche, F.; Wang, T.; Torchilin, V. P. Matrix Metalloproteinase 2-sensitive Multifunctional Polymeric Micelles for Tumor-specific Co-delivery of siRNA and Hydrophobic Drugs. *Biomaterials* **2014**, *35*, 4213–4222.

CHAPTER 8

NANOGELS: GENERAL CHARACTERISTICS, MATERIALS, AND APPLICATIONS IN DRUG DELIVERY

P. R. SARIKA[1] and NIRMALA RACHEL JAMES[2*]

[1]Department of Pharmaceutics and Pharmaceutical Chemistry, The University of Utah, Skaggs Research Building, Rm 2760, 30S, 2000E, Salt Lake City, UT 84112, United States of America

[2]Department of Chemistry, Indian Institute of Space Science and Technology (IIST), Valiamala, Thiruvananthapuram 695547, Kerala, India

*Corresponding author. E-mail: nirmala@iist.ac.in

ABSTRACT

Among the variety of polymeric nanoparticles designed for drug delivery, nanogels have captured special attention. These three-dimensional, cross-linked polymeric networks garner lot of special features in their tiny volume. High loading efficiency and colloidal stability facilitate encapsulation of various drugs and biomolecules. Nanogels can be tailor-made for specific applications. Size, shape, responsive behavior, surface chemistry, and morphology can be tuned for effective delivery applications. Nanogels have opened up vast opportunities in biomedical field, especially in drug delivery.

8.1 INTRODUCTION

Bringing a new drug to the market is a time-consuming and expensive process, since it involves several years of discovery, development, clinical testing, and obtaining regulatory approval. However, promising developments in biotechnology and pharmacogenomics have opened up possibilities of designing and realizing new drugs for specific applications. In the past two decades, research on drug delivery systems has achieved tremendous advancements. New drug discovery will be successful, only if we develop distinctive drug carriers simultaneously. Drug carrier should be highly stable in plasma with low toxicity and be able to overcome clearance by the reticuloendothelial system (RES). For the safe and successful delivery of the payload, the carrier must be able to vanquish various intracellular barriers such as plasma membrane, lysosomal enzymatic degradation, endosome entrapment, and nuclear membrane (Bildstein et al., 2011; Juliano et al., 2011; Reischl and Zimmer, 2009; Toita et al., 2011; Torchilin, 2008). An ideal carrier should possess all of the above-mentioned properties.

The boom in nanotechnology attracted scientists involved in the development of new drug delivery systems and the era of nanosized drug delivery systems thus started. Polymeric nanoparticles, lipid nanoparticles, carbon nanotubes, dendrimers, polymeric micelles, and nanogels are some of the nanoparticles developed so far for drug delivery (Fig. 8.1).

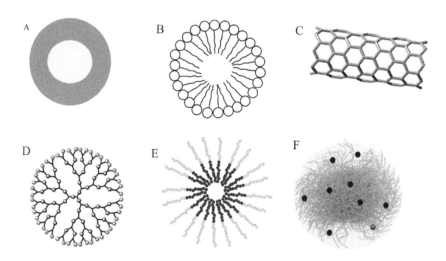

FIGURE 8.1 (A) Polymeric nanoparticles, (B) lipid nanoparticles, (C) carbon nanotubes, (D) dendrimers, (E) polymer micelles, and (F) nanogels.

Among different nanosized drug delivery systems, nanogels bagged more attention because of the unique properties offered by their nanosize (Oh et al., 2008). Nanogels are physically or chemically cross-linked, three-dimensional polymeric networks, confined to nanoscale dimensions (Kabanov and

Vinogradov, 2009b; Raemdonck et al., 2009a; Rejinold et al., 2012; Shah et al., 2012) Similar to their macroscopic counterparts (hydrogels), nanogels also have the capacity to possess large amount of water without dissolving in it. Nanogels have good stability in biological fluids (Yallapu et al., 2011) due to the negligible van der Waals forces of attraction between the particles. In nanogel dispersions, the Hamaker constants for the solvent and nanogel nanoparticles are equal and therefore, the chances for aggregation are trivial (Vincent, 2006). Nanogels also can attain biocompatibility and high aqueous dispersibility (Kabanov and Vinogradov, 2009a; Oh et al., 2008) since their nano size offers unique properties such as high drug-loading ability, easy-penetration ability (Choi et al., 2012), colloidal stability, and long circulation time (Beningo and Wang, 2002) in biological fluids. These advantages make them ideal candidates for drug delivery applications.

Drugs can be encapsulated in nanogels either by covalent conjugation or by physical entrapment. Encapsulated drug is released from nanogels by diffusion or degradation of the polymer networks. Size of the nanogels can be easily tuned to make them suitable for a particular application. Small size of the nanogels makes them appropriate choice for intracellular drug delivery since they can be easily taken up by the cells. Nano size also aids them to avoid renal clearance and uptake by the reticuloendothelial system. Nanogels are widely used for targeted drug delivery, owing to their ability to conjugate many directing ligands on surface due to their large surface area (Vasir and Labhasetwar, 2007).

8.2   METHODS OF PREPARATION

Nanogels have been synthesized by physical self-assembly of polymers, polymerization of monomers, chemical cross-linking of polymers and template-assisted nanofabrication.

Based on the interactions between the polymer chains in nanogels, they are classified as physically cross-linked and chemically cross-linked.

8.2.1   PHYSICALLY CROSS-LINKED NANOGELS

Physically cross-linked nanogels are prepared by exploiting noncovalent interactions such as hydrogen bonding, van der Waals forces, electrostatic and hydrophobic interactions between polymer chains. Hydrophobic and electrostatic interactions lead to the formation of different polymeric nanoparticles such as polymer micelles (Kataoka et al., 2001, polymer-soms (Lee and Feijen, 2012) and layer-by-layer nanoparticles (Ariga et al., 2011). However, all of these nanoparticles cannot be considered as nanogels unless they satisfy certain criteria. A nanogel should be cross-linked polymer network with size less than 100 nm and must possess an ability to retain large amount of water. Physically cross-linked nanogels are less stable because of the weak interactions. Moreover, it is difficult to control the size of nano-gels by physical interactions. Even though, these nanogels have stability issues, research on their applications have escalated in the last decade. This is attributed to the noninvolvement of toxic cross-linking agents/catalyst during their preparation. Sometimes, simple mixing of the polymer solution can result in physically cross-linked nanogels. The stability and size of these nanogels can be tuned by varying polymer composition and concentration, temperature, and ionic strength of the medium.

The most common method for preparation of physically cross-linked nanogels is self-assembly of amphiphilic polymers. Hydrophilic polymers are modified with hydrophobic moieties and controlled association of the resulting amphiphilic polymers in aqueous solution leads to physically cross-linked nanogels. Akiyoshi and coworkers prepared a variety of physi-cally cross-linked nanogels from hydrophobically modified polysaccharides (Akiyama et al., 2007; Hirakura et al., 2004; Sasaki and Akiyoshi, 2010). They prepared cholesterol-modified pullulan with a diameter of 20–30 nm which is used for protein encapsulation (Shimizu et al., 2008). Hydrophobic interactions between cholesterol groups on the pullulan provided cross-linking points. K. Akiyoshi and coworkers, also synthesized cholesterol-substituted cyclic dextrin (CH-Cdex) nanogels. Monodisperse and stable nanogels with a hydrodynamic radii of 10 nm is formed by the self-assembly of 4–6 CH-CDex macromolecules in water. The nanogel is projected to have application in delivery of fluoresce in isothiocyanate-labeled insulin (FITC-Ins) (Ozawa et al., 2009). Heparin and pluronic-based self-assembled nanogel has been prepared by conjugating carboxylated pluronic F127 to low-molecular-weight heparin. Nanogel administration hinders liver fibrosis via inhibition of the TGF-β/Smad pathway (Choi et al., 2011). A number of self-assembled nanogels are prepared based on similar methodologies. Biocompatible, naturally occurring polysaccharides such as chitosan, dextran, dextrin, mannan, hyaluronic acid, and heparin have been modified with hydrophobic moieties such as deoxycholic acid, cholesterol, and bile acid for nanogel preparation (Braeckmans et al., 2010; Kim et al., 2006;

Ozawa et al., 2009; Park et al., 2004; Sasaki and Akiyoshi, 2012; Toita et al., 2009).

Physically cross-linked nanogels can also be prepared through ionotropic gelation of multivalent cations and anions. Natural polymers such as chitosan, alginate, gellan gum, pectin, and arabic gum contain cations/anions on their structure. These polymers can form network structure by combining with the counterions and induce gelation by cross-linking. Ionotropic gelation depends on polymer and counterion concentration, temperature, pH of cross-linking solution, etc. (Patil et al., 2012). This simple and mild, physical cross-linking method by electrostatic interaction avoids possible toxicity of reagents and other undesirable effects. During ionotropic gelation method, polysaccharide (alginate, gellan, and pectin) solutions are added dropwise under constant stirring to the solutions containing counterions. Polysaccharides undergo ionic gelation due to the complexation between oppositely charged species. Low-molecular-weight (e.g., $CaCl_2$, $BaCl_2$, $MgCl_2$, $CuCl_2$, $ZnCl_2$, $CoCl_2$, pyrophosphate, tripolyphosphate, tetrapolyphosphate, etc.) and high-molecular-weight counterions (e.g., octyl sulphate, lauryl sulphate, hexadecyl sulphate, cetylstearyl sulphate, etc.) are used for ionotropic gelation method (Racoviță et al., 2009). Jonassen et al. developed chitosan nanogels by ionic cross-linking of cationic chitosan chains (CTS) with anionic tripolyphosphate (TPP). Physical stability of CTS/TPP nanogels can be varied by changing ionic strength, CTS concentration, and CTS/TPP ratio (Jonassen et al., 2012). Chitosan–TPP nanogels have been used for the encapsulation of metal ions such as Ag^+, Cu^{2+}, Zn^{2+}, Mn^{2+}, or Fe^{2+} and their antibacterial activity has been evaluated (Du et al., 2009). Mohammad Reza et al. developed chitosan–arabic gum nanoparticles by ionic gelation method for insulin delivery (Avadi et al., 2010). Alginate nanogels are also prepared by similar cross-linking with divalent calcium ions and the resulting nanoparticles are again complexed with chitosan. The prepared nanogels find application in insulin encapsulation (Sarmento et al., 2007).

Host–guest interactions are another well-studied way of preparing physically cross-linked nanogels. These nanogels are formed by the association of a molecule containing cavity (e.g., cyclodextrin) with another guest molecule. Host–guest interactions are relatively weak and lead to inclusion complexes. Host–guest systems are formed by various kinds of intermolecular interactions such as hydrogen bonding, hydrophobic interactions, π–π stacking, and metal–ligand binding (Cram, 1988). Cyclodextrins (CD) are the most widely employed host molecule in the construction of nanoassemblies owing to their excellent biocompatibility and low-toxicity. Gref and coworkers developed self-assembled nanogels of hydrophobically modified dextran and β-cyclodextrin by host–guest interactions. These nanogels are used for the encapsulation of hydrophobic molecules, benzophenone, and tamoxifen (Daoud-Mahammed et al., 2007). Zan et al. constructed polymeric nanogels via host–guest interactions for dual pH-triggered multistage drug delivery. This nanogel is suitable for deep tissue penetration, high-efficiency cellular uptake, and intracellular endo-lysosomal pH-responsive drug release (Zan et al., 2014).

8.2.2 CHEMICALLY CROSS-LINKED NANOGELS

Chemically cross-linked nanogels are formed by the introduction of covalent bonds between the polymer chains. Stable nanogels with desired size can be prepared by chemical crosslinking. Chemically cross-linked nanogels are developed either by the introduction of cross-linkable functional groups on the polymer chain or by the addition of external cross-linking agents. Properties such as swelling, degradation, and release kinetics of a nanogel can be tuned by varying the degree of cross-linking. Degradable nanogels can be prepared by inserting degradable covalent linkages such as amide, ester, carbonate, anhydride, phosphazene, and phosphate esters on polymers. These nanogels undergo degradation through hydrolysis or enzymatic cleavages (Yallapu et al., 2011).

Miniemulsion polymerization is one of the widely used techniques to prepare chemically cross-linked nanogels with controlled size. In miniemulsion, fine droplets of the discontinuous phase are dispersed in a continuous phase. Water-in-oil (W/O) or oil-in-water (O/W) miniemulsions are prepared by the application of high shear forces such as ultrasound and high-pressure homogenization. The resulting droplets are stabilized by suitable surfactants. Each droplet can act as an individual nanoreactor and the resulting nanogels will be monodispersed (Weiss and Landfester, 2010). Figure 8.2 explains the steps involved in miniemulsion technique.

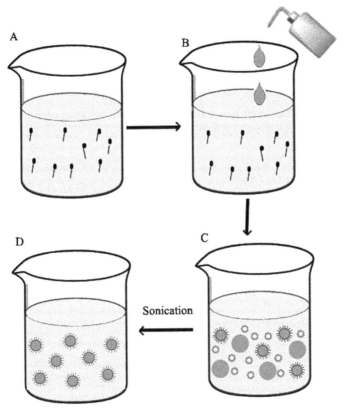

FIGURE 8.2   Steps involved in miniemulsion technique for the preparation of nanogels. (A) Surfactant-dissolved continuous phase, (B) addition of discontinuous phase to continuous phase, (C) surfactant molecules surround and stabilizes the droplets, and (D) monodispersed nanoparticles after sonication.

Nanogels prepared by atom transfer radical polymerization (ATRP) in W/O miniemulsion has been used to incorporate anticancer drugs and bioactive macromolecules (Oh et al., 2009; Oh et al., 2007). Siegwart et al. prepared uniformly cross-linked poly(ethylene oxide)-based nanogels by inverse miniemulsion ATRP. Gold nanoparticles, bovine serum albumin, and rhodamine B isothiocyanate-dextran are encapsulated into these nanogels to show their capability to encapsulate and deliver organic, inorganic, and biological molecules (Siegwart et al., 2009). Reversible addition-fragmentation chain transfer (RAFT) inverse miniemulsion polymerization of 2-(dimethylamino) ethyl methacrylate monomers yield biodegradable nanogels. These nanogels are cross-linked by disulphide cross-linker and the linkages undergo degradation by exposure to a reductive environment (Oliveira et al., 2011). Maciel et al. developed alginate/cystamine nanogels for delivery of an anticancer drug doxorubicin (Maciel et al., 2013). Sarika et al. developed gelatin–gum arabic aldehyde nanogels via fusion of two separate inverse miniemulsions of gelatin (gel) and gum arabic aldehyde (GAA). Cross-linking between GAA and gel occurs during the fusion of inverse miniemulsions (Sarika and James, 2015). These nanogels have been used for the delivery of curcumin to breast cancer cells (Sarika and Nirmala, 2016).

Redox responsive nanogels of thiol-modified poly(aspartic acid) have been prepared by W/O miniemulsion technique. Reducing environment of the cancer cells facilitates drug release from these nanogels (Krisch et al., 2016). Elaissari and coworkers prepared thermally sensitive poly(N-vinyl-caprolactam) nanogels by inverse miniemulsion polymerization (W/O) using span 80 as an oil-soluble surfactant and NaCl as a lipophobe (Medeiros et al., 2012).

Chemically cross-linked nanogels are also prepared from polymeric micelles. Physically cross-linked polymeric micelles are unstable and do not have control on drug release. Stability of micelles can be improved by covalent chemical cross-linking of either core or shell of the micelle. Carbodiimide-mediated amide bond cross-linking (Hawker and Wooley, 2005), photo-cross-linking methods (Ding et al., 2011b), "click" chemistry (Mortisen et al., 2010), and quarterisation of amino groups (Tamura et al., 2010) are used to prepare polymeric micelle nanogels.

Photolithography technique has been used to fabricate nanoparticles with uniform size, shape, and composition. Photolithography involves several steps including coating of low surface energy carrying UV-cross-linkable polymer on a photo-resist-coated wafer, molding the polymer into a pattern using a template and UV exposure, removal of template to reveal particles with thin residual layer, etching by plasma containing oxygen to remove residual layer, and finally collection of the fabricated particles by dissolution of the substrate in water or buffer (Glangchai et al., 2008; Sasaki and Akiyoshi, 2010). Schematic representation of the steps involved in photolithography is shown in Figure 8.3.

Good control over nanogel size is essential for *in vivo* drug delivery. Photolithographic fabrication of nanoparticles using a technique called PRINT (particle replication in nonwetting templates) is a very good fabrication method to develop monodispersed particles. Recently, Dunn et al. developed reductively responsive interfering RNA (siRNA)-conjugated hydrogel nanoparticles via PRINT technology for gene silencing (Dunn et al., 2012). Highly monodispersed particles of poly(lactic-co-glycolic acid) (PLGA) nanogels are prepared by PRINT technique. Scanning electron microscopy (SEM) images confirm the uniform size of PLGA nanogels. These nanogels shows 40% docetaxel loading and elevated plasma exposure and tumor delivery compared to free docetaxel (Chu et al., 2013).

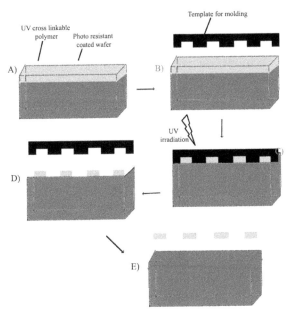

FIGURE 8.3   (A) Coating of low surface energy carrying UV-cross-linkable polymer on a photo-resist-coated wafer, (B) template for molding the polymer into a pattern, (C) UV exposure, (D) removal of the template to reveal particles, and (E) removal of the residual layer and collection of the fabricated particles by dissolution of the substrate in water or buffer.

8.3   BIOPOLYMER-BASED NANOGELS

Nanogels are prepared from synthetic as well as natural polymers (Fig. 8.4). PLGA is one of the widely used synthetic polymers for nanoparticle development (Mukerjee and Vishwanatha, 2009). Yallappau et al. developed PLGA (biodegradable polymer) nanoparticles, in presence of poly(vinyl alcohol) and poly(l-lysine) stabilizers, using a nano-precipitation technique and utilized for curcumin (Yallapu et al., 2010) delivery. Although synthetic polymers such as poly(ethylene glycol) (Ding et al., 2011a), poly(lactic acid)

(Lee et al., 2006), and poly(ε-caprolactone) (Musyanovych et al., 2008) proved their ability as suitable materials for the development of nanogels; natural polymers such as gum arabic (Sarika and James, 2015), chitosan (Brunel et al., 2009), dextran (Van Thienen et al., 2005), and hyaluronic acid (Lee et al., 2007) received more importance owing to their nontoxicity, biocompatibility, and biodegradability. Biopolymers possess all the properties of synthetic counterparts as well as many other properties including biodegradability, abundance in nature, renewability, and nontoxicity. In addition, biopolymers are decorated with large number of functional groups including hydroxyl, amino, and carboxylic acid groups. These functional groups enable further conjugation with targeting agents. Some of the biopolymers which have already been used for the fabrication of nanoparticles are alginate, chitosan, fibrinogen, hyaluronic acid, pullulan, gelatin, and dextran (Li et al., 2015).

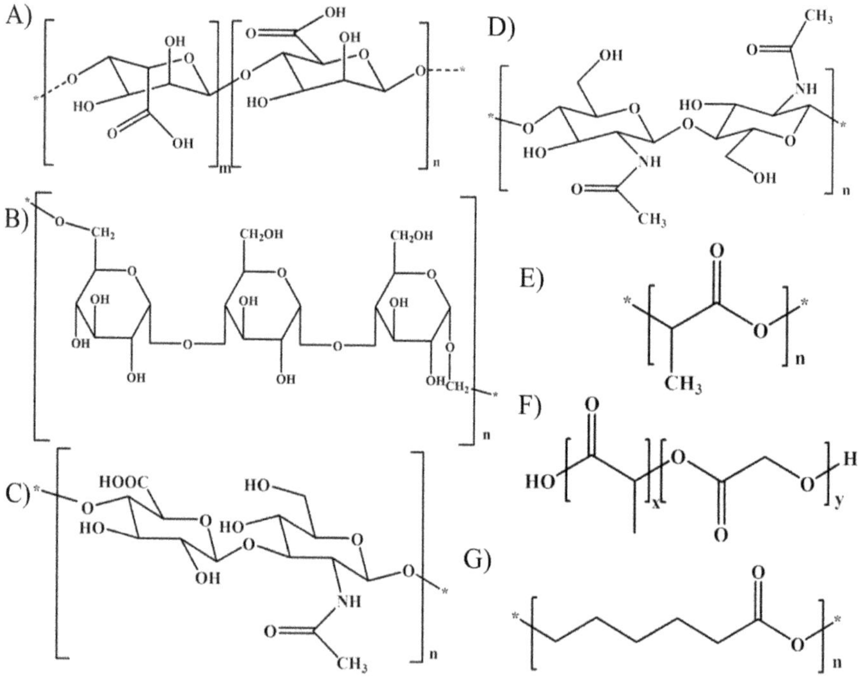

FIGURE 8.4 Structure of natural and synthetic polymers used for nanogel preparation. (A) Alginate, (B) pullulan, (C) hyaluronic acid, (D) chitosan, (E) polylactic acid, (F) poly(lactic-co-glycolic acid) (PLGA), and (G) poly(ε-caprolactone).

8.3.1   ALGINATE-BASED NANOGELS

Alginate is a polyanionic and biodegradable polysaccharide composed of β-D-mannuronic acid (M) and α-L-guluronic acid (G) units extracted from brown algae. The ratio of guluronate to mannuronate varies depending on the source it is derived from (Lee et al., 2012). Alginate has been explored extensively for various biomaterial preparations. Carboxylate functional groups on alginate can be cross-linked with divalent cations. Several alginate nanogels are prepared by ionic cross-linking with divalent cations (Ca^{2+}, Cu^{2+}, Zn^{2+}, or Mn^{2+}) (Hong et al., 2008). Bioreducible alginate nanogels are prepared by the ionic cross-linking of alginate and polyethylenimine, followed by disulfide cross-linking (Li et al., 2013). Gonçalves et al. developed dual-cross-linked dendrimer/alginate nanogels (AG/G5), using $CaCl_2$ as cross-linker and amine-terminated generation 5 dendrimer (G5) as a cross-linker. Dual cross-linking provides extrastability and doxorubicin release in a sustained way (Gonçalves et al., 2014). Alginate nanogels find applications in anticancer and nucleic acid delivery (Cai et al., 2012; Douglas et al., 2006; Parveen et al., 2013). Sarika et al. prepared alginic aldehyde-gelatin nanogels by inverse emulsion technique (Sarika et al., 2015) and used for curcumin drug delivery (Sarika et al., 2016).

8.3.2   PULLULAN-BASED NANOGELS

Pullulan is a linear, water-soluble extracellular polysaccharide produced by polymorphic fungus *Aureobasidium pullulans*. It is a nonionic polysaccharide with blood compatibility, biodegradability, nontoxicity, nonimmunogenicity, nonmutagenicity, and noncarcinogenicity. Its structure consists of α-1,4-linked glucose units, that are included in a α-1,6-linked maltotriose unit. Pullulan and its derivatives have applications in pharmaceutical, food, and biomedical field because of their biocompatibility and biodegradability. The enzyme pullulanase, hydrolyses the α-1-6 linkages of pullulan and converts the polymer into maltotriose units (Leathers, 2003). Pullulan nanogels are prepared by chemical cross-linking or hydrophobic modification and are employed to deliver anticancer drugs, proteins (Morimoto et al., 2012), growth factors (Fujioka-Kobayashi et al., 2012; Scomparin et al.,

2011), etc. Temperature- and pH-sensitive groups are introduced on pullulan chain and are used for stimuli-responsive drug delivery (Na et al., 2004). Seo et al. grafted pullulan with poly(l-lactide) and investigated its ability in temperature-responsive doxorubicin delivery (Seo et al., 2012).

8.3.3 DEXTRAN-BASED NANOGELS

Dextran is another biodegradable, biocompatible neutral polysaccharide with broad biomedical applications. Similar to pullulan nanogels, various dextran nanogels are prepared by ionic and chemical cross-linking methods. Hydroxyl groups on dextran are modified with different cross-linkable groups such as methacrylate, maleimide, and hydroxyethyl methacrylate. These modified dextran nanogels can be prepared by reverse miniemulsion or radical polymerization technique (Klinger et al., 2012; Peng et al., 2010; Van Thienen et al., 2007). Inverse emulsion polymerization technique has been used for cationically modified dextran nanogel preparation and their potential as siRNA carrier has been evaluated (Raemdonck et al., 2009b).

8.3.4 HYALURONIC ACID-BASED NANOGELS

Hyaluronic acid (HA) is an anionic, nonsulfated glycosaminoglycan present in the extracellular matrix of connective, epithelial, and neural tissues. HA plays vital role in many biological processes. HA is synthesized by hyaluronan synthase at the inner wall of the plasma membrane (Xu et al., 2012). It is made up of alternating units of D-glucuronic acid and N-acetyl-D-glucosamine, linked together via alternating β-1,4- and β-1,3-glycosidic bonds (Garg and Hales, 2004). HA is a well-exploited polysaccharide in various biomedical applications. It has been extensively used in pharmacology, drug delivery, surgery, wound healing, ophthalmology, dermatology, and plastic surgery (Kogan et al., 2007). HA-based biomaterials such as hydrogels (Isa et al., 2015), nanogels (Liang et al., 2016), and HA-drug conjugates (Camacho et al., 2015) attains application in tissue engineering, drug delivery, and wound healing.

Water et al. designed and developed HA-based nanogels for the delivery of peptide Novicidin. These nanogels are produced by microfluidics technique, and by tailoring the preparation parameters, encapsulation efficiency and zeta potential can be varied (Water et al., 2015). Yang et al. modified HA with methacrylate groups and used it for nanogel preparation by radical copolymerization. Doxorubicin-loaded HA nanogels could efficiently accumulate and penetrate the tumor matrix, and provides superior anti-tumor efficacy in H22 tumor-bearing mice. Bioresponsive and fluorescent HA-iodixanol nanogels are developed for targeted X-ray computed tomography imaging and chemotherapy of MCF-7 human breast tumor. Peng Liu and coworkers developed the ranostic nanogels for tumor diagnosis and chemotherapy. They have cross-linked functionalized HA with carbon dots. These nanogels offer fluorescence property and dual receptor-mediated targeted drug delivery with tumor microenvironment-responsive controlled release (Jia et al., 2016).

8.3.5   CHITOSAN-BASED NANOGELS

Chitosan is a linear cationic polysaccharide, prepared by the deacety-lation of chitin and is composed of randomly distributed β-(1-4)-linked D-glucosamine and N-acetyl-D-glucosamine units. Chitosan is soluble in aqueous acidic media, by protonation of the $-NH_2$ functional group on the C-2 position of the D-glucosamine repeating unit. Solubility of chitosan is related to the degree of deacetylation, ionic concentration, pH, nature of the acid used for protonation, and the distribution of acetyl groups along the chain. Biodegradability, nontoxicity, and biocompatibility benefited chitosan, to be used extensively in biotechnology, biomedicine, waste water treatment, and in cosmetics. Distinct material formulations such as hydrogels (Sudheesh Kumar et al., 2012), beads (Nguyen et al., 2016), microspheres (Fang et al., 2014), films (Perdones et al., 2014), composites (Chen et al., 2013), and nanoparticles (Ragelle et al., 2014) have been developed from chitosan for the delivery of drugs, proteins, and genes.

Pujana et al. developed genipin cross-linked chitosan nanogels by reverse microemulsion method (Pujana et al., 2013) and folate functionalization improves the solubility of the nanogels (Pujana et al., 2014b). Folate chitosan nanogels exhibited pH-responsive delivery of 5-fluorouracil. They also developed polyethylene glycol (PEG) diacids or tartaric acid cross-linked chitosan and functionalized with folic acid for cancer treatment purposes (Pujana et al., 2014a). Drug delivery to the brain is challenging due to poor drug permeability through the blood–brain barrier. Targeted delivery using nanoparticles opened up new possibilities. Azadi et al. developed methotrexate-loaded chitosan nanogels by ionic gelation method for drug delivery to brain. Intravenous administration of the drug-loaded nanogels exhibited high methotrexate concentration in brain (Azadi et al., 2013).

Inhibition of endocytic pathways by siRNA of flotillin-1 and Cdc42 provided information regarding the uptake mechanisms and intracellular fate of folate-functionalized glycol chitosan nanogels. Results also suggest the involvement of the actin cytoskeleton in nanogel uptake via macropino-cytosis (Pereira et al., 2015).

8.3.6 CYCLODEXTRIN-BASED NANOGELS

Cyclodextrins (CD) are nonreducing cyclic oligosaccharides composed of α-(1,4)-linked glucopyranose subunits with cage-like supramolecular structure. CD can form inclusion complexes with other molecules by intramolecular interactions. They are broadly used in many industrial products, technologies, and analytical methods because of their capability to form host–guest complexes. Among the three different CD (α, β, and γ), β-CD has been studied enormously because of its easy availability and low price (Del Valle, 2004). CD-based nanogels can be prepared by different cross-linking methods such as key–lock interactions (Gref et al., 2006), polymerization of CD monomers containing vinyl or acrylic functional groups (Kettel et al., 2011) and covalent cross-linking condensation with suitable bi/multifunctional agents.

γ-CD nanogels are prepared by emulsification/solvent evaporation process and dexamethasone is loaded into the nanogels. These nanogel eye drops showed constant dexamethasone concentration for at least 6 h in the tear fluid, whereas the concentration of the commercial product fell rapidly (Moya-Ortega et al., 2013). Kettel et al. demonstrated another interesting application of β-CD nanogels as a protective coating for textiles. The insecticide permethrin is loaded by inclusion complexation mechanism. Irradiation experiments revealed that, permethrin containing β-CD nanogels applied onto wool fabrics, reduces photobleaching when exposed to light. These nanogels can provide protection to wool (Kettel et al., 2014).

Incorporation of cyclodextrin into polymethylmethacrylate nanogels enhanced uptake and release rate of antiseptic and bacteriostatic agent chlorhexidine (Moya-Ortega et al., 2012). The nanogels are coated onto glass surfaces and exhibited antimicrobial activity. Dual responsive supramolecular nanogels developed from dextran-grafted-benzimidazole (Dex-g-BM) and thiol-β-cyclodextrin exhibited special supramolecular pH-sensitivity under acidic conditions. These nanogels showed higher doxorubicin (DOX) release under lower pH (pH <6) or higher glutathione (GSH) concentration conditions (Chen et al., 2014).

Several other natural polymers have also been explored for nanogel preparation. Chondroitin sulfate (Park et al., 2010), cycloamylose (Toita et al., 2009), curdlan (Na et al., 2000), heparin (Bae et al., 2008), and dextrin (Carvalho et al., 2010) are some of them.

8.4   STIMULI-RESPONSIVE NANOGELS

Stimuli-responsive nanogels are able to change their physicochemical properties such as volume, water content, hydrophobicity, and hydrophilicity in response to external stimuli. These nanogels can offer special features such as prolonged circulation time with high stability in blood stream, site-specific drug delivery, and controlled drug release. Hence, they are superior to conventional polymeric nanogels. Stimuli including temperature, light, magnetic field, pH, and ionic strength changes have been employed to produce physicochemical variations in nanogels (Bawa et al., 2009; de las Heras Alarcón et al., 2005). External stimuli can trigger release of drugs/biomolecules, encapsulated in these types of nanogels (Fig. 8.5).

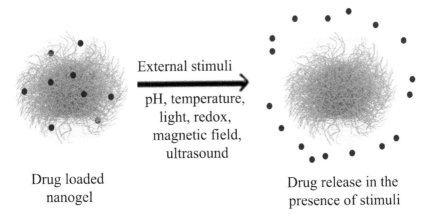

External stimuli

pH, temperature,
light, redox,
magnetic field,
ultrasound

Drug loaded
nanogel

Drug release in the
presence of stimuli

FIGURE 8.5   External stimuli such as pH, temperature, redox, light, and magnetic field can trigger drug release from stimuli-responsive nanogels.

8.4.1 pH-RESPONSIVE NANOGELS

pH-responsive nanogels are created by exploiting the pH variations in body followed by inflammation, infection, and cancer. In addition to that, there are pH differences in various body compartments. Digestive tract shows varied pH from stomach (pH 2) to colon (pH 7) and it is again different in cellular components such as lysosome (pH 4.5–5), endosomes (pH 5.5–6), and cytosol (pH 7.4) (Karimi et al., 2016). Targeted drug delivery to the infected sites with stimuli-responsive nanogels is highly effective because of their ability to provide maximum therapeutic effect with minimum side effects (Alvarez-Lorenzo and Concheiro, 2013, 2014) For example, inflammatory reaction drops pH of the region from 7.4 to 6.5 (Hunt et al., 1986). pH of cancer tissue is relatively lower than that of normal tissue. Due to high proliferation rate, vasculature of tumor is insufficient to provide enough nutritional and oxygen supply. Less oxygen supply leads to hypoxia, high production of lactic acid, and hydrolysis of adenosine triphosphate in tumor tissues. Subsequently, an acidic environment will be created in tumor tissues. This pH changes may be exploited for the delivery of drugs to tumor that are weak electrolytes with appropriate pKa (Gerweck and Seetharaman, 1996).

Zhao et al. developed pH-responsive nanogels of poly(N-2-hydroxyethyl acrylamide) (polyHEAA) and acrylic acid (AA) with controlled size and morphology by inverse-microemulsion free-radical polymerization. At high acidic pH and ionic strength, poly(HEAA/AA) nanogels exhibited high R6G (model drug) release due to the diminished interaction between nanogels and drug molecule (Zhao et al., 2012). Nonspecific delivery of many anticancer drugs can cause severe side effects. pH-responsive dextrin nanogels, loaded with doxorubicin provide controlled drug release to colorectal cancer cells (Manchun et al., 2015). pH-sensitive nanogels are also developed from lysosome and pectin by self-assembly and they exhibit accelerated release of methotrexate under acidic conditions (Lin et al., 2015). pH-sensitive nanogels are also used for protein delivery. Zeng et al. fabricated core–shell nanogels in situ through enzyme-catalyzed reactions. At endosomal pH conditions, these nanogels displayed fast FITC-BSA release in HeLa cells (Zeng et al., 2016). Oishi et al. developed pH-sensitive PEGylated nanogels by the emulsion copolymerization of 2-(N,N-diethylamino) ethylmethacrylate with heterobifunctional poly(ethylene glycol). Anticancer drug DOX is encapsulated into the nanogels by solvent vaporization. These nanogels exhibited no burst release of DOX under physiological pH, whereas significant release is observed at the endosomal pH (Oishi et al., 2007).

8.4.2   TEMPERATURE-SENSITIVE NANOGELS

Temperature-sensitive nanogels are prepared from polymers which are able to change volume in response to temperature. The temperature at which volume transition occurs is called volume phase transition temperature (VPTT). Depending on the changes in volume in response to temperature, these polymers are classified as negative and positive temperature responsive polymers. Negative-responsive polymers carry the lowest critical solution temperature (LCST) and shrink when environmental temperature increases above their VPTT. Poly(N-isopropyl acrylamide) (PNIPAM) is a widely used negative temperature responsive polymer with LCST at 32°C in aqueous solution. For temperature-induced drug delivery, polymeric materials with high VPTT are preferred. In this aspect, PNIPAM is not suitable for controlled drug release. Environmental conditions such as pH or polymer concentration has no influence on the critical solution temperature of PNIPAM. VPTT of PNIPAM has been modulated with several hydrophilic polymers, for temperature-mediated drug release. Yallappu et al. modified PNIPAM with poly(ethylene glycol)-maleic anhydride (PEG-MA) and vinyl pyrrolidone. NG-1 (PNIPAM nanogels) showed LCST of 38°C, whereas the modified nanogels NG-2 (PNIPAM with PEG-MA) and NG-3 (PNIPAM and vinyl pyrrolidone (VP) with PEG-MA) showed 39°C and 55°C, respectively. These nanogels can be used for intravascular delivery of rapamycin (Yallapu et al., 2008). Cationic core–shell thermosensitive nanogels are prepared from PNIPAM-g-PEI and the resulting nanogels are used for gene delivery (Cao et al., 2015). Recently, PNIPAM nanogels are used for topical delivery of protein to treat genetic skin diseases. Protein-loaded thermoresponsive poly(N-isopropylacrylamide)-polyglycerol-based nanogels with LCST 35°C showed 93% protein release when exposed to transglutaminase 1-deficient skin models (Witting et al., 2015). Abreu et al. grafted natural polymer cashew gum with poly(N-isopropyl acrylamide) and the resulting nanoparticles showed thermosensitive release of epirubicin (Abreu et al., 2016). Another important negative responsive polymer is poly(N-vinylcaprolactam) (PVCL). LCST of PVCL can be modulated by changing polymer chain length and concentration (Beija et al., 2011). Aguirre et al. developed dual responsive core–shell nanogels from temperature-responsive PVCL and pH-responsive PDEAEMA (poly(2-diethylaminoethyl) methacrylate). These core–shell nanogels are able to form complexes with siRNA through electrostatic interactions. *In vitro* release profiles proved the ability of dual responsive nanogels as a gene delivery system. siRNA release from the nanogel is controlled by the pH of the medium as well as by the cross-linking density of the PVCL-based shell (Aguirre et al., 2016). DOX-loaded self-assembled nanogels of pullulan-g-PLLA have been prepared for thermo-sensitive long-term drug release (Seo et al., 2012). Hydrophilic pullulan is grafted with hydrophobic poly(L-lactide) and the resulting poly(L-lactide)-g-pullulan (PLP) copolymers self-organized to nanogels in aqueous environment. Phase transition temperature of the nanogels can be modulated by changing the lactide content. Cumulative doxorubicin release is observed from the PLP nanogels when the temperature is increased from 37°C to 42°C.

Positive responsive nanogels with upper critical solution temperature (UCST) increase their volume at VPTT. Nanogels from UCST polymers are preferable for controlled drug release than LCST polymer-based nanogels. Drugs entrapped within the UCST polymer chains show more interaction with the polymer as compared to the interaction with LCST polymer thus premature drug release from these nanogels is less, compared to LCST nanogels. In addition to that, temperature-induced drug release from swollen nanogels may be more efficient than the squeezing release caused by nanogel deswelling (Li et al., 1998; Wang and Wu, 1997). Although positive temperature-responsive nanogels are more efficient for drug delivery, little research has been done to explore the use of these polymers (Ghugare et al., 2009).

8.4.3   LIGHT-SENSITIVE NANOGELS

Light-responsive nanogels are developed from polymers containing photo-active groups such as azobenzene, cinnamoyl, spirobenzopyran, and triphenylmethane. These groups are able to change size and shape, or form ionic or zwitterionic species, when exposed to light (Li and Keller, 2009). Even though UV radiation has been extensively used to trigger drug release from nanogels, it is harmful to tissues and cannot be used for deep tissue penetration. Near-infrared (NIR) radiation can provide deep penetration with low detrimental effects and hence, it is a promising alternative to UV irradiation (Weissleder, 2001). Indocyanine green incorporated diselenide bond-cross-linked poly(methacrylic acid) nanogels are developed for the NIR-mediated doxorubicin drug delivery. Indocyanine green trigger the cleavage of diselenide bond through generating reactive oxygen species. Upon NIR irradiation, these drug-loaded nanogels show targeted drug release (Tian et al., 2015). Khatun et al. developed multifunctional HA nanogels for light-glutathione-responsive controlled drug release. Doxorubicin is conjugated to light-responsive graphene by ester linkages, which is easily cleavable under acidic environment. Doxorubicin-graphene conjugate is coated with disulphide cross-linked HA. These nanogels can offer pH-, glutathione-, and light-responsive drug release. Nanogels are administered intravenously to tumor-bearing mice and irradiated with NIR laser (670 nm, 1 W cm^{-2}). Nanogels accumulate in the tumors passively and actively through both an enhanced permeability and retention effect and guidance of the receptor-mediated HA receptor. Heat generated upon irradiation provides photo-abla-tion effect sufficient to kill cancer cells (Khatun et al., 2015). Multiresponsive nanogels are prepared by encompassing different stimulus-responsive polymer components into the network. In addition to the above-mentioned responsive nanogels, magnetic field-responsive (Wang et al., 2014), redox-responsive (Liu et al., 2014), and biomolecule-responsive nanogels (Ye et al., 2014) have been developed.

8.5   NANOGEL DRUG DELIVERY ROUTES

Nanogels are mainly used for the administration of therapeutic agents. Common routes of administration are oral, nasal, transdermal, intravenous, etc. Oral drug delivery is the most convenient route of drug administration. Additionally, it is noninvasive and patient compliant administration method (des Rieux et al., 2006). Durán-Lobato et al. developed mannan-modified pH-responsive (HEMA-co-MAA) nanogel and evaluated its ability as a carrier for oral vaccine delivery. Triggered protein release at intestinal pH indicates the efficacy of (HEMA-co-MAA) nanogels in oral vaccine delivery (Durán-Lobato et al., 2014). Biodegradable, acetylated polyeth-yleneimine/polyethylene glycol (PEI/PEG) is complexed with interferon (IFN) for its oral delivery to norovirus infection. Rat experiments suggested no side effects of the nanogel–drug complex and significantly enhanced IFN activities against noroviruses (Kim et al., 2011). Oral administration of cancer chemotherapeutics is very difficult and development of oral nano-formulations are challenging. Vinogradov and coworkers developed poly-vinyl alcohol and dextrin-based nanogel conjugates with nucleotide analogs and demonstrated their enhanced activity against regular and drug-resistant cancers (Senanayake et al., 2011). This group used the same methodology for the oral administration of gemcitabine by conjugating its protected form with nanogel. The conjugate shows significantly higher activity than free drug in various cancer cells (Senanayake et al., 2013).

Transdermal delivery is the preferred route of drug administration, next to oral route. This noninvasive and easy mode of drug administration has many advantages. The barrier properties of the skin limit passive transdermal delivery of drugs. Presently, corticosteroid therapy used for skin diseases is not preferred due to its side effects. In order to achieve effective transdermal delivery, chemical penetration boosters or colloidal delivery systems have to be used. Several research attempts have been made to develop polymeric formulations for the treatment of skin diseases. Nanoformulations such as solid–lipid nanoparticles and lipid vesicles have been developed to improve transdermal drug delivery (Pierre and Costa, 2011; Schäfer-Korting et al., 2007). Polymeric nanoparticles have high drug-loading capacity and stability, compared to lipid nanoparticles. More than that, they offer controlled and sustained drug release over time. Mangalathillam et al. developed chitin nanogels for transdermal delivery of curcumin for skin cancer treatment (Mangalathillam et al., 2012). Two anti-inflammatory drugs, spantide II (SP) and ketoprofen (KP) have been delivered transdermally using poly(lactide-co-glycolic acid) and chitosan nanogels (Shah et al., 2012). Mavuso et al. have reviewed different polymeric colloidal nanogels used in transdermal drug delivery (Mavuso et al., 2015). Recently, Zabihi et al. conjugated nanogels with peptides and temoporfin (m-THPC) is encapsulated in it. Peptide conjugation to nanogel is found to enhance skin penetration of temoporfin (Zabihi et al., 2016).

Pulmonary delivery of drugs attracted tremendous scientific and biomedical interest in the healthcare research area. This is because of the high permeability, large absorptive surface area and good blood supply of lungs. More than that, the drug delivered by this route will experience low enzymatic activity and rapid absorption. Stocke et al. developed responsive nanogels for pulmonary drug delivery (Stocke et al., 2015). Palmitylacyl-ated exendin-4 (Ex4-C16)-loaded decholic acid-modified glycol chitosan nanogel offers a long-acting inhalation delivery system for treating type 2 diabetes (Lee et al., 2012).

Another route of drug administration using nanogel is intravenous (IV). IV medications have their own advantages and disadvantages. IV injection eliminates immediate elevation of drug levels in serum and in vital organs. IV injections are preferred in emergent situations such as cardiac arrest or narcotic overdose for an immediate, fast-acting therapeutic effect. Drugs given by IV injections are completely absorbed compared with other routes of medications. IV injections may damage surrounding tissues especially small peripheral veins. In addition to that, IV administration can cause significant harm or death if they are given incorrectly.

Surface-modified nanogels are used for the delivery of 5-flurouracil derivative to brain tissue through IV route (Soni et al., 2006). Shimoda et al. prepared dual cross-linked nanogels by the reaction of a thiol-modified poly(ethylene glycol) with acryloyl-modified cholesterol-bearing pullulan. These nanogels can be utilized as injectable nanocarriers for the controlled release of proteins (Shimoda et al., 2012). Hamidi et al. used Taguchi method to optimize conditions for developing efficient nanogels for IV delivery. They have studied effect of five factors such as chitosan concentration, pentasodium tripolyphosphate (TPP) concentration, CS-to-TPP volume ratio, addition time of the TPP solution to the CS solution, and temperature on the size of the nanogels particles (Hamidi et al., 2012).

8.6   CLINICAL TRIALS OF NANOGELS

Nanogels found applications in diagnostics, imaging and therapy. Even though, they have been explored for many biomedical applications, only few of the nanogel formulations have reached clinical trials (Kageyama et al., 2013; Saito et al., 2014). During the design of nanogel carriers for clinical applications, several critical factors have to be considered. Only 5–10% of the drug loaded in nanogels reaches the targeted sites and major portion gets accumulated in clearance organs such as kidney, liver, and spleen. This drawback pulls back its application in clinical trials (Vinogradov, 2010). Size, shape, surface properties, and composition of the nanogels play crucial role in their distribution in the tissues. Size and shape of the nanogels affect its clearance rate. Nanogels with size less than 200 nm can escape splenic filtration and very small nanogels (20 nm) get excreted through renal filtration (Elsabahy and Wooley, 2012; Zhang et al., 2012). Rod or filamentous nanogel shows longer circulation time than spherical nanogels (Geng et al., 2007). Nanogels with neutral surface charges generally exhibit long circulation time. In the case of stimuli-responsive nanogels, the responsiveness comes from the charged groups. Hence, it is difficult to maintain a proper balance between charge-depended responsive behavior and evasion of the nonspecific interactions of the nanogels with proteins or other biological molecules in the *in vivo* environment. Therefore, utmost care should be taken during the design of the nanogels for *in vivo* application. Abovementioned requirements should be considered for better *in vivo* efficacy.

Nanogels have extensively been used for targeted drug delivery to cancer cells. In reality, most of the nanogels cannot reach the core of the tumor, even if they are small enough. This is attributed to the high interstitial fluid pressure (Heldin et al., 2004), poor vasculature of the tumors. In order to achieve high accumulation in tumors, nanogel surfaces are decorated with targeting ligands. Folate molecules are one of the widely used targeting ligands in site-specific drug delivery owing to their over expression in many tumors. Folate receptors are also expressed in high levels on other normal organs such as small intestine, placenta, and kidneys (Desmoulin et al., 2012). Hence, active targeting is also challenging for efficient site-specific drug delivery. Even though, modern fabrication techniques help to control nanogel size and shape to a great extent, these techniques cannot offer control over the stoichiometry of the biomolecule conjugation. Targeting ligand introduction on the nanogel surface can alter surface charge and hydrophobicity, which may lead to increased opsonization, aggregation, and clearance (Phillips et al., 2010). Degradation kinetics and elimination profile of the nanogels designed for long-term clinical use is very important. Biodegradable natural or synthetic polymers are preferred for nanogel synthesis. Chemical modification can alter the degradation profile of the nanogels. Hence, clear knowledge about the trait of the nanogels in *in vivo* conditions is essential for long-term clinical applications.

8.7 CONCLUSIONS

Nanogels are special class of drug delivery systems with peculiar properties for encompassing drugs and small biological molecules. Research on nanogels has made significant progress in the past few decades. Physical properties of this delivery vehicle can be easily modulated by changing preparation methods. The recent developments allow researchers to control properties of nanogels such as size, stability, and biodegradability. Nanogels exhibit high-loading capacity, stability, biocompatibility, and biodegradability. Current research on nanogels is focused on the strategies to use them for the delivery of highly unstable molecules such as proteins, antibodies, or nucleic acids. Responsive nanogels offer stimuli-controlled release of the cargo at specific sites by changing their volume. In addition to that, by incorporating specific ligands, nanogels can provide targeted drug delivery. Of late, researchers are interested to develop multifunctional nanogels, using nanogel engineering technique. Nanogels developed from stimuli-responsive polymers, incorporated with drugs and quantum dots can perform responsive drug release along with imaging, simultaneously. Since multifunctional nanogels can provide many functions at a time, they are more preferable than single nanogels.

Inspite of the advancements in nanogel design and preparation, only a few of them have been cleared for clinical trials. In order to achieve the expected results in the clinical trials, careful engineering techniques are required. Problems in large-scale productions and batch-to-batch variability have to be addressed. Other challenges experienced by the nanogels in *in vivo* studies are efficient clearance and desired delivery on expected sites. More studies on the behavior of nanogels *in vivo* will help to ease their journey from bench side to bed side. Much more clinical trials are required for their applications in human beings.

KEYWORDS

- nanogels
- self-assembly
- drug delivery
- stimuli responsive
- cross-linked
- biopolymers

REFERENCES

Abreu, C. M.; Paula, H. C.; Seabra, V.; Feitosa, J. P.; Sarmento, B.; de Paula, R. C. Synthesis and Characterization of Non-toxic and Thermo-sensitive Poly(N-isopropyl Acrylamide)-grafted Cashew Gum Nanoparticles as a Potential Epirubicin Delivery Matrix. *Carbohydr. Polym.* **2016,** *154,* 77–85.

Aguirre, G.; Ramos, J.; Forcada, J. Advanced Design of t and pH Dual-responsive PDEAEMA–PVCL Core-shell Nanogels for siRNA Delivery. *J. Polym. Sci. Polym. Chem.* **2016,** *54,* 3203–3217.

Akiyama, E.; Morimoto, N.; Kujawa, P.; Ozawa, Y.; Winnik, F. M.; Akiyoshi, K. Self-assembled Nanogels of Cholesteryl-modified Polysaccharides: Effect of the Polysaccharide Structure on Their Association Characteristics in the Dilute and Semidilute Regimes. *Biomacromolecules* **2007,** *8,* 2366–2373.

Alvarez-Lorenzo, C.; Concheiro, A. *Smart Materials for Drug Delivery;* RSC: London: 2013; Vol. 1, p 1.

Alvarez-Lorenzo, C.; Concheiro, A. Smart Drug Delivery Systems: From Fundamentals to the Clinic. *Chem. Commun.* **2014,** *50,* 7743–7765.

Ariga, K.; Lvov, Y. M.; Kawakami, K.; Ji, Q.; Hill, J. P. Layer-by-layer Self-Assembled Shells for Drug Delivery. *Adv. Drug Deliv. Rev.* **2011,** *63,* 762–771.

Avadi, M. R.; Sadeghi, A. M. M.; Mohammadpour, N.; Abedin, S.; Atyabi, F.; Dinarvand, R.; Rafiee-Tehrani, M. Preparation and Characterization of Insulin Nanoparticles Using Chitosan and Arabic Gum with Ionic Gelation Method. *Nanomed. Nanotech. Biol. Med.* **2010,** *6,* 58–63.

Azadi, A.; Hamidi, M.; Rouini, M. R. Methotrexate-loaded Chitosan Nanogels as 'Trojan Horses' for Drug Delivery to Brain: Preparation and *In vitro/In vivo* Characterization. *Int. J. Biol. Macromol.* **2013,** *62,* 523–530.

Bae, K. H.; Mok, H.; Park, T. G. Synthesis, Characterization, and Intracellular Delivery of Reducible Heparin Nanogels for Apoptotic Cell Death. *Biomaterials* **2008,** *29,* 3376–3383.

Bawa, P.; Pillay, V.; Choonara, Y. E.; Du Toit, L. C. Stimuli-responsive Polymers and Their Applications in Drug Delivery. *Biomed. Mater.* **2009,** *4,* 022001.

Beija, M.; Marty, J. D.; Destarac, M. Thermoresponsive Poly(N-vinyl Caprolactam)-coated Gold Nanoparticles: Sharp Reversible Response and Easy Tunability. *Chem. Commun.* **2011,** *47,* 2826–2828.

Beningo, K. A.; Wang, Y.L. Fc-receptor-mediated Phagocytosis Is Regulated by Mechanical Properties of the Target. *J. Cell Sci.* **2002,** *115,* 849–856.

Bildstein, L.; Dubernet, C.; Couvreur, P. Prodrug-based Intracellular Delivery of Anticancer Agents. *Adv. Drug Deliv. Rev.* **2011,** *63,* 3–23.

Braeckmans, K.; Buyens, K.; Naeye, B.; Vercauteren, D.; Deschout, H.; Raemdonck, K.; Remaut, K.; Sanders, N. N.; Demeester, J.; De Smedt, S. C. Advanced Fluorescence Microscopy Methods Illuminate the Transfection Pathway of Nucleic Acid Nanoparticles. *J. Control. Release* **2010,** *148,* 69–74.

Brunel, F.; Véron, L.; Ladavière, C.; David, L.; Domard, A.; Delair, T. Synthesis and Structural Characterization of Chitosan Nanogels. *Langmuir* **2009,** *25,* 8935–8943.

Cai, H.; Ni, C.; Zhang, L. Preparation of Complex Nano-particles Based on Alginic acid/ Poly[(2-dimethylamino) Ethyl Methacrylate] and a Drug Vehicle for Doxorubicin Release Controlled by Ionic Strength. *Eur. J. Pharm. Sci.* **2012,** *45,* 43–49.

Camacho, K. M.; Kumar, S.; Menegatti, S.; Vogus, D. R.; Anselmo, A. C.; Mitragotri, S. Synergistic Antitumor Activity of Camptothecin–Doxorubicin Combinations and Their Conjugates with Hyaluronic Acid. *J. Control. Release* **2015,** *210,* 198–207.

Cao, P.; Sun, X.; Liang, Y.; Gao, X.; Li, X.; Li, W.; Song, Z.; Liang, G. Gene Delivery by a Cationic and Thermosensitive Nanogel Promoted Established Tumor Growth Inhibition. *Nanomedicine* **2015,** 10, 1585–1597.

Carvalho, V.; Castanheira, P.; Faria, T. Q.; Gonçalves, C.; Madureira, P.; Faro, C.; Domingues, L.; Brito, R. M.; Vilanova, M.; Gama, M.. Biological Activity of Heterologous Murine Interleukin-10 and Preliminary Studies on the Use of a Dextrin Nanogel as a Delivery System. *Int. J. Pharm.* **2010,** *400,* 234–242.

Chen, Y.; Chen, L.; Bai, H.; Li, L. Graphene Oxide–Chitosan Composite Hydrogels as Broad-spectrum Adsorbents for Water Purification. *J. Mater. Chem. A* **2013,** *1,* 1992–2001.

Chen, X.; Chen, L.; Yao, X.; Zhang, Z.; He, C.; Zhang, J.; Chen, X. Dual Responsive Supramolecular Nanogels for Intracellular Drug Delivery. *Chem. Commun.* **2014,** *50,* 3789–3791.

Choi, J. H.; Joung, Y. K.; Bae, J. W.; Choi, J. W.; Quyen, T. N.; Park, K. D.. Self-assembled Nanogel of Pluronic-conjugated Heparin as a Versatile Drug Nanocarrier. *Macromo. Res.* **2011,** *19,* 180–188.

Choi, W. I.; Lee, J. H.; Kim, J.-Y.; Kim, J.-C.;Kim, Y. H.; Tae, G. Efficient Skin Permeation of Soluble Proteins via Flexible and Functional Nano-carrier. *J. Control Release* **2012,** *157,* 272–278.

Chu, K. S.; Hasan, W.; Rawal, S.; Walsh, M. D.; Enlow, E. M.; Luft, J. C.; Bridges, A. S.; Kuijer, J. L.; Napier, M. E.; Zamboni, W. C. Plasma, Tumor and Tissue Pharmacokinetics of Docetaxel Delivered via Nanoparticles of Different Sizes and Shapes in Mice Bearing

SKOV-3 Human Ovarian Carcinoma Xenograft. *Nanomed. Nanotech. Biol. Med.* **2013**, *9*, 686–693.

Cram, D. J. The Design of Molecular Hosts, Guests, and Their Complexes (Nobel Lecture). *Angew. Chem. Int. Ed.* **1988**, *27*, 1009–1020.

Daoud-Mahammed, S.; Couvreur, P.; Gref, R. Novel Self-assembling Nanogels: Stability and Lyophilisation Studies. *Int. J. Pharm.* **2007**, *332*, 185–191.

de las Heras Alarcón, C.; Pennadam, S.; Alexander, C. Stimuli Responsive Polymers for Biomedical Applications. *Chem. Soc. Rev.* **2005**, *34*, 276–285.

Del Valle, E. M. Cyclodextrins and Their Uses: A Review. *Process Biochem* **2004**, *39*, 1033–1046.

des Rieux, A.; Fievez, V.; Garinot, M.; Schneider, Y.-J.; Préat, V. Nanoparticles as Potential Oral Delivery Systems of Proteins and Vaccines: A Mechanistic Approach. *J. Control Release* **2006**, *116*, 1–27.

Desmoulin, S. K.; Hou, Z.; Gangjee, A.; Matherly, L. H. The Human Proton-coupled Folate Transporter: Biology and Therapeutic Applications to Cancer. *Cancer Biol. Ther.* **2012**, *13*, 1355–1373.

Ding, J.; Shi, F.; Xiao, C.; Lin, L.; Chen, L.; He, C.; Zhuang, X.; Chen, X. One-step Preparation of Reduction-responsive Poly(Ethylene Glycol)-poly(Amino Acid)s Nanogels as Efficient Intracellular Drug Delivery Platforms. *Polym. Chem.* **2011a**, *2*, 2857–2864.

Ding, J.; Zhuang, X.; Xiao, C.; Cheng, Y.; Zhao, L.; He, C.; Tang, Z.; Chen, X. Preparation of Photo-cross-linked pH-responsive Polypeptide Nanogels as Potential Carriers for Controlled Drug Delivery. *J. Mater. Chem.* **2011b**, *21*, 11383–11391.

Douglas, K. L.; Piccirillo, C. A.; Tabrizian, M. Effects of Alginate Inclusion on the Vector Properties of Chitosan-based Nanoparticles. *J. Control Release* **2006**, *115*, 354–361.

Du, W. L.; Niu, S. S.; Xu, Y. L.; Xu, Z. R.; Fan, C. L. Antibacterial Activity of Chitosan Tripolyphosphate Nanoparticles Loaded with Various Metal Ions. *Carbohydr. Polym.* **2009**, *75*, 385–389.

Dunn, S. S.; Tian, S.; Blake, S.; Wang, J.; Galloway, A. L.; Murphy, A.; Pohlhaus, P. D.; Rolland, J. P.; Napier, M. E.; De Simone, J. M. Reductively Responsive siRNA-conjugated Hydrogel Nanoparticles for Gene Silencing. *J. Am. Chem. Soc.* **2012**, *134*, 7423–7430.

Durán-Lobato, M.; Carrillo-Conde, B.; Khairandish, Y.; Peppas, N. A. Surface-modified P (HEMA-co-MAA) Nanogel Carriers for Oral Vaccine Delivery: Design, Characterization, and *In vitro* Targeting Evaluation. *Biomacromolecules* **2014**, *15*, 2725–2734.

Elsabahy, M.; Wooley, K. L. Design of Polymeric Nanoparticles for Biomedical Delivery Applications. *Chem. Soc. Rev.* **2012**, *41*, 2545–2561.

Fang, J.; Zhang, Y.; Yan, S.; Liu, Z.; He, S.; Cui, L.; Yin, J. Poly(L-glutamic Acid)/Chitosan Polyelectrolyte Complex Porous Microspheres as Cell Microcarriers for Cartilage Regeneration. *Acta Biomater.* **2014**, *10*, 276–288.

Fujioka-Kobayashi, M.; Ota, M. S.; Shimoda, A.; Nakahama, K.I.; Akiyoshi, K.; Miyamoto, Y.; Iseki, S. Cholesteryl Group- and Acryloyl Group-bearing Pullulan Nanogel to Deliver BMP2 and FGF18 for Bone Tissue Engineering. *Biomaterials* **2012**, *33*, 7613–7620.

Garg, H. G; Hales, C. A. *Chemistry and Biology of Hyaluronan*, 1st ed.; Elsevier: Amsterdam, 2004.

Geng, Y.; Dalhaimer, P.; Cai, S.; Tsai, R.; Tewari, M.; Minko, T.; Discher, D. E. Shape Effects of Filaments Versus Spherical Particles in Flow and Drug Delivery. *Nat. Nanotechnol.* **2007**, *2*, 249–255.

Gerweck, L. E.; Seetharaman, K. Cellular pH Gradient in Tumor Versus Normal Tissue: Potential Exploitation for the Treatment of Cancer. *Cancer Res.* **1996**, *56*, 1194–1198.

Ghugare, S. V.; Mozetic, P.; Paradossi, G. Temperature-sensitive Poly(Vinyl Alcohol)/Poly(Methacrylate-co-N-isopropyl Acrylamide) Microgels for Doxorubicin Delivery. *Biomacromolecules* **2009**, *10*, 1589–1596.

Glangchai, L. C.; Caldorera-Moore, M.; Shi, L.; Roy, K. Nanoimprint Lithography Based Fabrication of Shape-specific, Enzymatically-triggered Smart Nanoparticles. *J. Control Release* **2008**, *125*, 263–272.

Gonçalves, M.; Maciel, D.; Capelo, D. B.; Xiao, S.; Sun, W.; Shi, X.; Rodrigues, J. O.; Tomás, H.; Li, Y. Dendrimer-assisted Formation of Fluorescent Nanogels for Drug Delivery and Intracellular Imaging. *Biomacromolecules* **2014**, *15*, 492–499.

Gref, R.; Amiel, C.; Molinard, K.; Daoud-Mahammed, S.; Sébille, B.; Gillet, B.; Beloeil, J. C.; Ringard, C.; Rosilio, V.; Poupaert, J. New Self-assembled Nanogels Based on Host–Guest Interactions: Characterization and Drug Loading. *J. Control. Release* **2006**, *111*, 316–324.

Hamidi, M.; Azadi, A.; Ashrafi, H.; Rafiei, P.; Mohamadi-Samani, S. Taguchi Orthogonal Array Design for the Optimization of Hydrogel Nanoparticles for the Intravenous Delivery of Small-molecule Drugs. *J. Appl. Polym. Sci.* **2012**, *126*, 1714–1724.

Hawker, C. J.; Wooley, K. L. The Convergence of Synthetic Organic and Polymer Chemistries. *Science* **2005**, *309*, 1200–1205.

Heldin, C.-H.; Rubin, K.; Pietras, K.; Östman, A. High Interstitial Fluid Pressure: An Obstacle in Cancer Therapy. *Nat. Rev. Cancer* **2004**, *4*, 806–813.

Hirakura, T.; Nomura, Y.; Aoyama, Y.; Akiyoshi, K. Photoresponsive Nanogels Formed by the Self-assembly of Spiropyrane-bearing Pullulan That Act as Artificial Molecular Chaperones. *Biomacromolecules* **2004**, *5*, 1804–1809.

Hong, J. S.; Vreeland, W. N.; De Paoli Lacerda, S. H.; Locascio, L. E.; Gaitan, M.; Raghavan, S. R. Liposome-templated Supramolecular Assembly of Responsive Alginate Nanogels. *Langmuir* **2008**, *24*, 4092–4096.

Hunt, C. A.; MacGregor, R. D.; Siegel, R. A. Engineering Targeted *In vivo* Drug Delivery. I. The Physiological and Physicochemical Principles Governing Opportunities and Limitations. *Pharm. Res.* **1986**, *3*, 333–344.

Isa, I. L. M.; Srivastava, A.; Tiernan, D.; Owens, P.; Rooney, P.; Dockery, P.; Pandit, A. Hyaluronic Acid Based Hydrogels Attenuate Inflammatory Receptors and Neurotrophins in Interleukin-1β Induced Inflammation Model of Nucleus Pulposus Cells. *Biomacromolecules* **2015**, *16*, 1714–1725.

Jia, X.; Han, Y.; Pei, M.; Zhao, X.; Tian, K.; Zhou, T.; Liu, P. Multi-functionalized Hyaluronic Acid Nanogels Crosslinked with Carbon Dots as Dual Receptor-mediated Targeting Tumor Theranostics. *Carbohydr. Polym.* **2016**, *152*, 391–397.

Jonassen, H.; Kjøniksen, A. L.; Hiorth, M. Stability of Chitosan Nanoparticles Cross-linked with Tripolyphosphate. *Biomacromolecules* **2012**, *13*, 3747–3756.

Juliano, R. L.; Ming, X.; Nakagawa, O. Cellular Uptake and Intracellular Trafficking of Antisense and siRNA Oligonucleotides. *Bioconjugate Chem.* **2011**, *23*, 147–157.

Kabanov, A. V.; Vinogradov, S. V. Nanogels as Pharmaceutical Carriers: Finite Networks of Infinite Capabilities. *Angew. Chem. Int. Ed.* **2009a**, *48*, 5418–5429.

Kabanov, A. V.; Vinogradov, S. V. Nanogels as Pharmaceutical Carriers: Finite Networks of Infinite Capabilities. *Angew. Chem. Int. Ed. Eng.* **2009b**, *48*, 5418–5429.

Kageyama, S.; Wada, H.; Muro, K.; Niwa, Y.; Ueda, S.; Miyata, H.; Takiguchi, S.; Sugino, S. H.; Miyahara, Y.; Ikeda, H. Dose-dependent Effects of NY-ESO-1 Protein Vaccine Complexed with Cholesteryl Pullulan (CHP-NY-ESO-1) on Immune Responses and Survival Benefits of Esophageal Cancer Patients. *J. Transl. Med.* **2013**, *11*, 1.

Karimi, M.; Eslami, M.; Sahandi-Zangabad, P.; Mirab, F.; Farajisafiloo, N.; Shafaei, Z.; Ghosh, D.; Bozorgomid, M.; Dashkhaneh, F.; Hamblin, M. R. pH-Sensitive Stimulus-responsive Nanocarriers for Targeted Delivery of Therapeutic Agents. *Wiley Interdiscip. Rev. Nanomed. Nanobiotechnol.* **2016**, *8*, 696–716.

Kataoka, K.; Harada, A.; Nagasaki, Y. Block Copolymer Micelles for Drug Delivery: Design, Characterization and Biological Significance. *Adv. Drug Deliv. Rev.* **2001**, *47*, 113–131.

Kettel, M. J.; Dierkes, F.; Schaefer, K.; Moeller, M.; Pich, A. Aqueous Nanogels Modified with Cyclodextrin. *Polymer* **2011**, *52*, 1917–1924.

Kettel, M. J.; Schaefer, K.; Groll, J.; Moeller, M. Nanogels with High Active β-cyclodextrin Content as Physical Coating System with Sustained Release Properties. *ACS Appl. Mater. Interfaces* **2014**, *6*, 2300–2311.

Khatun, Z.; Nurunnabi, M.; Nafiujjaman, M.; Reeck, G. R.; Khan, H. A.; Cho, K. J.; Lee, Y. K. A Hyaluronic Acid Nanogel for Photo–Chemo Theranostics of Lung Cancer with Simultaneous Light-responsive Controlled Release of Doxorubicin. *Nanoscale* **2015**, *7*, 10680–10689.

Kim, J. H.; Kim, Y. S.; Kim, S.; Park, J. H.; Kim, K.; Choi, K.; Chung, H.; Jeong, S. Y.; Park, R. W.; Kim, I. S. Hydrophobically Modified Glycol Chitosan Nanoparticles as Carriers for Paclitaxel. *J. Control. Release* **2006**, *111*, 228–234.

Kim, Y.; Thapa, M.; Hua, D. H.; Chang, K. O. Biodegradable Nanogels for Oral Delivery of Interferon for Norovirus Infection. *Antiviral Res.* **2011**, *89*, 165–173.

Klinger, D.; Aschenbrenner, E. M.; Weiss, C. K.; Landfester, K. Enzymatically Degradable Nanogels by Inverse Miniemulsion Copolymerization of Acrylamide with Dextran Methacrylates as Crosslinkers. *Polym. Chem.* **2012**, *3*, 204–216.

Kogan, G.; Šoltés, L.; Stern, R.; Gemeiner, P. Hyaluronic Acid: A Natural Biopolymer with a Broad Range of Biomedical and Industrial Applications. *Biotechnol. Lett.* **2007**, *29*, 17–25.

Krisch, E.; Messager, L.; Gyarmati, B.; Ravaine, V.; Szilágyi, A. Redox-and pH-Responsive Nanogels Based on Thiolated Poly(Aspartic Acid). *Macromol. Mater. Eng.* **2016**, *301*, 260–266.

Leathers, T. D. Biotechnological Production and Applications of Pullulan. *Appl. Microbiol. Biotechnol.* **2003**, *62*, 468–473.

Lee, J. S.; Feijen, J. Polymersomes for Drug Delivery: Design, Formation and Characterization. *J. Control Release* **2012**, *161*, 473–483.

Lee, K. Y.; Mooney, D. J. Alginate: Properties and Biomedical Applications. *Prog. Polym. Sci.* **2012**, *37*, 106–126.

Lee, W. C.; Li, Y. C.; Chu, I. Amphiphilic Poly(D, L-lactic Acid)/Poly(Ethylene Glycol)/Poly(D, L-lactic Acid) Nanogels for Controlled Release of Hydrophobic Drugs. *Macromol. Biosci.* **2006**, *6*, 846–854.

Lee, H.; Mok, H.; Lee, S.; Oh, Y. K.; Park, T. G. Target-specific Intracellular Delivery of siRNA Using Degradable Hyaluronic Acid Nanogels. *J. Control Release* **2007**, *119*, 245–252.

Lee, J.; Lee, C.; Kim, T. H.; Lee, E. S.; Shin, B. S.; Chi, S. C.; Park, E. S.; Lee, K. C.; Youn, Y. S. Self-assembled Glycol Chitosan Nanogels Containing Palmityl-acylated Exendin-4

Peptide as a Long-acting Anti-diabetic Inhalation System. *J. Control Release* **2012**, *161*, 728–734.

Li, J. K.; Wang, N.; Wu, X. S. Poly(Vinyl Alcohol) Nanoparticles Prepared by Freezing–Thawing Process for Protein/Peptide Drug Delivery. *J. Control Release* **1998**, *56*, 117–126.

Li, M. H.; Keller, P. Stimuli-responsive Polymer Vesicles. *Soft Matter* **2009**, *5*, 927–937.

Li, P.; Luo, Z.; Liu, P.; Gao, N.; Zhang, Y.; Pan, H.; Liu, L.; Wang, C.; Cai, L.; Ma, Y. Bioreducible Alginate-poly(Ethylenimine) Nanogels as an Antigen-delivery System Robustly Enhance Vaccine-elicited Humoral and Cellular Immune Responses. *J. Control Release* **2013**, *168*, 271–279.

Li, Y.; Maciel, D.; Rodrigues, J. o.; Shi, X.; Tomás, H. Biodegradable Polymer Nanogels for Drug/Nucleic Acid Delivery. *Chem. Rev.* **2015**, *115*, 8564–8608.

Liang, K.; Ng, S.; Lee, F.; Lim, J.; Chung, J. E.; Lee, S. S.; Kurisawa, M. Targeted Intracellular Protein Delivery Based on Hyaluronic Acid–Green Tea Catechin Nanogels. *Acta Biomater.* **2016**, *33*, 142–152.

Lin, L.; Xu, W.; Liang, H.; He, L.; Liu, S.; Li, Y.; Li, B.; Chen, Y. Construction of pH-sensitive Lysozyme/Pectin Nanogel for Tumor Methotrexate Delivery. *Colloids Surf. B* **2015**, *126*, 459–466.

Liu, J.; Detrembleur, C.; Hurtgen, M.; Debuigne, A.; De Pauw-Gillet, M. C.; Mornet, S.; Duguet, E.; Jérôme, C. Reversibly Crosslinked Thermo- and Redox-responsive Nanogels for Controlled Drug Release. *Polym. Chem.* **2014**, *5*, 77–88.

Maciel, D.; Figueira, P.; Xiao, S.; Hu, D.; Shi, X.; Rodrigues, J. o.; Tomás, H.; Li, Y. Redox-responsive Alginate Nanogels with Enhanced Anticancer Cytotoxicity. *Biomacromolecules* **2013**, *14*, 3140–3146.

Manchun, S.; Dass, C. R.; Cheewatanakornkool, K.; Sriamornsak, P. Enhanced Anti-tumor Effect of pH-responsive Dextrin Nanogels Delivering Doxorubicin on Colorectal Cancer. *Carbohydr. Polym.* **2015**, *126*, 222–230.

Mangalathillam, S.; Rejinold, N. S.; Nair, A.; Lakshmanan, V. K.; Nair, S. V.; Jayakumar, R. Curcumin Loaded Chitin Nanogels for Skin Cancer Treatment via the Transdermal Route. *Nanoscale* **2012**, *4*, 239–250.

Mavuso, S.; Marimuthu, T.; E Choonara, Y.; Kumar, P.; C du Toit, L.; Pillay, V. A Review of Polymeric Colloidal Nanogels in Transdermal Drug Delivery. *Curr. Pharm. Des.* **2015**, *21*, 2801–2813.

Medeiros, S. F.; Santos, A. M.; Fessi, H.; Elaissari, A. Thermally-sensitive and Magnetic Poly(N-vinylcaprolactam)-based Nanogels by Inverse Miniemulsion Polymerization. *J. Colloid Sci. Biotechnol.* **2012**, *1*, 99–112.

Morimoto, N.; Hirano, S.; Takahashi, H.; Loethen, S.; Thompson, D. H.; Akiyoshi, K. Self-assembled pH-sensitive Cholesteryl Pullulan Nanogel as a Protein Delivery Vehicle. *Biomacromolecules* **2012**, *14*, 56–63.

Mortisen, D.; Peroglio, M.; Alini, M.; Eglin, D Tailoring Thermoreversible Hyaluronan Hydrogels by Click Chemistry and RAFT Polymerization for Cell and Drug Therapy. *Biomacromolecules* **2010**, *11*, 1261–1272.

Moya-Ortega, M. D.; Alvarez-Lorenzo, C.; Sigurdsson, H. H.; Concheiro, A.; Loftsson, T. Cross-linked Hydroxypropyl-β-cyclodextrin and γ-cyclodextrin Nanogels for Drug Delivery: Physicochemical and Loading/Release Properties. *Carbohydr. Polym.* **2012**, *87*, 2344–2351.

Moya-Ortega, M. D.; Alves, T. F.; Alvarez-Lorenzo, C.; Concheiro, A.; Stefánsson, E.; Thor-steinsdóttir, M.; Loftsson, T. Dexamethasone Eye Drops Containing γ-cyclodextrin-Based Nanogels. *Int. J. Pharm.* **2013,** *441,* 507–515.

Mukerjee, A.; Vishwanatha, J. K. Formulation, Characterization and Evaluation of Curcumin-loaded PLGA Nanospheres for Cancer Therapy. *Anticancer Res.* **2009,** *29,* 3867–3875.

Musyanovych, A.; Schmitz-Wienke, J.; Mailänder, V.; Walther, P.; Landfester, K. Prepara-tion of Biodegradable Polymer Nanoparticles by Miniemulsion Technique and Their Cell Interactions. *Macromol. Biosci.* **2008,** *8,* 127–139.

Na, K.; Park, K. H.; Kim, S. W.; Bae, Y. H. Self-assembled Hydrogel Nanoparticles from Curdlan Derivatives: Characterization, Anti-cancer Drug Release and Interaction with a Hepatoma Cell Line (HepG2). *J. Control Release* **2000,** *69,* 225–236.

Na, K.; Lee, K. H.; Bae, Y. H. pH-sensitivity and pH-dependent Interior Structural Change of Self-assembled Hydrogel Nanoparticles of Pullulan Acetate/Oligo-sulfonamide Conjugate. *J. Control Release* **2004,** *97,* 513–525.

Nguyen, T. A.; Fu, C.-C.; Juang, R.-S. Effective Removal of Sulfur Dyes from Water by Biosorption and Subsequent Immobilized Laccase Degradation on Crosslinked Chitosan Beads. *Chem. Eng. J.* **2016,** *304,* 313–324.

Oh, J. K.; Siegwart, D. J.; Lee, H. I.; Sherwood, G.; Peteanu, L.; Hollinger, J. O.; Kataoka, K.; Matyjaszewski, K. Biodegradable Nanogels Prepared by Atom Transfer Radical Polymer-ization as Potential Drug Delivery Carriers: Synthesis, Biodegradation, *In vitro* Release, and Bioconjugation. *J. Am. Chem. Soc.* **2007,** *129,* 5939–5945.

Oh, J. K.; Drumright, R.; Siegwart, D. J.; Matyjaszewski, K. The Development of Microgels/Nanogels for Drug Delivery Applications. *Prog. Polym. Sci.* **2008,** *33,* 448–477.

Oh, J. K.; Bencherif, S. A.; Matyjaszewski, K. Atom Transfer Radical Polymerization in Inverse Miniemulsion: A Versatile Route Toward Preparation and Functionalization of Microgels/Nanogels for Targeted Drug Delivery Applications. *Polymer* **2009,** *50,* 4407–4423.

Oishi, M.; Hayashi, H.; Iijima, M.; Nagasaki, Y. Endosomal Release and Intracellular Delivery of Anticancer Drugs Using pH-sensitive PEGylated Nanogels. *J. Mater. Chem.* **2007,** *17,* 3720–3725.

Oliveira, M. A. M.; Boyer, C.; Nele, M.; Pinto, J. C.; Zetterlund, P. B.; Davis, T. P. Synthesis of Biodegradable Hydrogel Nanoparticles for Bioapplications Using Inverse Miniemulsion RAFT Polymerization. *Macromolecules* **2011,** *44,* 7167–7175.

Ozawa, Y.; Sawada, S. i.; Morimoto, N.; Akiyoshi, K. Self-Assembled Nanogel of Hydropho-bized Dendritic Dextrin for Protein Delivery. *Macromol. Biosci.* **2009,** *9,* 694–701.

Park, K.; Kim, K; Kwon, I. C.; Kim, S. K.; Lee, S.; Lee, D. Y.; Byun, Y. Preparation and Characterization of Self-assembled Nanoparticles of Heparin-deoxycholic Acid Conju-gates. *Langmuir* **2004,** *20,* 11726–11731.

Park, W.; Park, S.-j.; Na, K. Potential of Self-organizing Nanogel with Acetylated Chon-droitin Sulfate as an Anti-cancer Drug Carrier. *Colloids Surf. B* **2010,** *79,* 501–508.

Parveen, S.; Mitra, M.; Krishnakumar, S.; Sahoo, S. K. Retraction Notice to Enhanced Antip-roliferative Activity of Carboplatin Loaded Chitosan-alginate Nanoparticles in Retinoblas-toma Cell Line *Acta Biomaterialia* **2013,** *6,* 7075.

Patil, P.; Chavanke, D.; Wagh, M. A Review on Ionotropic Gelation Method: Novel Approach for Controlled Gastroretentive Gelispheres. *Int. J. Pharm. Pharm. Sci.* **2012,** *4,* 27–32.

Peng, K.; Cui, C.; Tomatsu, I.; Porta, F.; Meijer, A. H.; Spaink, H. P.; Kros, A. Cyclodextrin/ Dextran Based Drug Carriers for a Controlled Release of Hydrophobic Drugs in Zebrafish Embryos. *Soft Matter* **2010**, *6*, 3778–3783.

Perdones, Á.; Vargas, M.; Atarés, L.; Chiralt, A. Physical, Antioxidant and Antimicrobial Properties of Chitosan–Cinnamon Leaf Oil Films as Affected by Oleic Acid. *Food Hydrocoll.* **2014**, *36*, 256–264.

Pereira, P.; Pedrosa, S. S.; Wymant, J. M.; Sayers, E.; Correia, A.; Vilanova, M.; Jones, A. T.; Gama, F. M. siRNA Inhibition of Endocytic Pathways to Characterize the Cellular Uptake Mechanisms of Folate-functionalized Glycol Chitosan Nanogels. *Mol. Pharm.* **2015**, *12*, 1970–1979.

Phillips, M. A.; Gran, M. L.; Peppas, N. A. Targeted Nanodelivery of Drugs and Diagnostics. *Nano Today* **2010**, *5*, 143–159.

Pierre, M. B. R.; Costa, I. d. S. M. Liposomal Systems as Drug Delivery Vehicles for Dermal and Transdermal Applications. *Arch. Dermatol. Res.* **2011**, *303*, 607–621.

Pujana, M. A.; Pérez-Álvarez, L.; Iturbe, L. C. C.; Katime, I. Biodegradable Chitosan Nanogels Crosslinked with Genipin. *Carbohydr. Polym.* **2013**, *94*, 836–842.

Pujana, M. A.; Pérez-Álvarez, L.; Iturbe, L. C. C.; Katime, I. pH-sensitive Chitosan-Folate Nanogels Crosslinked with Biocompatible Dicarboxylic Acids. *Eur. Polym. J.* **2014a**, *61*, 215–225.

Pujana, M. A.; Pérez-Álvarez, L.; Iturbe, L. C. C.; Katime, I. Water Soluble Folate–Chitosan Nanogels Crosslinked by Genipin. *Carbohydr. Polym.* **2014b**, *101*, 113–120.

Racoviță, S.; Vasiliu, S.; Popa, M.; Luca, C. Polysaccharides Based on Micro- and Nanoparticles Obtained by Ionic Gelation and Their Applications as Drug Delivery Systems. *Rev. Roum. Chim.* **2009**, *54*, 709–718.

Raemdonck, K.; Demeester, J.; De Smedt, S. Advanced Nanogel Engineering for Drug Delivery. *Soft Matter* **2009a**, *5*, 707–715.

Raemdonck, K.; Naeye, B.; Buyens, K.; Vandenbroucke, R. E.; Høgset, A.; Demeester, J.; De Smedt, S. C. Biodegradable Dextran Nanogels for RNA Interference: Focusing on Endosomal Escape and Intracellular siRNA Delivery. *Adv. Funct. Mater.* **2009b**, *19*, 1406–1415.

Ragelle, H.; Riva, R.; Vandermeulen, G.; Naeye, B.; Pourcelle, V.; Le Duff, C. S.; D'Haese, C.; Nysten, B.; Braeckmans, K.; De Smedt, S. C. Chitosan Nanoparticles for siRNA Delivery: Optimizing Formulation to Increase Stability and Efficiency. *J. Control Release* **2014**, *176*, 54–63.

Reischl, D.; Zimmer, A. Drug Delivery of siRNA Therapeutics: Potentials and Limits of Nanosystems. *Nanomed. Nanotech. Biol. Med.* **2009**, *5*, 8–20.

Rejinold, N. S.; Nair, A.; Sabitha, M.; Chennazhi, K.; Tamura, H.; Nair, S. V.; Jayakumar, R. Synthesis, Characterization and *In vitro* Cytocompatibility Studies of Chitin Nanogels for Biomedical Applications. *Carbohydr. Polym.* **2012**, *87*, 943–949.

Saito, T.; Wada, H.; Yamasaki, M.; Miyata, H.; Nishikawa, H.; Sato, E.; Kageyama, S.; Shiku, H.; Mori, M.; Doki, Y. High Expression of MAGE-A4 and MHC Class I Antigens in Tumor Cells and Induction of MAGE-A4 Immune Responses are Prognostic Markers of CHP-MAGE-A4 Cancer Vaccine. *Vaccine* **2014**, *32*, 5901–5907.

Sarika, P.; James, N. R. Preparation and Characterisation of Gelatin–Gum Arabic Aldehyde Nanogels via Inverse Miniemulsion Technique. *Int. J. Biol. Macromol.* **2015**, *76*, 181–187.

Sarika, P.; Kumar, P. A.; Raj, D. K.; James, N. R. Nanogels Based on Alginic Aldehyde and Gelatin by Inverse Miniemulsion Technique: Synthesis and Characterization. *Carbohydr. Polym.* **2015**, *119*, 118–125.

Sarika, P.; Nirmala, R. J. Curcumin Loaded Gum Arabic Aldehyde-gelatin Nanogels for Breast Cancer Therapy. *Mater. Sci. Eng. C* **2016**, *65*, 331–337.

Sarika, P.; James, N. R.; Raj, D. K. Preparation, Characterization and Biological Evaluation of Curcumin Loaded Alginate Aldehyde–Gelatin Nanogels. *Mater. Sci. Eng. C* **2016**, *68*, 251–257.

Sarmento, B.; Ribeiro, A.; Veiga, F.; Ferreira, D.; Neufeld, R. Insulin-loaded Nanoparticles are Prepared by Alginate Ionotropic Pre-gelation Followed by Chitosan Polyelectrolyte Complexation. *J. Nanosci. Nanotechnol.* **2007**, *7*, 2833–2841.

Sasaki, Y.; Akiyoshi, K. Nanogel Engineering for New Nanobiomaterials: From Chaperoning Engineering to Biomedical Applications. *Chem. Rec.* **2010**, *10*, 366–376.

Sasaki, Y.; Akiyoshi, K. Self-assembled Nanogel Engineering for Advanced Biomedical Technology. *Chem. Lett.* **2012**, *41*, 202–208.

Schäfer-Korting, M.; Mehnert, W.; Korting, H.-C. Lipid Nanoparticles for Improved Topical Application of Drugs for Skin Diseases. *Adv. Drug Deliv. Rev.* **2007**, *59*, 427–443.

Scomparin, A.; Salmaso, S.; Bersani, S.; Satchi-Fainaro, R.; Caliceti, P. Novel folated and Non-folated Pullulan Bioconjugates for Anticancer Drug Delivery. *Eur. J. Pharm. Sci.* **2011**, *42*, 547–558.

Senanayake, T. H.; Warren, G.; Vinogradov, S. V. Novel anticancer Polymeric Conjugates of Activated Nucleoside Analogues. *Bioconjugate Chem.* **2011**, *22*, 1983–1993.

Senanayake, T. H.; Warren, G.; Wei, X.; Vinogradov, S. V. Application of Activated Nucleoside Analogs for the Treatment of Drug-resistant Tumors by Oral Delivery of Nanogel-drug Conjugates. *J. Control. Release* **2013**, *167*, 200–209.

Seo, S.; Lee, C.-S.; Jung, Y.-S.; Na, K. Thermo-sensitivity and Triggered Drug Release of Polysaccharide Nanogels Derived from Pullulan-g-poly(l-lactide) Copolymers. *Carbohydr. Polym.* **2012**, *87*, 1105–1111.

Shah, P. P.; Desai, P. R.; Patel, A. R.; Singh, M. S. Skin Permeating Nanogel for the Cutaneous Co-delivery of Two Anti-inflammatory Drugs. *Biomaterials* **2012**, *33*, 1607–1617.

Shimizu, T.; Kishida, T.; Hasegawa, U.; Ueda, Y.; Imanishi, J.; Yamagishi, H.; Akiyoshi, K.; Otsuji, E.; Mazda, O. Nanogel DDS Enables Sustained Release of IL-12 for Tumor Immunotherapy. *Biochem. Biophys. Res. Commun.* **2008**, *367*, 330–335.

Shimoda, A.; Sawada, S. I.; Kano, A.; Maruyama, A.; Moquin, A.; Winnik, F. M.; Akiyoshi, K. Dual Crosslinked Hydrogel Nanoparticles by Nanogel Bottom-up Method for Sustained-release Delivery. *Colloids Surf. B* **2012**, *99*, 38–44.

Siegwart, D. J.; Srinivasan, A.; Bencherif, S. A.; Karunanidhi, A.; Oh, J. K.; Vaidya, S.; Jin, R.; Hollinger, J. O.; Matyjaszewski, K. Cellular Uptake of functional Nanogels Prepared by Inverse Miniemulsion ATRP with Encapsulated Proteins, Carbohydrates, and Gold Nanoparticles. *Biomacromolecules* **2009**, *10*, 2300–2309.

Soni, S.; Babbar, A. k.; Sharma, R. k.; Maitra, A. Delivery of Hydrophobised 5-fluorouracil Derivative to Brain Tissue Through Intravenous Route Using Surface Modified Nanogels. *J. Drug Target.* **2006**, *14*, 87–95.

Stocke, N. A.; Arnold, S. M.; Hilt, J. Z. Responsive Hydrogel Nanoparticles for Pulmonary Delivery. *J. Drug. Deliv. Sci. Tec.* **2015**, *29*, 143–151.

Sudheesh Kumar, P.; Lakshmanan, V.-K.; Anilkumar, T.; Ramya, C.; Reshmi, P.; Unnikrishnan, A.; Nair, S. V.; Jayakumar, R. Flexible and Microporous Chitosan Hydrogel/Nano ZnO Composite Bandages for Wound Dressing: *In vitro* and *In vivo* Evaluation. *ACS Appl. Mater. Interfaces* **2012**, *4*, 2618–2629.

Tamura, A.; Oishi, M.; Nagasaki, Y. Efficient siRNA Delivery Based on PEGylated and Partially Quaternized Polyamine Nanogels: Enhanced Gene Silencing Activity by the Cooperative Effect of Tertiary and Quaternary Amino Groups in the Core. *J. Control Release* **2010**, *146*, 378–387.

Tian, Y.; Zheng, J.; Tang, X.; Ren, Q.; Wang, Y.; Yang, W. Near-infrared Light-responsive Nanogels with Diselenide-cross-linkers for On-demand Degradation and Triggered Drug Release. *Part. Part. Syst. Character.* **2015**, *32*, 547–551.

Toita, S.; Soma, Y.; Morimoto, N.; Akiyoshi, K. Cycloamylose-based Biomaterial: Nanogel of Cholesterol-bearing Cationic Cycloamylose for siRNA Delivery. *Chem. Lett.* **2009**, *38*, 1114–1115.

Toita, S.; Sawada, S.-i.; Akiyoshi, K. Polysaccharide Nanogel Gene Delivery System with Endosome-escaping Function: Co-delivery of Plasmid DNA and Phospholipase A 2. *J. Control Release* **2011**, *155*, 54–59.

Torchilin, V. P. Cell Penetrating Peptide-modified Pharmaceutical Nanocarriers for Intracellular Drug and Gene Delivery. *Pept. Sci.* **2008**, *90*, 604–610.

Van Thienen, T.; Lucas, B.; Flesch, F.; Van Nostrum, C.; Demeester, J.; De Smedt, S. On the Synthesis and Characterization of Biodegradable Dextran Nanogels with Tunable Degradation Properties. *Macromolecules* **2005**, *38*, 8503–8511.

Van Thienen, T.; Raemdonck, K.; Demeester, J.; De Smedt, S. Protein Release from Biodegradable Dextran Nanogels. *Langmuir* **2007**, *23*, 9794–9801.

Vasir, J. K.; Labhasetwar, V. Biodegradable Nanoparticles for Cytosolic Delivery of Therapeutics. *Adv. Drug Deliv. Rev.* **2007**, *59*, 718–728.

Vinogradov, S. V. Nanogels in the Race for Drug Delivery. *Nanomedicine* **2010**, *5*, 165–168.

Wang, N.; Wu, X. S. Preparation and Characterization of Agarose Hydrogel Nanoparticles for Protein and Peptide Drug Delivery. *Pharm. Dev. Technol.* **1997**, *2*, 135–142.

Wang, H.; Yi, J.; Mukherjee, S.; Banerjee, P.; Zhou, S. Magnetic/NIR-thermally Responsive Hybrid Nanogels for Optical Temperature Sensing, Tumor Cell Imaging and Triggered Drug Release. *Nanoscale* **2014**, *6*, 13001–13011.

Water, J. J.; Kim, Y.; Maltesen, M. J.; Franzyk, H.; Foged, C.; Nielsen, H. M. Hyaluronic Acid-based Nanogels Produced by Microfluidics-facilitated Self-assembly Improves the Safety Profile of the Cationic Host Defense Peptide Novicidin. *Pharm. Res.* **2015**, *32*, 2727–2735.

Weiss, C. K.; Landfester, K. *Hybrid Latex Particles Preparation with (Mini) Emulsion Polymerization*; Springer Verlag: New York, 2010; Vol. 233; p 185.

Weissleder, R. A Clearer Vision for *In vivo* Imaging. *Nature Biotechnol.* **2001**, *19*, 316–317.

Witting, M.; Molina, M.; Obst, K.; Plank, R.; Eckl, K. M.; Hennies, H. C.; Calderón, M.; Frieß, W.; Hedtrich, S. Thermosensitive Dendritic Polyglycerol-based Nanogels for Cutaneous Delivery of Biomacromolecules. *Nanomed. Nanotech. Biol. Med.* **2015**, *11*, 1179–1187.

Xu, X.; Jha, A. K.; Harrington, D. A.; Farach-Carson, M. C.; Jia, X. Hyaluronic Acid-based Hydrogels: From a Natural Polysaccharide to Complex Networks. *Soft Matter* **2012**, *8*, 3280–3294.

Yallapu, M. M.; Vasir, J. K.; Jain, T. K.; Vijayaraghavalu, S.; Labhasetwar, V. Synthesis, Characterization and Antiproliferative Activity of Rapamycin-loaded Poly(N-isopropylacrylamide)-based Nanogels in Vascular Smooth Muscle Cells. *J. Biomed. Nanotechnol.* **2008**, *4*, 16–24.

Yallapu, M. M.; Gupta, B. K.; Jaggi, M.; Chauhan, S. C. Fabrication of Curcumin Encapsulated PLGA Nanoparticles for Improved Therapeutic Effects in Metastatic Cancer Cells. *J. Colloid Interface Sci.* **2010,** *351,* 19–29.

Yallapu, M. M.; Jaggi, M.; Chauhan, S. C. Design and Engineering of Nanogels for Cancer Treatment. *Drug Discov. Today* **2011,** *16,* 457–463.

Ye, T.; Yan, S.; Hu, Y.; Ding, L.; Wu, W. Synthesis and Volume Phase Transition of Concanavalin A-based Glucose-responsive Nanogels. *Polym. Chem.* **2014,** *5,* 186–194.

Zabihi, F.; Wieczorek, S.; Dimde, M.; Hedtrich, S.; Börner, H. G.; Haag, R. Intradermal Drug Delivery by Nanogel-peptide Conjugates; Specific and Efficient Transport of Temoporfin. *J. Control Release* **2016,** *242,* 35–41.

Zan, M.; Li, J.; Luo, S.; Ge, Z. Dual pH-triggered Multistage Drug Delivery Systems Based on Host–Guest Interaction-associated Polymeric Nanogels. *Chem. Commun.* **2014,** *50,* 7824–7827.

Zeng, Z.; She, Y.; Peng, Z.; Wei, J.; He, X. Enzyme-mediated In Situ Formation of pH-sensitive Nanogels for Proteins Delivery. *RSC Adv.* **2016,** *6,* 8032–8042.

Zhang, L.; Cao, Z.; Li, Y.; Ella-Menye, J.-R.; Bai, T.; Jiang, S. Softer Zwitterionic Nanogels for Longer Circulation and Lower Splenic Accumulation. *ACS Nano.* **2012,** *6,* 6681–6686.

Zhao, C.; Chen, Q.; Patel, K.; Li, L.; Li, X.; Wang, Q.; Zhang, G.; Zheng, J. Synthesis and Characterization of pH-sensitive Poly(N-2-hydroxyethyl Acrylamide)–Acrylic Acid (Poly(HEAA/AA)) Nanogels with Antifouling Protection for Controlled Release. *Soft Matter* **2012,** *8,* 7848–7857.

NANOGELS: PREPARATION, APPLICATIONS, AND CLINICAL CONSIDERATIONS

MARCOS LUCIANO BRUSCHI*,
FERNANDA BELINCANTA BORGHI-PANGONI,
JÉSSICA BASSI DA SILVA, MARIANA VOLPATO JUNQUEIRA,
and SABRINA BARBOSA DE SOUZA FERREIRA

Postgraduate Program in Pharmaceutical Sciences, Laboratory of Research and Development of Drug Delivery Systems, Department of Pharmacy, State University of Maringá, Maringá, Paraná, Brazil

Corresponding author. E-mail: mlbruschi@uem.br; mlbruschi@gmail.com

ABSTRACT

Normally prepared with hydrophilic polymers, the hydrogels have emerged as a versatile drug delivery system. This system can be developed in many physical forms, when confined to a nanoscopic dimensions, the nanogels present good permeation capabilities. They have presented properties, which can overcome the limitations from the conventional hydrogels. Commonly, this recent category of drug delivery systems has been classified based on the cross-linking method as either chemically or physically or (covalently) cross-linked nanogels. Furthermore, the formulations could be prepared without employing energy or adverse conditions. There are many methods of synthesis of nanogels and the most used ones are the preparation using polymer precursors and via heterogeneous polymerization of monomers. Mostly, nanogel systems have demonstrated low toxicity, since their composition is basically water and biocompatible polymers. Moreover, the incorporation of chemical moieties can make the nanogels responsive to some stimulus. Nanogels can be administered by many routes. Recent studies reveal the nanogel as a proposal system for many diseases such as cancer and diabetes treatment. However, they can also be used for diagnostics and imaging. Therefore, in this chapter we have assessed and discussed strategies for preparation, applications, characterization, as well as some obstacles to the clinical use of this novel approach as potentially therapeutic modality.

9.1 INTRODUCTION

In the last two decades, a great deal of interest has emerged in hydrogels as the most versatile platform for targeted drug delivery system and sustained release (Jiang et al., 2014; Soni and Yadav, 2016). Hydrogels are frequently prepared with hydrophilic polymer matrices (Oh et al., 2008), which present a microporous structure containing pore sizes and tunable porosities with dimensions covering from cells, human organs to viruses (Jiang et al., 2014). Recently, there has been a growing interest in nanogels, a hydrogel form confined to a nanoscopic dimensions (Soni and Yadav, 2016). The term nanogels is usually characterized by gel particles less than 100 nm in size (Kabanov and Vinogradov, 2009); nanogels typically range in size of 20–200 nm in diameter (Sultana et al., 2013). Because of these small-size characteristics, it is effective in avoiding the rapid renal exclusion, present good permeation capabilities, and can cross the blood brain barrier (Sultana et al., 2013).

The term "nanogels" is defined by nanosized particles shaped by cross-linked polymer, based on chemical or physical networks, which swell in an appropriate solvent (Sultana et al., 2013). Nanogels have three-dimensional networks that are hydrophilic and capable of retaining a high content of water (Gota et al., 2009).

Nanogels and nanosized drug delivery systems have many advantages especially the controlled release of drugs. Nanosized delivery systems can be useful in systems that require achieving finer temporal control due to their large surface area, also systems that need a burst release or sustained protein release. Furthermore, nanogels can be designed to facilitate the encapsulation of many classes of bioactive compounds (Chacko et al., 2012) and respond to intracellular signals (Yu and Ding, 2008). On the other hand, nanogels have some disadvantages such as monomer or surfactant traces may remain that can impart toxicity and hence, need expensive technique to remove completely the surfactants and solvents at the end of the whole preparation process (Sultana et al., 2013).

The formulations could be prepared without employing energy or adverse conditions such as sonication or homogenization (Sultana et al., 2013).

Hydrogels can be prepared with any water-soluble polymer, using a wide range of chemical compositions. Furthermore, hydrogels can be developed in many physical forms, including coatings, slabs, films, microparticles, and nanoparticles (Hoare and Kolane, 2008). Traditionally, nanogels have been placed and classified based on the cross-linking method as either chemically or physically or (covalently) cross-linked nanogels (Soni et al., 2016). There are many methods of synthesis of nanogels, the most used ones are preparation from polymer precursors and via heterogeneous polymerization of monomers (Chacko et al. 2012).

These systems are considered excellent drug delivery systems because particle properties can be prepared to prevent the fast clearance by phagocytic cells and allow the drug targeting. Furthermore, they can provide controlled and sustained drug release at the exact target site, with ability to reach the slimmest capillary vessel (Sultana et al., 2013).

The main properties of nanogel are biocompatibility and degradability, besides this it can presents swelling property in aqueous media, higher drug-loading capacity and nonimmunologic response among others. Nanocomposite hydrogel presented exceptional swelling/deswelling, optical, and mechanical properties, which can overcome the limitations from the conventional hydrogels that are chemically cross-linked (Okada and Usuki, 2006; Sultana et al., 2013). A special property of nanogels is that they are able to solubilize hydrophobic drugs. Furthermore, nanogels can be administered by many routes such as oral, pulmonary, nasal, topical, parenteral, and intraocular (Sultana et al., 2013).

These nanostructured gels can be designed to a wide range of applications, as potential nanomedicine carriers (Soni and Yadav, 2016), but also as diagnostic agents carrier in their core (Sultana et al., 2013). Nanosized gels have found usage in several fields with a great number of applications (Chacko et al., 2012; Tahara and Akiyoshi, 2015), in pharmaceutical, biomedical engineering, and biomaterials science, with enormous impact in drug delivery area (Oh et al., 2008). Nanomaterials can be used in many fields, especially in biomaterials science because of their biocompatibility, high water content, good mechanical properties, and tunable chemical and three-dimensional (3D) physical structure (Oh et al., 2008). Hydrogels hybridized with nanosized materials have been developed and studied using polyaniline nanosticks (Meng et al., 2007), carbon nanotubes (Wang and Chen, 2007), and inorganic clays (Haraguchi et al., 2006). After being injected, nanomaterials can circulate in the body, and they have the ability to target diseases at the site of disorder. Nanogels are well known as one of the most effective immunological drug carriers (Tahara and Akiyoshi, 2015). Nanomaterials are especially useful in cancer therapy, where the nanosized delivery systems target cancers by the increased effect of retention and permeability (Maeda et al., 2000).

9.2 STRATEGIES FOR PREPARATION OF NANOGELS

Present and future of nanogel applications require a high degree of control over diverse properties such as stability, bioconjugation, dimension, and biodegradation. The stability displays a huge importance, since it can ensure prolonged circulation in the blood stream without premature release with possible adverse side effects. Moreover, their particle size should be lower than 200 nm in diameter to facilitate the cellular uptake of them and decrease nanoparticle uptake by mononuclear phagocyte system, increasing their circulation time in blood. In addition, these nanogels should demonstrate biodegradability in order to modulate the release of drugs and remove the empty device after drug release (Oh et al., 2008). Despite, there are a large number of strategies to produce nanogels, these methods should be well chosen in order to guarantee these requirements. For example, to control the particle size it is necessary to optimize the polymer concentration, as well as, the environmental parameters, such as temperature, pH, and ionic strength. Thus, the advances in polymer chemistry lead to high diversity and control over their composition, architecture, and functionality, then, providing the development of nanogels with optimized characteristics (Soni et al., 2016).

9.2.1 PHOTOLITHOGRAPHIC TECHNIQUES

This method is useful to prepare tridimensional microgel rings for drug delivery (Oh et al., 2008) (Fig. 9.1). Photolithographic technique is based on the release of molded gels from stamps or replica molds, previously developed. Poly(dimethylsiloxane) (PDMS) stamps are used to mold, release, and stack the gel into tridimensional structures (Oh et al., 2008; Sultana et al., 2013).

A top-down method named "particle replication in nonwetting templates" (PRINT) was developed to produce PEG-based gels with submicron size and controlled particle size, shape, and composition. This method can be used to incorporate DNA, proteins and low-molecular-weight drugs. In this case, the molding material used a photocurable perfluoropolyether (PFPE) replica. Moreover, PRINT have been considered as GMP-compliant (good manufacturing practice) platform for particle fabrication on a large scale (Oh et al., 2008; Sultana et al., 2013).

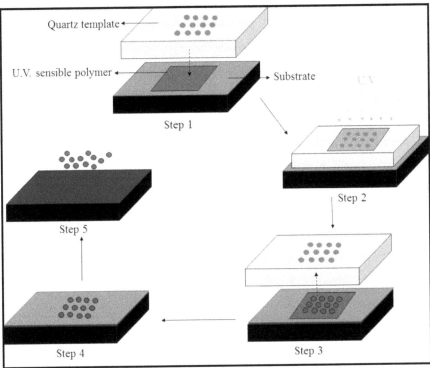

FIGURE 9.1   Schematic illustration of five steps in the photolithography nanogel synthesis.

9.2.2 MICROMOLDING TECHNIQUE

This technique is considered as a potential method to fabricate submicron-sized microgels. It is very similar to lithographic method; however, it does not use the lithographic equipment and cleanroom facilities. The features of the mold stamp can be controlled and consequently, their size and shape are precise. In this process, cells were suspended in a hydrogel precursor solution composed by methacrylated hyaluronic acid (MeHA). The resulting solution was deposited onto plasma-cleaned hydrophilic PDMS patterns and photo-cross-linked via exposure to UV light. The resulting cell-laden microgels are removed, hydrated, and then harvested. They can be molded into many shapes such as square, prisms, disks, and strings (Oh et al., 2008; Sultana et al., 2013).

9.2.3 MICROFLUIDIC PREPARATION

For this technique, it is necessary the fabrication of microfluidic devices by soft lithography using elastomeric materials as building blocks (Fig. 9.2). These devices are comprised of inlets for monomers and continuous liquids, and microchannel with a tapered junction where two immiscible phases are mixed. Consequently, these monomers are emulsified by breaking liquid into droplets, which are *in situ* cross-linked by photopolymerization. The achievement of monodisperse particles with wide range of shapes and morphologies can be optimized by the variation of flow rates of liquids and precise control of reaction time (Oh et al., 2008).

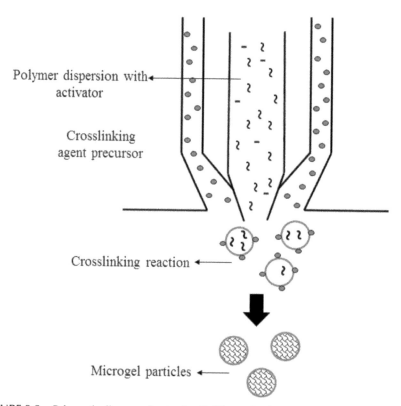

FIGURE 9.2   Schematic diagram about microfluidic production of microgel particles.

Diverse approaches have been used to obtain nanogels by microfluidic method. The gelation methods in microchannel are chemical gelation, physical gelation by temperature change, reversible shear thinning, and ionic cross-linking and coalescence-induced gelation. Chemical gelation, mostly used for synthetic polymers, uses photoinduced polymerization to cross-link low viscosity monomer droplets dispersed in a continuous phase with surfactants. On the other hand, the physical gelation by ionic cross-linking can be classified in internal and external cross-linking and this method was employed to prepare microgels of alginate cross-linked with calcium ions (Ca^+). Thus, the external gelation occurs when the aqueous sodium alginate solution is emulsified in soybean oil composed by sodium acetate, while, the internal resulted in the preparation of stable monodisperse spherical alginate microgels (Oh et al., 2008).

9.2.4 FABRICATION OF BIOPOLYMERS

Natural polymers such as, chitosan (CS), hyaluronic acid (HA), and dextran are carbohydrate-based materials with interesting properties that make them useful for diverse applications. Among these properties, they are known to be nontoxic, highly water soluble, biocompatible, and biodegradable. Moreover, these materials have demonstrated high concentration of functional groups such as, amino groups in CS and carboxylic acid groups in HA, which can be cross-linked with additional functional cross-links, resulting in nanogels. These functional groups can be used with bioconjugation for drug targeting. There are diverse methods developed for preparation of nanogels containing biopolymers: water-in-oil (W/O) heterogeneous emulsion, aqueous homogeneous gelation, spray-drying method, and chemical cross-linking of dextran (Oh et al., 2008; Sultana et al., 2013).

9.2.4.1 WATER-IN-OIL HETEROGENEOUS EMULSION METHODS

These methods involve two steps: the emulsification of aqueous droplets in continuous oil phase using oil-soluble surfactants, followed by cross-linking of biopolymers with hydrophilic cross-links. These techniques display some variations and another classification including inverse (mini)emulsion (also called as emulsion cross-linking technique), reverse micelle and membrane emulsification (Oh et al., 2008; Sultana et al., 2013).

9.2.4.1.1 Inverse (Mini) Emulsion Method

The W/O emulsion is obtained from a continuous hydrophobic phase and aqueous biopolymer droplets, using a homogenizer or a high-speed mechanical stirrer. Subsequently, these droplets are cross-linked, purified, and dried by lyophilization process, where the size of these nanogels can be controlled by the content of surfactants and cross-linking agents as well as the speed of agitation. The organic solvents commonly used are mineral oil and hexane, while the oil-soluble surfactant is Aerosol-OT (sodium bis(2-ethylhaxyl) sulfosuccinate) that provides colloidal stability for these formulations. Some CS-based nanogels were formed with the incorporation of many drugs, such as 5-fluorouracil, phenobarbitone, theophylline, pentazocine, and progesterone (Oh et al., 2008; Sultana et al., 2013).

9.2.4.1.2 Reverse Micellar Method

This technique is also based in W/O dispersion with the difference that a large number of oil-soluble surfactants are used to form a thermodynamically stable micellar solution comprising aqueous droplets dispersed in the continuous oil phase. The size of these micellar droplets range from 10 to 100 nm in diameter and it has been applied to anticancer drugs (Oh et al., 2008; Sultana et al., 2013).

9.2.4.1.3 Membrane Emulsification

This method provides spherical particles with highly uniform size distribution, since it involves the use of a membrane with uniform pore size ranging from 0.1 to 18 μm. In this sense, a liquid pass under an adequate pressure. The droplets resulted in W/O and O/W emulsions, multiple emulsions such as O/W/O and solid/O/W dispersions, and the particles display diverse shapes such as microspheres, hollow spheres, core-shell microcapsules, and organic–inorganic hybrid materials. This technique can be combined to a cross-linking method to form particles with uniform size distribution. Thus, their size is controlled by the membrane pore size, velocity of the continuous phase, and pressure of the transmembrane (Oh et al., 2008; Sultana et al., 2013).

9.2.4.2   AQUEOUS HOMOGENEOUS GELATION METHOD

Biopolymer-based nanoparticles in water are prepared by covalent chemical cross-linking. This strategy has been used mainly for CS-based nanogels in the treatment of cancer with a higher uniform nanogel size. Furthermore, this method is advantageous, since it avoids a possible toxicity of cross-links and reagents of chemical cross-linking method (Oh et al., 2008).

9.2.4.3   SPRAY-DRYING METHOD

This technique is based on the use of a spray dryer, composed of atomizer and drying chamber, where solutions and suspensions of drugs and polymers are atomized into fine droplets. Then, the solvent from the droplets is quickly evaporated by a stream of hot air in the drying chamber, resulting in the formation of microspheres or microgels. In this sense, the size of nanogels is determined by nozzle size, spray flow rate, atomization speed, and rate of cross-linking. The most common biopolymers used by this technique are CS and HA, that have been employed to carry various drugs such as cimetidine, sodium diclofenac, vitamin D, ampicillin, and betamethasone disodium phosphate (Oh et al., 2008).

9.2.4.4   CHEMICAL CROSS-LINKING OF DEXTRAN

There are diverse methods related to the preparation of biodegradable dextran (DEX)-based nanogels, such as carbodiimide coupling, Michael addition reaction, and free radical polymerization (Oh et al., 2008; Sultana et al., 2013).

9.2.4.4.1   Michael Addition Reaction

In this technique, hydroxyl groups of DEX are modified with either thiols or acrylates. This reaction has been also used in the preparation of HA-based hydrogels (Oh et al., 2008).

9.2.4.4.2 Free Radical Polymerization

Many methacrylated dextrans can be synthesized using hydroxyl groups of DEX that react with different methacrylated precursors, such as methacrylated DEX derivatives of glycidyl methacrylate, 2-hydroxyethyl methacrylate (DEX-HEMA), and hydroxyl-terminated HEMA-lactate (DEX-LA-HEMA) (Oh et al., 2008). The preparation of well-defined synthetic microgels occurs by many heterogeneous polymerization reactions of hydrophilic or water-soluble monomers in presence of either dysfunctional or multifunctional cross-links. These reactions include precipitation, inverse miniemulsion, inverse microemulsion, and dispersion polymerization with an uncontrolled free radical polymerization process (Oh et al., 2008; Sultana et al., 2013).

9.2.4.5   DISPERSION AND POLYMERIZATION

It allows the preparation of micron-sized particles with narrow size distribution containing monomers, polymeric stabilizers, and initiators soluble in organic solvent as continuous phase. In the beginning of this process, the polymerization occurs in a homogeneous reaction mixture, then, the resultant polymers become insoluble in the continuous medium, making a stable dispersion of polymeric particles with colloidal stabilizers. Drugs and magnetic nanoparticles have been applied to these systems by physical or chemical incorporation (Oh et al., 2008; Sultana et al., 2013).

9.2.4.6   PRECIPITATION AND POLYMERIZATION

This method also involves the formation of a homogeneous mixture with initiation and polymerization in a homogeneous solution. To provide the isolation of particles, it is required to use the cross-linker. Consequently, the cross-linked nanogels have irregular shape with high polydispersity (Oh et al., 2008; Sultana et al., 2013).

9.2.4.7   INVERSE (MINI) EMULSION POLYMERIZATION

It is a W/O polymerization process which is composed of aqueous droplets stably dispersed with hydrophobic surfactant in a continuous organic medium. By the inverse emulsion process, the stable dispersions are stirred by the mechanical stirrer, at the same time the sonication process is used for inverse miniemulsion polymerization (Oh et al., 2008; Sultana et al., 2013).

9.2.4.8   INVERSE MICROEMULSION POLYMERIZATION

This technique produces thermodynamically stable systems with the addition of emulsifier above the critical micellar concentration (CMC), where aqueous droplets are dispersed with the help of a large content of oil-soluble surfactants. This process has yielded systems useful for gene delivery, as well as, for photodynamic therapy that could kill rat glioma cells (Oh et al., 2008; Sultana et al., 2013).

9.2.5   HETEROGENEOUS CONTROLLED/LIVING RADICAL POLYMERIZATION

It is considered as a versatile tool for the preparation of polymer-protein/peptide bioconjugates with controlled molecular weight, narrow molecular weight distribution, designed architecture, and many end-functionalities. Diverse techniques have been reported, such as atom transfer radical polymerization (ATRP), stable free radical polymerization (SFRP), and reversible addition fragmentation chain transfer (RAFT) polymerization. ATRP have been used to incorporate peptide sequences, biotin, and streptavidin (Oh et al., 2008; Sultana et al., 2013).

Moreover, these methods provide biomaterials with diverse applications and many advantages (Oh et al., 2008; Sultana et al., 2013):

- Resulting particles preserve high degree of halide end-functionality to enable chain extension to form functional block copolymers;
- These systems enable the incorporation of high concentration of drugs after polymerization;
- Degradable in a reducing environment to individual polymeric chains;
- Their properties such as, swelling ratio, degradation behavior, and colloidal stability of nanogels prepared by ATRP are superior to other methods such as miniemulsion polymerization;
- Nanogels with enhanced circulation time in the blood, due to groups that prevent nanoparticle uptake by reticuloendothelial system (RES).

9.2.6 CONVERSION OF MACROSCOPIC GELS TO NANOGELS

Hydrogels are easy to prepare, since it is not necessary to control the synthetic parameters to control size. Thus, these gels are formed by bulk polymerization with solid and the network structure in macroporous blocks, which are crushed, ground, and sieved to form gels in the desired particle size. However, this process displays some disadvantages such as it is time and energy consuming and there is significant loss of material as well as particles with different shapes and sizes (Sultana et al., 2013).

9.3 POLYMERIC PRECURSORS OF NANOGELS

Nanogels can be prepared from polymer precursors. Polymers are macromolecules formed from smaller structural units (the monomers). This led to the recognition of these polymers as a promising material for drug delivery (Hamidi et al., 2008). Papers have been recently published developing controlled-release nanogels as a delivery system using different polymeric compositions (Table 9.1).

TABLE 9.1 Composition of Polymeric Nanogels from Recent Publications.

Polymeric nanogels composites	References
Alginate	Ruel-Gariépy and Leroux (2004)
Carboxymethyl chitosan–linoleic acid	Asadi and Khoee (2016)
CS	Ruel-Gariépy and Leroux (2004)
CS and CMCS	Fan and Wang (2016)
Cholesterol-modified polysaccharides (e.g., dextran, mannan, pullulan, amylopectin, and cholesteryl pullulan (Ls-CHP)	Feng et al. (2016)
Copolymer composed by OEG and PDS	Gessner et al. (2000)
Dextran	Giulbudagian et al. (2016)
Gelatin	Gota et al. (2009)
Hydroxypropyl cellulose–poly(acrylic acid)	Asadi and Khoee (2016)
HPMC	Hamidi et al. (2008)
PEG	Abd El-Rehim et al. (2013)
PEG–PEI	Haraguchi (2007)
Pullulan	Haraguchi et al. (2006); Hoare and Kohane (2008)

CS, chitosan; CMCS, carboxymethyl chitosan; HPMC, hydroxypropyl methylcellulose; PDS, pyridyl disulfide; PEG, polyethylene glycol; PEG–PEI, polyethylene glycol and polyethylenimine; OEG, oligoethyleneglycol.

PEGylated nanogels are nanoparticles obtained by the modification of a particle surface by covalently grafting adsorbing or entrapping polyethylene glycol (PEG). PEGylated nanogels improve the circulation time making the drug available for longer time. Furthermore, they can deliver their drug load into tumor cells, following intravenous injection (Soni et al., 2016).

Chitosan (α(1-4)-2-amino-2-deoxy β-D-glucan) is a deacetylated form of chitin, it is a polysaccharide that is generously present in crustacean shells (Hamidi et al., 2008). Chitosan presents a characteristic of adhering to the mucosal surfaces, caused by interactive forces, which results in sol–gel transition stages. This characteristic has been exploited in mucosal drug delivery (Shutava and Lvov, 2006).

The bioavailability of some drugs encapsulated in alginate can be significantly higher than those with free drugs (Hamidi et al., 2008). Alginic acid is an anionic biopolymer consisting of linear chains of α-L-glucuronic acid and β-D-mannuronic acid. The alginate polymer presents huge properties, such as a wide degree of aqueous solubility, with gelation tendency in appropriate conditions. Besides, it has biocompatibility, nontoxicity, and a great porosity of the resulting gels (Tønnesen and Karlsen, 2002). Many drugs were encapsulated in alginate against many diseases as (Hamidi et al., 2008), for example, alginate-encapsulated azol antifungal and antitubercular drugs for murine tuberculosis (Ahmad et al., 2007).

Some studies have demonstrated the formation of nanogel using polymer amphiphilic in an aqueous medium such as pullulan, hydrophobized polysaccharides, and hydrophobized pullulan. Investigation was performed using cholesterol-bearing pullulan nanogels prepared by physical cross-linking (Fig. 9.3). The self-assembly process is denominated as the autonomous organization of components into aggregates that are structurally defined (Akiyoshi et al., 1999; Akiyoshi et al., 2000; Sultana et al., 2013).

Nanogel drug delivery formulations improve the safety and effectiveness of certain anticancer drugs. Nanogel technology ensures that cancer treatment presents targeted delivery of drugs with expected high therapeutic efficacy with low toxicities to surrounding tissues. Some nanogel examples are polyethyleneimine nanogels with size-dependent property nanogel based on suicide gene hTERT–CD-TK delivered for lung cancer. Another example is polyacrylamide with novel core–shell magnetic nanogel for radiopharmaceutical carrier for cancer radiotherapy (Sultana et al., 2013).

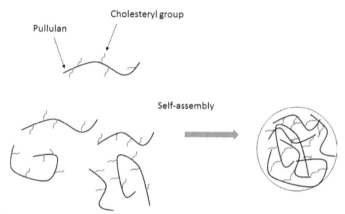

FIGURE 9.3 Schematic representation of nanogel preparation by physical cross-linking (self-assembly).

Polymeric-based nanogels can be used in ophthalmic purposes. Nanogels prepared by pH-sensitive polyvinyl pyrrolidone-poly(acrylic acid) (PVP/PAAc) using γ-radiation-induced polymerization of acrylic acid (AAc) in an aqueous solution of polyvinyl pyrrolidone (PVP) were used to encapsulate pilocarpine. This system was developed to maintain for prolonged period of time in an appropriate pilocarpine concentration at the site of action (Abd El-Rehim et al., 2013).

Polymeric nanogels have also been studied for diabetes. An injectable nanogel, produced with a mixture of oppositely charged nanoparticles to respond to glucose and releases developed to make the nanogel respond to increased acidity dextran. In this system a modified polysaccharide was used, which the nanoparticle in the gel holds spheres of dextran loaded with an enzyme that converts glucose into glucuronic acid and insulin. Then, glucose molecules are able to enter and diffuse through the gel, readily (Sultana et al., 2005).

Poly-(lactide-co-glycolic acid) and chitosan were studied to prepare bilayer nanoparticles for anti-inflammatory purposes. The surface of the nanoparticles was modified with oleic acid. Hydroxypropyl methylcellulose (HPMC) and Carbopol were used to prepare the nanogels in order to achieve the desired viscosity (Shah et al., 2012).

In this way, nanogels can be prepared using different types of polymers for numerous applications in the health area.

9.4   LOADING NANOGELS WITH ACTIVE AGENTS

An excellent carrier should demonstrate a high drug-loading capacity, while reducing the required amount of carrier. There are three incorporation methods to load a drug into a nanogel, such as physical entrapment, covalent conjugation, and self-assembly (Sultana et al., 2013).

The physical entrapment has been employed to incorporate proteins in cholesterol-modified pullulan and siRNA in hyaluronic acid nanogels (Akiyoshi et al., 2000; Lee et al., 2009). Moreover, hydrophobic molecules can be loaded in nonpolar domains; however, the hydrophobic interaction of the drug molecules with the nanogel has resulted in low drug loading (no more than 10%). As examples of hydrophobic interaction, prostaglandin E2 has been solubilized cholesterol-modified pullulan, while N-hexylcarbamoil-5-fluoroacil was incorporated in cross-linked nanogels of N-isopropylacrylamide (NIPAAm) and N-vinylpirrolidone (VP) copolymers. Moreover, doxorubicin (DOX) have been loaded in nanogels of pluronic F127 (Kabanov and Vinogradov, 2009; Sultana et al., 2013).

Another way to incorporate drugs into nanogels is the covalent attachment during nanogel synthesis or with preformed nanogels. For example, cisplatin has been loaded in PEG-b-PMA nanogels. Furthermore, enzyme molecules were incorporated into nanogels in order to increase their shelf life and thermostability, such as N-succinidomido acrylate polymerized with polyacrylamide nanogels (Kabanov and Vinogradov, 2009; Sultana et al., 2013).

The self-assembly, known as an autonomous organization of components into structurally well-defined aggregates has diverse advantages such as cost-effective, versatile and facile, stable and robust structures. This process is characterized by the diffusion, followed by the binding of molecules through noncovalent interactions, such as electrostatic and/or hydrophobic associations (Kabanov and Vinogradov, 2009; Sultana et al., 2013).

9.5   MECHANISM OF DRUG RELEASE FROM NANOGELS

There are diverse mechanisms that provide drug release from nanogels, this knowledge is very important in order to get modified release and therapeutic efficacy. There are diverse classification types such as, simple diffusion, degradation of the nanogel, a shift in the pH value, displacement by counter ions present in the environment, or transitions induced by an external energy source (Kabanov and Vinogradov, 2009; Sultana et al., 2013). On the other hand, these release mechanisms can be classified into diffusion-controlled, swelling-controlled, and chemical-controlled. In this item, the last approach will be deeper exposed, since the other classification is related to the stimuli involved, which will be demonstrated in item of stimuli-responsive nanogels (Hamidi et al., 2008).

The diffusion is considered as a simple mechanism release, successfully employed in diverse nanomedicines. In this mechanism, molecules of different sizes and characteristics diffuse into/out of nanogel matrix during the loading and storage periods. The drug diffusion out of a nanogel matrix is dependent on the mesh sizes of the matrix of the gel, which is affected by the degree of cross-linking and chemical structure of the monomers (Hamidi et al., 2008), for example, the release of DOX from Pluronic blocks copolymer nanogels. Moreover, this mechanism is very common in stimuli-responsive systems, such as pH, temperature, redox, and magnetic (Kabanov and Vinogradov, 2009; Sultana et al., 2013).

Another mechanism reported in the literature is the swelling-controlled release is faster than nanogel distention; whereas, the chemical-controlled release is due to the chemical reactions between the polymer network and the drug such as polymeric chain cleavage via hydrolytic or enzymatic degradation, or reversible/irreversible reactions. Furthermore, two different mechanisms reported as controlling the rate of drug release are the surface or bulk erosion of nanogels or the binding equilibrium among the drug-binding moieties incorporated in the nanogels (Hamidi et al., 2008).

9.6   STIMULI-RESPONSIVE NANOGELS

The incorporation of chemical moieties into the polymers can make the nanogels responsive to some stimulus (Fig. 9.4). In this sense, these carriers alter their hydrophilicity and/or hydrophobicity in response to specific environment within the body (pH, temperature, redox conditions, or enzyme concentration) or externally applied stimuli (light or magnetic field). These changes are expressed by the swelling or deswelling of the nanogel network, leading to the release of drugs due to solvent penetration into free spaces. Because of their nanosized dimensions, nanogels display a very rapid response to changes in the environmental conditions when compared to hydrogels. Thus, the method of preparation of these formulations have demonstrated a huge importance, since it can control the structure of the materials and consequently, their responsiveness to stimuli (Chacko et al., 2012; Soni and Yadav, 2016; Soni et al., 2016).

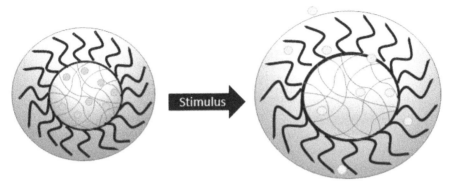

FIGURE 9.4 Stimuli-responsive core-cross-linked nanogel with drug physically trapped and drug release after stimulus exposure.

These stimuli can be classified into two types: chemical or physical. A chemical response is required to cleave the linking bond between carrier and drug, when drugs are covalently bound to the nanogels, such as an enzymatic reaction, a reaction with biological chemicals, changes in pH, redox, or light. On the other hand, drug encapsulated through physical or supramolecular interactions can release the drug by chemical response or physical changes in the conformation, such as temperature or pH (Chacko et al., 2012).

9.6.1 pH-RESPONSIVE NANOGEL SYSTEMS

Changes in pH values due to physiological conditions or diseases such as cancer can be used as a trigger to the delivery of drugs from nanogels. The mechanism involved in this process is related to the increase of the osmotic pressure inside the carrier and consequently, swelling of the nanogel (Soni et al., 2016). This strategy has been broadly used in cancer mainly with DOX, since the differences in pH between healthy tissue and tumor tissue can be employed as stimuli for the triggered drug release (Oerlemans et al., 2010). It can be explained by the high rate of aerobic and anaerobic glycolysis in cancer cells, that decrease the pH of tumor tissue to 6.8 and the endosomal/lysosomal compartments of tumor cells (pH 4.0–6.5), while the healthy tissues display a pH value of 7.4 (Chen et al., 2016; Soni and Yadav, 2016).

A wide range of polymers has been used to develop this strategy in cancer treatment. Among them, the acetal bond has been broadly used to form pH-sensitive carriers and networks for intracellular drug delivery. These groups are rapidly hydrolyzed at mildly acidic pH from tumor cells and relatively stable under physiological conditions. Chen et al. (2016) developed pH-degradable nanogels containing acetal-linked polyvinyl alcohol (PVA) for encapsulation of paclitaxel (PTX) and intracellular release. Consequently, these nanogels could degrade fast at a mildly acidic pH, while the degradation products are PVA, poly(hydroxyethyl acrylate) (PHEA), and acetaldehyde without any toxic byproducts (Chen et al., 2016).

Cuggino et al. (2016) developed different nanogels based on poly(N-isopropylacrylamide) (poly(NIPA)) and poly(acrylic acid) (poly(AAc)) as carriers to the delivery of DOX hydrochloride (DOX-HCl). These nanogels have demonstrated low cytotoxicity, excellent cellular penetration properties, and triggered sustained release of DOX-HCl at lysosomal conditions. Additionally, these nanogels exhibited drug-loading capacity and efficiency because of ionic interaction (Cuggino et al., 2016).

Another pH-responsive drug-carrying system composed by glycol chitosan (GCS) that recognizes tumor pH and release DOX was developed by Oh et al. (Oh et al., 2010). The pH sensitivity property is due to 3-diethylaminopropyl isothiocyanate (DEAP), which was grafted in GCS. It was found that the release of DOX was enhanced at pH 6.8 when compared to physiological pH, thus such nanogels would maximize the therapeutic activity of the drug for the treatment. The authors suggested that DOX-loaded GCS-g-DEAP could target DOX in cancer-associated acidic pH (6.8) and also be used for triggering release at endosomal pH (6.0) (Oh et al., 2010).

Additionally, DOX was also incorporated in nanogels composed of hydrophobized pullulan (PUL)-N_α-Boc-L-histidine (bHis) conjugates. As well as the other nanogels, DOX showed increased release in acidic pH of tumors. As well as the other nanogels, DOX showed increased release in acidic pH of tumors (Na et al., 2007).

DOX was loaded in chitin nanogels and displayed higher DOX release in acidic pH when compared to neutral pH. After 1 h, 32% the drug was released, that was similar in acidic and neutral pH. However, after 24 h, 60% was released in acidic condition, while only 40% was released in the physiological condition. This is due to the higher swelling of nanogel in acidic pH (Jayakumar et al., 2012).

Another chitosan-based nanogel was developed for pH-responsive strategy. This nanogel is composed of chitosan-g-poly(N-isopropylacrylamide) (CS-g-PNIPAm) for the delivery of oridonin (ORI) and displayed a pH-triggered fast drug release under acidic condition. Furthermore, ORI-loaded nanogels demonstrated increased cellular toxicity when compared to

ORI solution at the same pH. At acidic conditions (pH 6.5), the anticancer cytotoxicity of the carriers against HepG2 cells was significantly increased than at pH 7.4 (Duan et al., 2011).

9.6.2 TEMPERATURE-RESPONSIVE NANOGEL SYSTEMS

As well as the hydrogels, the nanogels can be formulated with thermoresponsive polymers. They exhibit a drastic and discontinuous change of their physical properties with temperature, especially near body temperature. This property is related to a phenomenon called sol–gel transition, that is, the temperature at which it occurs the transition from a solution to a gel (Klouda and Mikos, 2008). Nanogel allows a control of the release of the drugs, as well as the incorporation of drugs with poor water-solubility.

The abrupt change of solubility is governed by the balance between hydrophilic and hydrophobic groups. As a result of this balance, a little change in the temperature in the polymer aqueous solution is capable of creating a new adjustment of the hydrophilic and hydrophobic interactions between the polymer blocks and water molecules (Bajpai et al., 2008; Matanović et al., 2014).

Many polymers display a thermoresponsive phase transition property. These polymers can be divided into two major classes, which are focused on their origin: synthetic materials and naturally occurring polymers (Ruel-Gariépy and Leroux, 2004).

Gelatin can be used for the formulation of transdermal product, both nanogel and nanogel incorporated into the solid film. This polymer allows the permeation of anti-inflammatory drugs better than commercial formulation, for example, diclofenac sodium's permeation was 130 mg/cm^2 per hour to nanogel against 30 mg/cm^2 per hour to commercial topic gel (Carmona-Moran et al., 2016).

Chitosan is another natural polymer that presents thermoresponsive characteristics. Associated with poly(NIPAAm-co-acrylamide), the systems can be applied to delivery of paclitaxel to improve antitumor efficacy in mice bearing HT-29 colon carcinoma tumors after intravenous administration. On the other hand, dextran can be associated with lauryl and cyclodextrin to incorporation of benzophenone, a sunscreen agent, or with denatured lysozyme cores, in this case is also possible the association with antibodies directed to endothelial determinant to local treatment of pulmonary inflammation with dexamethasone. Besides that, nanogel of dextran and lysozyme is commonly used in the formation of in situ silver nanoparticles, displaying a bactericidal property from *Escherichia coli* and bacteriostatic properties from *Staphylococcus aureus* (Soni et al., 2016).

Metals, such as silver and gold, can also be incorporated in poly(2-[N,N-diethylamino]ethyl methacrylate) nanogels, in this case, this system improves the photothermal efficacy in response to the laser-irradiation resulting in selective cytotoxicity in cancer cells, in addition, they can increase cancer cells' radiosensitivity. Polystyrene helps in cell imaging as well as in photo-thermal therapy where their strong absorption in the NIR region is used for photothermal conversion (Maya et al., 2013; Soni et al., 2016).

Polystyrene and PEG improves the efficacy and cytotoxicity of mouse melanoma cells from curcumin. However, to date carcinoma studies are the most common applied small interfering RNAs, micro RNAs, oligonucleotides, and antisense oligodeoxynucleotides. For this purpose, it is important to use cationic polymers, as PEG–PEI nanogel, poly(ethylene glycol) and poly(ethyl ethylene phosphate) nanogel, cross-linking between oligo (L-lactic acid)-poly(ethylene oxide)-poly(propylene oxide)-poly(ethylene oxide)-oligo (L-lactic acid) and poly(ethylene glycol) grafted poly(lysine), and poly(lactide)-g-pullulan (PLP1 and 2) copolymers. These nanogels permit the complexation and stabilization drugs, allow to cross the blood–brain barrier, increase of cytotoxicity and the legality of cells (Maya et al., 2013; Soni et al., 2016).

9.6.3 REDOX-RESPONSIVE NANOGELS

Redox-responsive nanogels have drawn increasing attention because they comprise redox-responsive materials with a high redox potential difference (100–1000 fold) between intracellular space and the extracellular space. This is due to glutathione (GSH)/glutathione disulfide (GSSG), the major redox couple in animal cells, determining the antioxidative capacity. In this way, GSH have demonstrated higher concentration in the intracellular environment (about 2–10 mM) when compared to the extracellular fluids and circulation (about 2–20 µM). Moreover, tumor cells have higher GSH (about four times) concentration than healthy cells. So, GSH can be considered as an ideal internal stimulus for rapid destabilization of nanocarriers inside cells to accomplish efficient intracellular drug release (Maciel et al., 2013; Maya et al., 2013; Miao et al., 2016).

In addition, GSH is able to reduce disulfide bonds by serving as an electron donor. In this manner, disulfide bond is stable in the bloodstream because of the low concentrations of GSG. On the other hand, these disulfide cross-linked nanogels are destabilized and the drug is released after intracellular uptake (Maciel et al., 2013; Park et al., 2017). This strategy has been developed in diverse studies, that will be demonstrated in the next paragraphs (Kim et al., 2016; Lee et al., 2007; Maciel et al., 2013; Oh et al., 2009; Park et al., 2017; Ryu et al., 2010).

Nanogels containing hyaluronic acid (HA) was conjugated to thiol groups to form disulfide linkages and was used as a drug delivery system of small interfering RNA (siRNA). As expected, the release rates of siRNA were regulated by GSH concentrations in intracellular space and it was suggested that this formulation could be used to specific delivery of anticancer drugs (Lee et al., 2007).

Kwon et al. and Ryu et al. developed oligo ethylene glycol (OEG) and pyridyl sulfite (PDS) nanogels to encapsulate DOX or dyes. Varying the OEG and PDS units, the nanogel demonstrated stability and redox release. Besides, these systems displayed efficient cell uptake (Oh et al., 2009; Ryu et al., 2010).

Alginate sodium-based nanogel was developed by Maciel et al. to release DOX. Alginate sodium (AG) was in situ cross-linked with cystamine (Cys) to get disulfide linkages. Cys/AG-DOX demonstrated high drug-encapsulation efficiency, rapid release under conditions mimicking intracellular environment, enhanced DOX internalization for CAL-72 cells (an osteosarcoma cell line), and exhibited higher toxic activity than free DOX (Maciel et al., 2013).

Kim et al. synthesized a nanogel for photodynamic therapy cross-linked with disulfide bonds and the photosensitizer used was chlorin e6 (Ce6). These nanogels were easily encapsulated into HeLa cells and Ce6-demonstrated photoactivity under laser exposure (Kim et al., 2016).

Another disulfide nanogel was developed with poly(amino acid) to release DOX. This system provides a completely disintegrable drug carrier and demonstrated an efficient translocation of DOX to the nucleus of the cancer cells (Park et al., 2017).

9.6.4 MAGNETIC-RESPONSIVE NANOGELS SYSTEMS

The magnetic field-responsive polymers are introduced by implementation of iron oxide nanoparticles, which can be guided to the target site by magnetic field. However, the magnetic guiding of these systems to organ or tumor deeply placed can cause difficulties in this application (Hrubý et al., 2015; James et al., 2014).

Incorporation of magnetic particles into a nanogel is a promising alternative to obtain different biological functions to increase, especially, cancer therapeutics. These systems are applied in the field of magnetic resonance imaging (MRI) as contrast agent, near-infrared or multimodal imaging applications, magnetically targeted photodynamic therapy, visible targeting, and targeted thermosensitive chemotherapy and luminescence (Maya et al., 2013; Yallapu et al., 2011).

Nanogels are used as a drug delivery, so present lower release, besides that was a high magnetic content, high saturation magnetization and super paramagnetic behavior, good storage stability, and can be improve the efficacy therapeutic (Maya et al., 2013).

They are especially useful in chemotherapeutic drug delivery. PVP hydrogel with PVA, carboxymethyl chitosan cross-linked with epichlorohydrin, poly(ethylene glycol) methyl ether methacrylate, and N-2-(aminoethyl) methacrylamide hydrochloride are some of the polymer that allow the incorporation of iron oxide particles. The magnetic material comprises about 10% of structure, and consists of maghemite and magnetite, which displays ferro-ferrimagnetic behavior (Demarchi et al., 2014; Hamidi et al., 2008; Maya et al., 2013).

9.6.5 OTHER STIMULI-RESPONSIVE NANOGEL SYSTEMS

Besides the temperature, pH, redox, and magnetic-responsive nanogels, there are other stimuli less commonly used to release drugs from these carriers, such as light, enzymes, and glucose. The light is another strategy employed in nanogels, Kang et al. developed a core–shell nanogel with near infrared-light (NIR)-responsive. The core was made from an Au–Ag-based nanorods and the shell from a double-stranded oligonucleotide cross-linked polyacrylamide. Consequently, when the carrier is exposed to the NIR light the drug is released (Kang et al., 2011; Soni et al., 2016).

The activities of the bacterial lipases, that are abundant in the microbial flora, have been also used as a stimulus to the delivery of drugs. Xiong et al. constructed a nanogel to release antibiotics composed by a lipase-sensitive poly(ε-caprolactone) (PCL) interlayer between the polyphosphoester core and the PEG shell. Thus, the antibiotics are rapidly released in presence of lipase or lipase-secreting bacteria (Soni et al., 2016; Xiong et al., 2012).

Glucose has been employed as a stimulus to release insulin in amphoteric nanogels composed by poly(N-isopropylacrylamide) functionalized with aminophenylboronic acid (PBA). Consequently, the binding of glucose drives a gel-swelling response and, then, insulin release. These nanogels demonstrated high capacity of insulin uptake and selective release in an "on–off" manner, according to the glucose concentration in the blood (Hoare and Kohane, 2008; Soni et al., 2016).

Multistimuli-responsive nanogels, which can respond to a combination of more than one stimulus, can be more efficient and specific in some targeted therapy for some diseases such as cancer, in comparison to single responsive nanogels (Rijcken et al., 2007; Soni et al., 2016). According to the literature, multistimuli-responsive carriers were most commonly used for cancer and the drugs most commonly incorporated in these nanogels were paclitaxel (PTX) and DOX, as demonstrated in Table 9.2.

TABLE 9.2 Multi-stimuli-responsive Nanogels.

Stimuli	Drug	Polymers	References
Temperature pH Redox	PTX	Monomethyl oligo(ethylene glycol) acrylate; Ortho ester-containing acrylic monomer; Disulfide-containing cross-linker	Park et al. (2017)
Redox/pH	DOX	Oligoethyleneglycol; Pyridyldisulfide	Gessner et al. (2000)
pH/temperature	PTX	Poly(N-isopropylacryl amide co-acrylic acid) block; PCL block	Qiao et al. (2011)
pH/temperature	PTX	mPEG2000-Isopropylideneglycerol	Rachmawati et al. (2012)
pH/temperature	DOX	Poly(N-isopropylacrylamide-co-acrylic acid)/Fe_3O_4	Gota et al. (2009)
pH/magnetic	siRNA	2-vinylpyridine and divinylbenzene	Rijcken et al. (2007)

DOX, doxorubicin; PTX, paclitaxel; PCL, poly(ε-caprolactone); siRNA, small interfering RNA.

A triply responsive nanogels were prepared by miniemulsion radical copolymerization of ortho ester-containing acrylic monomer, 2-(5,5-dimethyl-1,3-dioxan-2-yloxy) ethyl acrylate and monomethyloligo(ethylene glycol) acrylate that was cross-linked by bis(2-acryloyloxyethyl) disulfide. The nanogels presented thermo/pH/redox responsive behavior and were dependent on the cross-linking and composition polymer. These new nanogels were able to encapsulate the hydrophobic drug PTX, had good stability at pH 7.4, and showed potential cytotoxicity to tumor cells (Qiao et al., 2011).

Nanosized polymeric delivery dual responsive systems that encapsulate intracellular DOX and release them in response to a specific intracellular stimulus were studied for cancer therapy (Asadi and Khoee, 2016).

In this way, many stimuli-responsive nanogel systems can be developed and studied, using different applications and purposes.

9.7 CHARACTERIZATION OF NANOGELS

Nanogels are characterized by their size, charge, structure, and cytotoxicity. The successful of the delivery drug mainly depends of the size and surface characteristics among other factors (Sarika et al., 2016a) (Table 9.3).

TABLE 9.3 Characterization of Nanogels.

Polymers	Drug	Characterization assays	References
Algynate aldehyde–gelatin	Curcumin	Size distribution and zeta potential measurement (DLS); scanning electron microscopy; thermal studies; encapsulation efficiency; *in vitro* drug release studies; cytotoxicity studies; cellular uptake studies; FTIR spectra; 1H NMR spectra; drug release; XRD; hemolysis test	Ryu et al. (2012)
poly(amino acid)	Doxorubicin	Cytotoxicity studies; cellular uptake studies	Maya et al. (2013)
Gelatin–dopamine	Doxorubicin	Emission scanning electron; size measurement; cytotoxicity studies; cellular uptake studies	Gota et al. (2009)
Hydroxypropyl methylcellulose	Insulin	Size measurement; transmission electron microscopy; entrapment efficiency; FTIR; drug release	Hamidi et al. (2008)
Ovalbumin–dextran	Curcumin	Size distribution and zeta potential measurement (DLS); FTIR spectra; thermal studies; XRD; transmission electron microscopy	Giulbudagian et al. (2016)

TABLE 9.3 *(Continued)*

Polymers	Drug	Characterization assays	References
Copolymer-izing ornithine methacrylamide	Folic acid	Transmission electron microscopy; size distribution and zeta potential measurement (DLS); cellular uptake studies; cytotoxicity studies	Ryu et al. (2010)
Copolymer that contains OEG and PDS unit	Doxorubicin	Size distribution and zeta potential measurement (DLS); transmission electron microscopy	Gessner et al. (2000)
Chitosan/β-glycerol phosphate	Myricetin	Size distribution and zeta potential measurement (DLS); Rhological properties; cellular uptake studies; determination of swelling ratios; cytotoxicity studies; erosing testing; drug release; Kinetic studies;	Sarika et al. (2016a)
Egg yolk LDL complex	Curcumin	size distribution and zeta potential measurement (DLS); FTIR spectra; scanning electron microscopy; stability on pH; encapsulation efficiency; drug release	Sarika et al. (2016b)
Chitosan	Insulin	Size distribution and zeta potential measurement (DLS); Encapsulation efficiency; FTIR spectra; drug release; cytotoxicity studies; permeation; mucoadhesive capacity	Schmitt et al. (2010)

DLS, dynamic light scattering; FTIR, Fourier–transform infrared; LDL, low-density lipoproteins; NMR, nuclear magnetic resonance; OEG, oligoethyleneglycol; PDS, pyridyldisulfide; XRD, X-ray diffraction.

Frequently, the hydrodynamic diameter of nanogels is measured by DLS assay and a small polydispersity index which means that the size distribution is relatively homogeneous in size (Fan and Wang, 2016). Despite, recent works have demonstrated that different preparations of this system can be the main reason of the large variation in the size and also that some types of nanogels can be transformed into the microgels when exposed to high-temperature condition (e.g., physiological temperature) (Cho et al., 2008; Sarika et al., 2016a; Yao et al., 2016). Moreover, the zeta potential is an index of the physical stability to a dispersion system because in this assay the electrostatic barriers which can prevent the nanoparticles are quantified to from aggregation and agglomeration (Rachmawati et al., 2012; Yao et al., 2016). Considering that, positive surface charge promotes protein adsorption as well as hydrophobic surface, the hydrophilic polymers carrying negative surface charges are demonstrate to be more suitable to the development of nanogel drug delivery systems (Gessner et al., 2000; Ryu et al., 2012; Schmitt et al., 2010).

The microscopy has also been a great method to do physical studies such as the analyses of the nanogel morphology (Cho et al., 2008; Sarika et al., 2016a). The scanning electron microscopy (SEM), for example, has been used to characterize the morphology of several nanogels. Moreover, using microscopy analyses the dispersion degree and the size of the nanoparticle can be observed. However, to measure the size by the SEM can offer a different value (generally smaller) when compare to the DLS method, since here the measurement is performed on the dried particles (Fan and Wang, 2016; Sarika et al., 2016a). By the confocal laser scanning microscopy it is possible to prove the *in vitro* cellular uptake of the nanogel, confirming effectively the distribution of the system into the sick cell (Park et al., 2017; Sarika et al., 2016a). Therefore, it can show if the targeting nanostructure is more distributed within the nucleus, in the cytoplasm, or in the perinuclear region as well (Park et al., 2017).

The cytotoxicity assay has been widely considered one quick and efficient approach to evaluating the biocompatibility of biomaterials (Fan and Wang, 2016). Furthermore, the majority nanogel systems have demonstrated low cytotoxicity, probably because the main composition of them is water and biocompatible polymers.

Thermal studies and X-ray diffraction (XRD) have also been used to the characterization of nanogels loaded with a specific drug in order to check the differences in the thermal stability between loaded and nonloaded nanoparticles, as well as to confirm the encapsulation of the active agent in the nanogel network. In addition, FTIR spectra can show the cross-linking network between the polymers of the nanogel structure and also demonstrate drug encapsulation, in the same way that ^{1}H NMR spectra (Feng et al., 2016; Sarika et al., 2016a).

On the other hand, there is an *in vitro* erosion testing to evaluate the erosion profile of the nanogels into different buffers, which can be related to the denser network structures of the nanogel (Yao et al., 2016). Moreover, the stability of nanogel in different pH conditions has been reported (Zhou et al., 2016). Another complementary test to characterize nanogel profile is the hemolytic test by hemocompatibility studies, which confirm the compatibility of nanogels with blood cells (Sarika et al., 2016a; Yu et al., 2004).

Gels prior to becoming a nanogel system can be evaluated for the rheology. In this way, the transition temperature and changes in the storage (G') and the loss (G") module can be measured allowing to know hydrophobic interactions or any elastic gel network formed (Yao et al., 2016). Moreover, swelling behavior also is a reported assay in the literature and the capability of a nanogel to absorb water. The high swelling degree suggests a great capacity for accessing and retaining water molecules within the polymer network. On the other hand, lower swelling ratio is probably due to weaker hydrogen bonding, since the most important governing forces during this phenomenon are hydrogen bonds between functional groups of water and the nanogel structure (Yao et al., 2016).

9.8 NANOGEL APPLICATIONS

Different kinds of nanogels have been extensively studied for the delivery of active agents as screenings and treatments in many diseases in many different applications areas. The majority of papers reveal the nanogel system as a proposal system for many diseases treatment such as anticancer drugs (Asadi and Khoee, 2016; Bhattarai et al., 2010; Fan and Wang, 2016; Kabanov and Vinogradov, 2009; Liang et al., 2016; Li et al., 2016; Sarika et al., 2016b; Yallapu et al., 2011) and diabetes (Saluma et al., 2013; Zhao et al., 2016; Wang et al., 2016). Furthermore, nanogels can be used to immunotherapy (Saluma et al., 2013; Tahara and Akiyoshi, 2015), antileishmaniasis delivery (Kabanov and Vinogradov 2009), oral delivery systems for lipophilic nutrients (Yao et al., 2016; Zhou et al., 2016), ophthalmic delivery (Bhattarai et al., 2010), fortify functional foods and beverages (Feng and Wang, 2016), skin barrier (Giulbudagian et al., 2016), and various shapes, as film, sheet, uneven sheet, hollow tube, bellows, that can be used to different propose

(Haraguchi, 2007). They, also, can be used in diagnostics and imaging (Soni et al., 2016).

Nanogels were being used to improve the therapeutic efficacy and to reduce unwanted side effects of anticancer drugs, as toxic effects on normal tissue, beside the a very narrow therapeutic window. These systems accumulate in the tumor after administration by enhanced permeability and retention effect. Compared to normal treatment, they presented low toxicities to surrounding tissues and high therapeutic efficacy (Asadi and Khoee, 2016; Saluma et al., 2013; Yallapu et al., 2011).

Another characteristics is the ability to enter cells via specific nucleoside transports, being activated intracellular by formation of nucleoside 5-diphosphate, synthesis of NATP, phosphorylation into nucleoside 5-phosphates by intracellular nucleoside kinases, and conversion of ribonucleotides into deoxyribonucleotides by nucleoside reductases (Kabanov and Vinogradov, 2009) (Fig. 9.5).

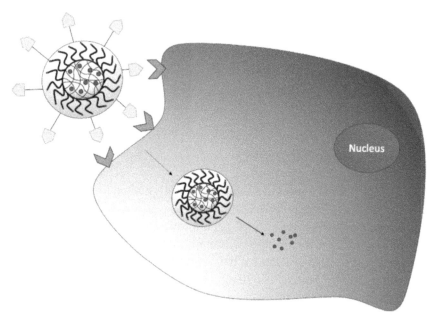

FIGURE 9.5   Cellular uptake of targeting nanogel system with cytoplasm drug release.

DOX can be incorporated into both gelatin nanogel and redox-responsive nanogels with oligoethyleneglycol (OEG) and pyridyl disulfide (PDS) units. Both systems presented excellent cytocompatibility and size distribution in range of 80–115 nm, shows that it can be easily internalized. Besides, gelatin nanogels can be effectively internalized and localized into cells, indicating the ability to overcome cellular barriers and serve as efficient intracellular drug delivery carrier. The redox-responsive system also show an effectively endocytosed by A2780CP ovarian cancer cells (Asadi and Khoee, 2016; Fan and Wang, 2016).

Studies with hyaluronic acid-epigallocatechin gallate conjugates (HA–EGCG) with granzyme B (GzmB) display a significant cytotoxic effect. Intracellular trafficking and caspase assay demonstrated that cell death occurs by apoptosis triggered by the delivery of drug to the cytosol of those cells. The most important point observed was the toxicity targeting against CD44-overexpressing HCT-116 cancer cells but not CD44-deficient cells (Liang et al., 2016).

Other polymers are being tested to be used as carrier. Polysorbate 80 demonstrates the ability to cross the blood–brain barrier, allow the 5-fluo-rouracil (5-FU) accumulation in rabbit brains (Kabanov and Vinogradov, 2009). Chitosan-based nanogel added to glutaraldehyde was tested to breast cancer, resulting in a reduction of the tumoral progression rate. Moreover, the breast cancer metastatic was prevented above 70% of recurrence (Bhattarai et al., 2010). Guam-arabic nanogel improves the bioavailability and therapeutic efficacy of curcumin toward cancer cells (Sarika et al., 2016b).

Other possibility is the incorporation of fluorescent cross-linkable carbon dots (CCD) into zwitterionic multifunctional nanogel drug delivery vehicle. In this type of systems the zwitterionic nanogel allows the addition of drug, while the CCD works as a marker, enable the real-time tracking and locating of the nanogel. These kind of multifunctional nanogels hold a great potential for targeted delivery and simultaneous imaging in cancer therapy (Li et al., 2016).

A controlled and prolonged release system is needed to reduce the injection frequency of insulin and increase the compliance of diabetic patient. In this sense, two polymers are commonly employed on insulin delivery: hydroxypropylmethyl cellulose (HPMC), and carboxymethyl chitosan (CMCS) (Saluma et al., 2013; Wang et al., 2016; Zhao et al., 2016). The HPMC nanogels are pH and temperature responsive so the release can be controlled by these two mechanisms, their encapsulation is around 96% (Zhao et al., 2016). On the other hand, CMCS gels can be charged positively or negatively, this opposite gels show different results to permeation and permeation in different intestinal segments, the negative nanogel was 1.7-folds (permeation) and 3.0-folds (adhesion) higher for rat jejunum.

Besides that, negative CMCS demonstrated a level of glucose in the blood 3 mM/L lower than the positive CMCS, showing more effect on the blood glucose level administration (Wang et al., 2016).

High water content, structural flexibility of nanogels, biocompatibility, and fluid-like transport properties make them ideal carriers for various imaging probes and contrast agents. Multiple functional groups were incorporated at both surface and interior of nanogels permit the incorporation of inorganic nanoparticles, multiple dye and/or reporter molecules. The administration of magnetic nanoparticles, for example, iron oxide, shows a colloidal stability and better sensitivity when compared with the same agents nonencapsulated. Besides that, nanogels are capable to incorporate a large rate of magnetic particles that generate stronger local magnetic fields. In the same context, contrast agents based on manganese and gadolinium are toxic to issues and are rapidly cleared from the body, so the conjugation to nanogels promote a safe administration of these composts, and also exhibit 5.0-fold more efficacies compared to clinically used Magnevist (Soni et al., 2016).

9.9   OBSTACLE TO CLINICAL USE

Nanogels are a recent category of drug delivery systems. Therefore, few studies have shown how these structures behave in the body. Initially, many nanogels were developed to aim in reduction biodegradation and toxicity, or to increase bioavailability, cell-selective accumulation, and encapsulation of drugs.

However, the *in vivo* behavior, after application of these nanocarriers, has displayed a huge gap when compared with *in vitro* properties. This occurs specially by interaction of surface of nanogels with serum components, which may accelerate the clearance of the structure from blood circulation (Vinogradov, 2010). Other factor that influences the deactivation is the capture by macrophages of lungs, liver, spleen, and endothelial reticulum. Besides that, only 5–10% of injected dose reaches the target site. All these issues are related with particle size, shape, charge, composition, and surface properties of nanogels (Soni et al., 2016; Vinogradov, 2010).

The obstacles can be classified into clearance, charge, biomolecule delivery, targeted delivery, degradation, and drug release. However, in general all difficulties are related to delivery of genes (Soni et al., 2016).

9.9.1 CLEARANCE

The spleen performs a primordial role in the clearance of inert, active or strange particles from the blood circulation. In this sense, it was demonstrated that the spleen bed filters particles larger than 150 nm more efficiently, small and rigid structure (<20 nm) are usually excreted via renal filtration, because these can easily pass through interendothelial cell-slits and gain access to the venous circulation (Moghimi et al., 1991; Soni et al., 2016). Size, flexibility, and shape also influence the circulation half-life of particles, so are extremely important and well-defined parameters during the development.

It is essential to also consider the opsonization process, which facilitates phagocytosis, and binding with serum proteins, which allow the uptake by the organs of mononuclear phagocytic system (MPS), such as spleen and liver (Tamura et al., 2011).

For nanogels composed of PEG, for example, the following characteristics need to be considered: an unit dose of these nanoparticles generates PEG-specific IgM antibodies, additionally, the acceleration of blood clearance compromises the efficacy of treatment (Ishihara et al., 2009; Shiraishi et al., 2013).

9.9.2 CHARGE

The occurrence of surface charge can increase the propensity of hemolysis, alter plasma circulation profile, opsonization profile, and recognition by cells in the MPS's organs. The greater the neutrality of these particles the longer is the circulation time. The major part of these systems is stimuli-responsive, and this characteristic is related with presence of charged groups incorporated into network. Consequently, it is difficult to achieve this balance between charge-related responsive behaviors and prevent interactions of nanogels and body components (Soni et al., 2016).

Nanogels of PEI are commonly used to delivery of nucleic acids. As the PEI is a cationic polymer and the DNA is acid, the complexation minimizes the charge-related toxicity of the carrier. However, when the nucleic acids are released intracellularly, the polymer returns to native cationic and toxicity state, forming of vacuoles and disturbance in the cell cycle. How the toxicity is dependent on the PEI molecular weight (MW), low MW–PEI conjugated to the development of carriers, has been used (Lv et al., 2006; Soni et al., 2016).

9.9.3 BIOMOLECULE DELIVERY

The common carriers for the delivery of genes are the viral vectors; although this strategy has a limited load capacity and risk of immune reactions. A good alternative is the nanogels, but they display a low efficiency relative to viral vector caused by degradation of nucleic acids in the endosomes. As already mentioned above, the carriers used are cationic in order to stable complex with negative charge of nucleic acid, protecting the genes from degradation and improve the circulation time. Many of these particles do not remain intact in circulation after the decomplexing, being quickly cleared by the kidney. As well as genes, therapeutic protein can also be incorporated into nanogels; in this case, it is important to remember the activity of proteins are highly dependent on their conformation, being sensitive to several steps into development, so the selection of process and equipment is essential to preserve protein function (Soni et al., 2016).

9.9.4 TARGETED DELIVERY

Nanogels can be used to delivery chemotherapeutics to tumor, but the probability of these systems reaching the core of tumor is limited which, among other factors, is due to the hampered interstitial transport and the high interstitial fluid pressure. Other challenge is the low vascularization, and development of hypoxia and necrosis makes inaccessible the core to the nanogel. Small and well-vascularized tumors can be potential target to these systems. The nonspecificity of conjugation to receptor ligands to only one tissue, and the nonhomogeneity of receptors on the cells, causes a nonuniform accumulation of the nanogels in the tissue. Other factors to be considered are the fact of biomolecules, as antibodies and peptides, are very sensitive to bioconjugation reactions, that can promote a lower reproducibility of nanogel–antibody conjugate between the batches, and the variation of stoichiometry of conjugation that alters the binding affinity of the antibody (Soni et al., 2016).

9.9.5 DEGRADATION

The nanogels have molecular weight above the renal threshold (approximately 40 kDa for copolymers), so they cannot be removed from the body by kidneys; so, if the polymers are nondegradable they would have a tendency to accumulate in the organism. Thus, it is preferred to use biodegradable (synthetic or natural) nanogels, even though chemical reaction in the body can alter the rate of degradation or transform to a nonbiodegradable polymer (Vercauteren et al., 1990).

9.9.6 DRUG RELEASE

Nanogels are complex systems, which can present an environmentally-responsive release. Moreover, they can make the degradation of the biodegradable material that compose them at the target site. In this way, it is difficult to control the rate of both stimuli-responsive drug release and degradation kinetics; thus, the results obtained in the *in vitro* tests may vary with those obtained in *in vivo* experiments. Besides that, a burst release can happen, resulting in substantial drug loss in the circulation upon intravenous administration, and an exposure to toxic drugs in healthy organs (Huang and Brazel, 2001).

9.10 CONCLUSION

Nanogels are nanosized particles designed by cross-linked polymer that swell in an appropriate solvent. In recent times, nanogels have been studied as a category of drug delivery systems. The nanogels have some advantages that make them ideal carriers for many purposes, as they present high water content, biocompatibility, fluid-like transport properties, and structural flexibility. Nanogels have also presented some obstacles mostly related to delivery of genes. These obstacles are usually related to charge, clearance, biomolecule delivery, degradation, drug release, and targeted delivery. Nanogels are normally characterized around their charge, structure, size, and cytotoxicity. Nanogels can be responsive to some stimulus depending on the incorporation of chemical moieties into the polymers. These stimuli-responsive can be triggered by reactions to specific environment within the body (pH, temperature, redox conditions, or enzyme concentration) or by externally stimuli (light or magnetic field). Nanogel applications require an extraordinary degree of control over different properties, such as biodegradation, stability, dimension, and bioconjugation. Different kinds of nanogels have been developed for the delivery of active agents as a screenings and treatments in many diseases in many different applications areas. Nanogels are being used to reduce the undesirable side effects and to improve the therapeutic efficacy of anticancer drugs, and can be moreover used to diagnostic and imaging. In this sense, nanogels are a potentially therapeutic modality that must be studied and can be developed, characterized and applied for many purposes.

KEYWORDS

- **drug delivery**
- **nanotechnology**
- **polymers**
- **hydrogels**
- **dispersion**
- **characterization**

REFERENCES

Abd El-Rehim, H. A.; Swilem, A. E.; Klingner, A.; Hegazy, E. S. A.; Hamed, A. A. Developing the Potential Ophthalmic Applications of Pilocarpine Entrapped into Polyvinylpyrrolidone-poly(acrylic Acid) Nanogel Dispersions Prepared by γ Radiation. *Biomacromolecules* **2013,** *14* (3), 688–698.

Ahmad, Z.; Sharma, S.; Khuller, G. K. Chemotherapeutic Evaluation of Alginate Nanoparticle-encapsulated Azole Antifungal and Antitubercular Drugs Against Murine Tuberculosis. *Nanomed. Nanotechnol. Biol. Med.* **2007,** *3* (3), 239–243.

Akiyoshi, K.; Sasaki, Y.; Sunamoto, J. Molecular Chaperone-like Activity of Hydrogel Nanoparticles of Hydrophobized Pullulan: Thermal Stabilization with Refolding of Carbonic Anhydrase B. *Bioconjug. Chem.* **1999,** *10* (3), 321–324.

Akiyoshi, K.; Kang, E. C.; Kurumada, S.; Sunamoto, J.; Principi, T.; Winnik, F. M. Controlled Association of Amphiphilic Polymers in Water: Thermosensitive Nanoparticles Formed by Self-assembly of Hydrophobically Modified Pullulans and Poly(N-isopropylacrylamides). *Macromolecules* **2000,** *33* (9), 3244–3249.

Asadi, H.; Khoee, S. Dual Responsive Nanogels for Intracellular Doxorubicin Delivery. *Int. J. Pharm.* **2016,** *511* (1), 424–435.

Bajpai, A. K.; Shukla, S. K.; Bhanu, S.; Kankane, S. Responsive Polymers in Controlled Drug Delivery. *Prog. Polym. Sci.* **2008,** *33* (11), 1088–1118.

Bhattarai, N.; Gunn, J.; Zhang, M. Chitosan-based Hydrogels for Controlled, Localized Drug Delivery. *Adv. Drug Deliv. Rev.* **2010,** *62* (1), 83–99.

Carmona-Moran, C. A.; Zavgorodnya, O.; Penman, A. D.; Kharlampieva, E.; Bridges, S. L.; Hergenrother, R. W.; Singh, J. A.; Wick, T. M. Development of Gellan Gum Containing Formulations for Transdermal Drug Delivery: Component Evaluation and Controlled

Drug Release Using Temperature Responsive Nanogels. *Int. J. Pharm.* **2016,** *509* (1–2), 465–476.

Chacko, R. T.; Ventura, J.; Zhuang, J.; Thayumanavan, S. Polymer Nanogels: A Versatile Nanoscopic Drug Delivery Platform. *Adv. Drug Deliv. Rev.* **2012,** *64* (9), 836–851.

Chen, W.; Hou, Y.; Tu, Z.; Gao, L.; Haag, R. pH-degradable PVA-based Nanogels via Photo-crosslinking of Thermo-preinduced Nanoaggregates for Controlled Drug Delivery. *J. Control. Release* **2016.**

Cho, K.; Wang, X.; Nie, S.; Chen, Z.; Shin, D. M. Therapeutic Nanoparticles for Drug Delivery in Cancer. *Clin. Cancer Res.* **2008,** *14* (5), 1310–1316.

Cuggino, J. C.; Molina, M.; Wedepohl, S.; Igarzabal, C. I. A.; Calderón, M.; Gugliotta, L. M. Responsive Nanogels for Application as Smart Carriers in Endocytic pH-triggered Drug Delivery Systems. *Eur. Polym. J.* **2016,** *78,* 14–24.

Demarchi, C. A.; Debrassi, A.; Buzzi, F. D. C.; Corrêa, R.; Filho, V. C.; Rodrigues, C. A.; Nedelko, N.; Demchenko, P.; Ślawska-Waniewska, A.; Dłużewski, P.; Greneche, J.-M. A Magnetic Nanogel Based on O-carboxymethylchitosan for Antitumor Drug Delivery: Synthesis, Characterization and *in vitro* Drug Release. *Soft Matter* **2014,** *10* (19), 3441–3450.

Duan, C.; Zhang, D.; Wang, F.; Zheng, D.; Jia, L.; Feng, F.; Liu, Y.; Wang, Y.; Tian, K.; Wang, F.; Zhang, Q. Chitosan-G-poly(N-isopropylacrylamide) Based Nanogels for Tumor Extra-cellular Targeting. *Int. J. Pharm.* **2011,** *409* (1–2), 252–259.

Fan, C.; Wang, D. A. Novel Gelatin-based Nano-gels with Coordination-induced Drug Loading for Intracellular Delivery. *J. Mater. Sci. Technol.* **2016,** *32* (9), 840–844.

Feng, J.; Wu, S.; Wang, H.; Liu, S. Improved Bioavailability of Curcumin in Ovalbumin-dextran Nanogels Prepared by Maillard Reaction. *J. Funct. Foods* **2016,** *27,* 55–68.

Gessner, A.; Waicz, R.; Lieske, A.; Paulke, B. R.; Mäder, K.; Müller, R. H. Nanoparticles with Decreasing Surface Hydrophobicities: Influence on Plasma Protein Adsorption. *Int. J. Pharm.* **2000,** *196* (2), 245–249.

Giulbudagian, M.; Rancan, F.; Klossek, A.; Yamamoto, K.; Jurisch, J.; Neto, V. C.; Schrade, P.; Bachmann, S.; Rühl, E.; Blume-Peytavi, U.; Vogt, A.; Calderón, M. Correlation Between the Chemical Composition of Thermoresponsive Nanogels and Their Interaction with the Skin Barrier. *J. Control. Release* **2016,** *243,* 323–332.

Gota, C.; Okabe, K.; Funatsu, T.; Harada, Y.; Uchiyama, S. Hydrophilic Fluorescent Nanogel Thermometer for Intracellular Thermometry. *J. Am. Chem. Soc.* **2009,** *131* (8), 2766–2767.

Hamidi, M.; Azadi, A.; Rafiei, P. Hydrogel Nanoparticles in Drug Delivery. *Adv. Drug Deliv. Rev.* **2008,** *60* (15), 1638–1649.

Haraguchi, K. Nanocomposite Hydrogels. *Curr. Opin. Solid State Mater. Sci.* **2007,** *11,* 47–54.

Haraguchi, K.; Takehisa, T.; Ebato, M. Control of Cell Cultivation and Cell Sheet Detachment on the Surface of Polymer/Clay Nanocomposite Hydrogels. *Biomacromolecules* **2006,** *7,* 3267–3275.

Hoare, T. R.; Kohane, D. S. Hydrogels in Drug Delivery: Progress and Challenges. *Polymer (Guildf)* **2008,** *49* (8), 1993–2007.

Hrubý, M.; Filippov, S. K.; Štěpánek, P. Smart Polymers in Drug Delivery Systems on Cross-roads: Which Way Deserves Following? *Eur. Polym. J.* **2015,** *65,* 82–97.

Huang, X.; Brazel, C. S. On the Importance and Mechanisms of Burst Release in Matrix-controlled Drug Delivery Systems. *J. Control. Release* **2001,** *73* (2–3), 121–136.

Ishihara, T.; Takeda, M.; Sakamoto, H.; Kimoto, A.; Kobayashi, C.; Takasaki, N.; Yuki, K.; Tanaka, K. I.; Takenaga, M.; Igarashi, R.; Maeda, T.; Yamakawa, N.; Okamoto, Y.; Otsuka, M.; Ishida, T.; Kiwada, H.; Mizushima, Y.; Mizushima, T. Accelerated Blood Clearance Phenomenon upon Repeated Injection of Peg-modified Pla-nanoparticles. *Pharm. Res.* **2009**, *26* (10), 2270–2279.

James, H. P.; John, R.; Alex, A. Smart Polymers for the Controlled Delivery of Drugs—A Concise Overview. *Acta Pharm. Sin. B* **2014**, *4* (2), 120–127.

Jayakumar, R.; Nair, A.; Rejinold, N. S.; Maya, S.; Nair, S. V. Doxorubicin-loaded pH-Responsive Chitin Nanogels for Drug Delivery to Cancer Cells. *Carbohydr. Polym.* **2012**, *87* (3), 2352–2356.

Jiang, Y.; Chen, J.; Deng, C.; Suuronen, E. J.; Zhong, Z. Click Hydrogels, Microgels and Nanogels: Emerging Platforms for Drug Delivery and Tissue Engineering. *Biomaterials* **2014**, *35* (18), 4969–4985.

Kabanov, A. V.; Vinogradov, S. V. Nanogels as Pharmaceutical Carriers: Finite Networks of Infinite Capabilities. *Angew. Chem. Int. Ed.* **2009**, *48* (30), 5418–5429.

Kang, H.; Trondoli, A. C.; Zhu, G.; Chen, Y.; Chang, Y.; Liu, H.; Huang, Y.; Zhang, X.; Tan, W. Near-infrared Light-responsive Core-shell Nanogels for Targeted Drug Delivery. *ACS Nano* **2011**, *5* (6), 5094–5099.

Kim, H.; Kim, B.; Lee, C.; Ryu, J. L.; Hong, S.; Kim, J.; Ha, E.; Paik, H. Redox-responsive Biodegradable Nanogels for Photodynamic Therapy Using Chlorin e6. *J. Mater. Sci.* **2016**, *51* (18), 8442–8451.

Klouda, L.; Mikos, A. G. Thermoresponsive Hydrogels in Biomedical Applications. *Eur. J. Pharm. Biopharm.* **2008**, *68* (1), 34–45.

Lee, H.; Mok, H.; Lee, S.; Oh, Y.; Gwan, T. Target-specific Intracellular Delivery of siRNA Using Degradable Hyaluronic Acid Nanogels. *J. Control. Release* **2007**, *119*, 245–252.

Lee, Y.; Park, S. Y.; Kim, C.; Park, T. G. Thermally Triggered Intracellular Explosion of Volume Transition Nanogels for Necrotic Cell Death. *J. Control. Release* **2009**, *135* (1), 89–95.

Li, W.; Liu, Q.; Zhang, P.; Liu, L. Zwitterionic Nanogels Crosslinked by Fluorescent Carbon Dots for Targeted Drug Delivery and Simultaneous Bioimaging. *Acta Biomater.* **2016**, *40*, 254–262.

Liang, K.; Ng, S.; Lee, F.; Lim, J.; Chung, J. E.; Lee, S. S.; Kurisawa, M. Targeted Intracellular Protein Delivery Based on Hyaluronic Acid-Green Tea Catechin Nanogels. *Acta Biomater.* **2016**, *33*, 142–152.

Lv, H.; Zhang, S.; Wang, B.; Cui, S.; Yan, J. Toxicity of Cationic Lipids and Cationic Polymers in Gene Delivery. *J. Control. Release* **2006**, *114* (1), 100–109.

Maciel, D.; Figueira, P.; Xiao, S.; Hu, D.; Shi, X.; Rodrigues, J. C.; Tomás, H.; Li, Y. Redox-Responsive Alginate Nanogels with Enhanced Anticancer Cytotoxicity. *Macromolecules* **2013**, *14* (9), 3140–3146.

Maeda, H.; Wu, J.; Sawa, T.; Matsumura, Y.; Hori, K. Tumor Vascular Permeability and the EPR Effect in Macromolecular Therapeutics: A Review. *J. Control. Release* **2000**, *65* (1–2), 271–284.

Matanović, M. R.; Kristl, J.; Grabnar, P. A. Thermoresponsive Polymers: Insights into Decisive Hydrogel Characteristics, Mechanisms of Gelation, and Promising Biomedical Applications. *Int. J. Pharm.* **2014**, *472* (1–2), 262–275.

Maya, S.; Sarmento, B.; Nair, A.; Rejinold, N. S.; Nair, S. V; Jayakumar, R. Smart Stimuli Sensitive Nanogels in Cancer Drug Delivery and Imaging : A Review. *Curr. Pharm. Design* **2013,** *19* (41),7203–7218.

Meng, L.; Lu, Y.; Wang, X.; Zhang, J.; Duan, Y.; Li, C. Facile Synthesis of Straight Polyaniline Nanostick in Hydrogel. *Macromolecules* **2007,** *40,* 2981–2983.

Miao, C.; Li, F.; Zuo, Y.; Wang, R.; Xiong, Y. Novel Redox-Responsive Nanogels Based on Poly(Ionic Liquid)s for the Triggered Loading and Release of Cargos. *RSC Adv.* **2016,** *6,* 3013–3019.

Moghimi, S. M.; Porter, C. J. H.; Muir, I. S.; Illum, L.; Davis, S. S. Non-phagocytic Uptake of Intravenously Injected Microspheres in Rat Spleen: Influence of Particle Size and Hydrophilic Coating. *Biochem. Biophys. Res. Commun.* **1991,** *177* (2), 861–866.

Na, K.; Eun, S. L.; You, H. B. Self-organized Nanogels Responding to Tumor Extracellular pH: pH-dependent Drug Release and *in vitro* Cytotoxicity Against MCF-7 Cells. *Bioconjug. Chem.* **2007,** *18* (5), 1568–1574.

Oerlemans, C.; Bult, W.; Bos, M.; Storm, G.; Nijsen, J. F. W.; Hennink, W. E. Polymeric Micelles in Anticancer Therapy: Targeting, Imaging and Triggered Release. *Pharm. Res.* **2010,** *27* (12), 2569–2589.

Oh, J. K.; Drumright, R.; Siegwart, D. J.; Matyjaszewski, K. The Development of Microgels/Nanogels for Drug Delivery Applications. *Prog. Polym. Sci.* **2008,** *33* (4), 448–477.

Oh, J. K.; Lee, D. I.; Park, J. M. Biopolymer-based Microgels/Nanogels for Drug Delivery Applications. *Prog. Polym. Sci.* **2009,** *34* (12), 1261–1282.

Oh, N. M.; Oh, K. T.; Baik, H. J.; Lee, B. R.; Lee, A. H.; Youn, Y. S.; Lee, E. S. A Self-organized 3-diethylaminopropyl-Bearing Glycol Chitosan Nanogel for Tumor Acidic pH Targeting: *In vitro* Evaluation. *Colloid. Surf. B Biointerfaces* **2010,** *78* (1), 120–126.

Okada, A.; Usuki, A. Twenty Years of Polymer-clay Nanocomposites. *Macromol. Mater. Eng.* **2006,** *291* (12), 1449–1476.

Park, C. W.; Yang, H. M.; Woo, M. A.; Lee, K. S.; Kim, J. D. Completely Disintegrable Redox-responsive Poly(Amino Acid) Nanogels for Intracellular Drug Delivery. *J. Ind. Eng. Chem.* **2017,** *45,* 182–188.

Qiao, Z. Y.; Zhang, R.; Du, F. S.; Liang, D. H.; Li, Z. C. Multi-responsive Nanogels Containing Motifs of Ortho Ester, Oligo(ethylene Glycol) and Disulfide Linkage as Carriers of Hydrophobic Anti-cancer Drugs. *J. Control. Release* **2011,** *152* (1), 57–66.

Rachmawati, H.; Shaal, L. A.; Muller, R. H..; Keck, C. M. Development of Curcumin Nanocrystal: Physical Aspects. *Pharm. Nanotechnol.* **2012,** *1* (2), 204–214.

Rijcken, C. J. F.; Soga, O.; Hennink, W. E.; van Nostrum, C. F. Triggered Destabilisation of Polymeric Micelles and Vesicles by Changing Polymers Polarity: An Attractive Tool for Drug Delivery. *J. Control. Release* **2007,** pp 131–148.

Ruel-Gariépy, E.; Leroux J.-C. In Situ-forming Hydrogels—Review of Temperature-sensitive Systems. *Eur. J. Pharm. Biopharm.* **2004,** *58* (2), 409–426.

Ryu, J-H; Chacko, R. T.; Jiwpanich, S.; Bickerton, S.; Babu, R. P.; Thayumanavan, S. Self-Cross-linked Polymer Nanogels : A Versatile Nanoscopic Drug Delivery Platform. *J. Am. Chem. Soc.* **2010,** *132* (48), 17227–17235.

Ryu, J. H.; Bickerton, S.; Zhuang, J.; Thayumanavan, S. Ligand-decorated Nanogels: Fast One-pot Synthesis and Cellular Targeting. *Biomacromolecules* **2012,** *13* (5), 1515–1522.

Sarika, P. R.; James, N. R.; Anil kumar, P. R.; Raj, D. K. Preparation, Characterization and Biological Evaluation of Curcumin Loaded Alginate Aldehyde–Gelatin Nanogels. *Mater. Sci. Eng. C* **2016a,** *68,* 251–257a.

Sarika, P. R.; Nirmala, R. J. Curcumin Loaded Gum Arabic Aldehyde–Gelatin Nanogels for Breast Cancer Therapy. *Mater. Sci. Eng. C* **2016b,** *65,* 331–337b.

Schmitt, F.; Lagopoulos, L.; Käuper, P.; Rossi, N.; Busso, N.; Barge, J.; Wagnières, G.; Laue, C.; Wandrey, C.; Juillerat-Jeanneret, L. Chitosan-based Nanogels for Selective Delivery of Photosensitizers to Macrophages and Improved Retention in and Therapy of Articular Joints. *J. Control. Release* **2010,** *144* (2), 242–250.

Shah, P. P.; Desai, P. R.; Patel, A. R.; Singh, M. S. Skin Permeating Nanogel for the Cutaneous Co-delivery of Two Anti-inflammatory Drugs. *Biomaterials* **2012,** *33* (5), 1607–1617.

Shiraishi, K.; Hamano, M.; Ma, H.; Kawano, K.; Maitani, Y.; Aoshi, T.; Ishii, K. J.; Yokoyama, M. Hydrophobic Blocks of PEG-conjugates Play a Significant Role in the Accelerated Blood Clearance (ABC) Phenomenon. *J. Control. Release* **2013,** *165* (3), 183–190.

Shutava, T. G.; Lvov, Y. M. Nano-engineered Microcapsules of Tannic Acid and Chitosan for Protein Encapsulation. *J. Nanosci. Nanotechnol.* **2006,** *6* (6), 1655–1661.

Soni, G.; Yadav, K. S. Nanogels as Potential Nanomedicine Carrier for Treatment of Cancer: A Mini Review of the State of the Art. *Saudi Pharm. J.* **2016,** *24* (2), 133–139.

Soni, K. S.; Desale, S. S.; Bronich, T. K. Nanogels: An Overview of Properties, Biomedical Applications and Obstacles to Clinical Translation. *J. Control. Release* **2016,** *240,* 109–126.

Sultana, F.; Arafat, M.; Sharmin, S. An Overview of Nanogel Drug Delivery System. *J. App. Pharm. Sci.* **2013,** *3,* 95–105.

Tahara, Y.; Akiyoshi, K. Current Advances in Self-assembled Nanogel Delivery Systems for Immunotherapy. *Adv. Drug Deliv. Rev.* **2015,** *95,* 65–76.

Tamura, M.; Ichinohe, S.; Tamura, A.; Ikeda, Y.; Nagasaki, Y. *In vitro* and *In vivo* Characteristics of Core-shell Type Nanogel Particles: Optimization of Core Cross-linking Density and Surface Poly(Ethylene Glycol) Density in PEGylated Nanogels. *Acta Biomater.* **2011,** *7* (9), 3354–3361.

Tønnesen, H. H.; Karlsen, J. Alginate in Drug Delivery Systems. *Drug Dev. Ind. Pharm.* **2002,** *28* (6), 621–630.

Vercauteren, R.; Bruneel, D.; Schacht, E.; Duncan, R. Effect of the Chemical Modification of Dextran on the Degradation by Dextranase. *J. Bioact. Compat. Polym.* **1990,** *5* (1), 4–15.

Vinogradov, S. V. Nanogels in the Race for Drug Delivery. *Nanomedicine* **2010,** *5* (2), 165–168.

Wang, Z.; Chen, Y. Supramolecular Hydrogels Hybridized with Single-walled Carbon Nanotubes. *Macromolecules* **2007,** *40* (9), 3402–3407.

Wang, J.; Xu, M.; Cheng, X.; Kong, M.; Liu, Y.; Feng, C.; Chen, X. Positive/Negative Surface Charge of Chitosan Based Nanogels and Its Potential Influence on Oral Insulin Delivery. *Carbohydr. Polym.* **2016,** *136,* 867–874.

Xiong, M. H.; Bao, Y.; Yang, X. Z.; Wang, Y. C.; Sun, B.; Wang, J. Lipase-sensitive Polymeric Triple-layered Nanogel For "On-demand" drug Delivery. *J. Am. Chem. Soc.* **2012,** *134* (9), 4355–4362.

Yallapu, M. M.; Jaggi, M.; Chauhan, S. C. Design and Engineering of Nanogels for Cancer Treatment. *Drug Discov. Today* **2011,** *16* (9–10), 457–463.

Yao, Y.; Xia, M.; Wang, H.; Li, G.; Shen, H.; Ji, G.; Meng, Q.; Xie, Y. Preparation and Evaluation of Chitosan-based Nanogels/Gels for Oral Delivery of Myricetin. *Eur. J. Pharm. Sci.* **2016,** *91,* 144–153.

Yu, L.; Ding, J. Injectable Hydrogels as Unique Biomedical Materials. *Chem. Soc. Rev.* **2008,** *37* (8) 1473–1481.

Yu, S-R.; Zhang, X-P.; He, Z-M.; Liu, Y-H.; Liu, Z- H. Effects of Ce on the Short-term Biocompatibility of Ti-Fe-Mo-Mn-Nb-Zr Alloy for Dental Materials. *J. Mater. Sci. Mater. Med.* **2004,** *15* (6), 687–691.

Zhang, L.; Guo, R.; Yang, M.; Jiang, X.; Liu, B.; Xiong, M. H. et al. Thermo and pH Dual-responsive Nanoparticles for Anti-cancer Drug Delivery. *Adv. Mater.* **2007,** *19,* 2988–2992.

Zhao, D.; Shi, X.; Liu, T.; Lu, X.; Qiu, G.; Shea, K. J. Synthesis of Surfactant-free Hydroxypropyl Methylcellulose Nanogels for Controlled Release of Insulin. *Carbohydr. Polym.* **2016,** *151,* 1006–1011.

Zhou, M.; Hu, Q.; Wang, T.; Xue, J.; Luo, Y. Effects of Different Polysaccharides on the Formation of Egg Yolk LDL Complex Nanogels for Nutrient Delivery. *Carbohydr. Polym.* **2016,** *153,* 336–344.

CHAPTER 10

POLYSACCHARIDE-BASED NANOGELS AS DRUG DELIVERY SYSTEMS

ZAHRA ESKANDARI[1,2], FATEMEH BAHADORI[3], and EBRU TOKSOY ONER[2*]

[1]*Department of Chemistry, Biochemistry Division, Faculty of Sciences and Arts, Yildiz Technical University, Istanbul, Turkey*

[2]*IBSB, Department of Bioengineering, Marmara University, Kadikoy 34722, Istanbul, Turkey*

[3]*Department of Pharmaceutical Biotechnology, Faculty of Pharmacy, Bezmialem Vakif University, Fatih 34093, Istanbul, Turkey*

[*]Corresponding author. E-mail: ebru.toksoy@marmara.edu.tr

ABSTRACT

Nanogels are ideal drug delivery carriers due to their responsiveness to a wide variety of environmental stimuli, flexibility, and versatility; ability for site-specific and time-controlled delivery of bioactive agents; excellent drug loading capacity; and extended circulation times. Hence, nanogels are also emerged in various fields such as biosensing and imaging, tissue engineering, and medical diagnostics. Among nanogels, polysaccharide-based ones have fascinated attention as a vesicle of different pharmaceutical agents. The main objective of this chapter is to offer an overview of various polysaccharide-based nanogels. After a brief introduction on the basic concepts of drug delivery systems, an overview of nanogels is presented and their advantages and classification is discussed. The succeeding parts focus on polysaccharides, polysaccharide-based nanogel drug delivery systems, and their methods of synthesis and applications. Finally, current challenges and future perspectives are discussed.

10.1 INTRODUCTION

Drug delivery systems (DDS) have been introduced to increase the effectiveness of a drug system through increasing the permeability, solubility, and metabolic stability of a drug molecule. DDS are based on interdisciplinary approaches which combine polymer science, bioconjugate chemistry, pharmaceutics, and molecular biology (Devi et al., 2010). The primary goal of DDS is to transport pharmaceutical agents to the systemic circulation by controlling the pharmacodynamics, pharmacokinetics, nonimmunogenicity, nonspecific toxicity, and biorecognition of target site to reach a desired pharmacological effect (Tiwari et al., 2012). The primary advantages of DDS over traditional systems are their tendency to deliver the pharmaceutical agents more selectively to a specific site, elimination of over- or underdosing, increase in patient compliance, prevention of side effects, and more consistent absorption within the cell of interest (Robinson and Mauger, 1991). Nanoparticles (NPs), within a size range from 10 to 1000 nm in diameter, have unique features that make them attractive tools in DDS. These features include their quantum properties, their high surface area, and their ability to carry and adsorb other compounds (Debele et al., 2016; Kozlowska et al., 2009). Due to their relatively large functional surface area, NPs can encapsulate or immobilize high amount of drugs either through covalent or noncovalent interactions (Luo et al., 2014). Hydrogel NPs or nanogels are an important group of nanosized carriers that are typically comprised of highly hydrated, cross-linked hydrophilic polymers (Ferrer et al., 2014). Polysaccharide-based nanogels have received a lot of attention as a vesicle of different pharmaceutical agents due to their unique multifunctional groups as well as their physicochemical properties, biodegradability and biocompatibility (Debele et al., 2016; Joung et al., 2012). The multifunctional groups of the polysaccharide backbone permit facile chemical modification to synthesize NPs with various structures (Saravanakumar et al., 2012). Some polysaccharides have the ability to distinguish specific cell types and make it possible to synthesize targeted DDS through receptor-mediated endocytosis while some others can be modified with other polymers and then used in the DDS (Debele et al., 2016).

The main objective of this chapter is to offer an overview of various polysaccharide-based nanogels as vesicles of pharmaceutical agents. After a brief introduction on the basic concepts of DDS, an overview of nanogels is presented. A brief account on nanogels, their advantages, and classification is discussed. The succeeding parts focus on polysaccharides, polysaccharide-based nanogel DDS, and their methods of synthesis and applications. Finally, current challenges and future perspectives are discussed.

10.2 DRUG DELIVERY SYSTEMS

The introduction of drug in human body may be achieved through several anatomic routes (Zempsky, 1998). To achieve the desired therapeutic effects, the choice of the most suitable administration route is of indisputable importance (Coelho et al., 2010). As a consequence, several factors such as the properties of the drug itself, the disease to be treated, and the desired therapeutic time must be taken into account when administrating a drug. Drugs can be administrated either directly to the target site or by systemic routes (Golan et al., 2011).

When traditional methods of pharmaceutical treatments, including oral administration of liquids and solid pills or injection of active chemical drugs, are applied, repeated administrations are necessary to maintain a certain drug dose in the body (Coelho et al., 2010). Although these treatments are effective, dose peaks at administration times are sometimes unavoidable. Accordingly, the impossibility of controlling the drug level over a long period is considered as one of the main disadvantages of traditional methods of treatments (Coelho et al., 2010). During the last two decades, new strategies and approaches have been developed to meet this issue and to control several parameters that are considered crucial for enhancing the treatment performance. Some of these parameters include the rate, period of time, and targeting of delivery. These attempts led to the emergence of the so called DDS (Jain, 2008). The primary purpose of using a DDS is not only to deliver a biologically active compound in a controlled way but also to maintain the desired level of drug in the body (Coelho et al., 2010). Furthermore, targeted drug delivery is also possible, in which one can direct the drug toward a specific tissue or organ (Bajpai et al., 2008).

Advantages of DDS over traditional formulations include the ability to deliver a drug more selectively to a specific site; more accurate, easier, less frequent dosing; decreased variability in systemic drug concentrations; reductions in toxic metabolites; and absorption that is more consistent with the mechanism of action (Robinson and Mauger, 1991). The capability of drugs to cure and relieve illnesses could be maximized if the drugs could reach the target site in sufficient amount over sufficient time. Developments in DDS are toward this objective and the following subparts will discuss three of the most important issues and processes that are of significance in the development of DDS, namely, mucus as an important biological barrier and the principles of adhesion, and the mechanisms of gelation.

10.2.1 MUCUS AS AN IMPORTANT BIOLOGICAL BARRIER TO DRUG DELIVERY

Mucus is a complex hydrogel which comprises glycoproteins (mucins), lipids, salts, enzymes, deoxyribonucleic acid (DNA), and cellular debris and covers various epithelial surfaces in the human body (Lai et al., 2009). Mucus behaves as a barrier to protect the underlying tissues against the extracellular environment (Sigurdsson et al., 2013). Therefore, mucus can adversely affect the action or absorption of drugs administered through oral, pulmonary, nasal, vaginal, or other routes. Mucus is a barrier to the penetration of foreign molecules but it must, nevertheless, permit the diffusion of some special molecules between the cell and the surface of the mucus layer (Round et al., 2002). The barrier function of mucus depends on the physiochemical characteristics of the drug and the glycoprotein (Sigurdsson et al., 2013). It also depends on the tertiary conformation of the glycoprotein, which in turn, depends on environmental factors at the target site, such as ionic strength, pH, and the existence of other agents (Khanvilkar et al., 2001).

There may be two major mechanisms that stop particles from diffusing through mucus gel. Either they stick to the mucin fibers or mucus components (i.e., interaction filtering) or they will be hindered by the size of the mesh spacing between the mucin fibers (i.e., size filtering) (Lieleg and Ribbeck, 2001; Olmsted et al., 2001). Size filtering will not affect drug molecules in solution since the mesh size is at least 100 times greater than most common drug molecules; however, interaction filtering can affect any drug molecule, particle, or protein, regardless of their size (Sigurdsson et al., 2013).

Some approaches have been introduced, so far, to overcome the mucus barrier. One strategy would be to increase the adhesion to the underlying epithelium. This method would increase the possibility of the carrier to be in close contact with the epithelium for longer periods to unload its payload (Sigurdsson et al., 2013). However, the problem remains because the carrier still needs to cross the mucus layer first. In spite of this, such carriers could be useful by considering the fact that the mucus blanket is not perfect but exhibits larger holes, very thin domains, and voids (Sims and Horne, 1997), which increases the chance for such particles to reach the epithelium. Another approach to boost particle contact with the epithelium is to design the particles that are able to penetrate through the mucus blanket (Sigurdsson et al., 2013). This feature is usually called mucopenetration.

Other, more recent, approaches to solve this problem are approaches and methods that include the application of external force fields, preferably magnetic forces. Here the magnetic forces could not only direct inhaled aerosols to a targeted region of the lungs (Dames et al., 2007) but could also be useful in terms of overcoming the mucus layer on mucosal tissues (Sanders et al., 2009). This approach is especially preferable when combined with previous strategies (Sigurdsson et al., 2013). Particles with mucolytic surface and magnetic core were demonstrated to penetrate through biological hydrogels with high efficiency (Kuhn et al., 2006). Furthermore, the application of motorized nanoobjects as introduced in previous research (Dhar et al., 2006) could be thought to be of success in this context (Sigurdsson et al., 2013).

To sum this up, mucus has been known as an important biological barrier to drug diffusion already for a long time. This is particularly true for lipophilic drugs which have special features and interact with the glycoproteins and lipids in the mucus, something which eventually leads to lower absorption and bioavailability (Sigurdsson et al., 2013). More innovative DDS are required in order to overcome the mucus barrier. Recent research studies on nanoparticulate drug carriers propose that it can be possible to render mucopenetrating properties by minimizing any mucoadhesive interactions via some appropriate surface chemistry (Sigurdsson et al., 2013). Demonstrating the ability of such merely mucoinert particles to be transported through mucus over physiologically relevant distances needs further research. It is possible to expand this concept by designing particles with mucolytic properties. Ideally such carriers would be capable to selectively enhance their own transport across mucus, but without disrupting its structure and biologically important barrier functions (Sigurdsson et al., 2013).

10.2.2 THE PRINCIPLES OF BIOADHESION AND MUCOADHESION

Mucoadhesive DDS are the delivery systems that utilize the bioadhesion property of certain polymers which become adhesive on hydration and hence can be used for targeted drug delivery for an extended period of time (Mahajan et al., 2013). Bioadhesion is defined as a phenomenon in which two materials, at least one of which is biological, are held together by means of interfacial forces. This attachment could be between an artificial material and biological substrate, such as adhesion between a polymer and a biological membrane. In the case of polymer attached to the mucin layer of a mucosal tissue, the term "mucoadhesion" is mainly used (Gandhi et al., 2011). Mucoadhesive DDS can be delivered by various routes: buccal, oral, vaginal, rectal, nasal, and ocular. There are six general theories of adhesion, which have been referred to in the investigation of mucoadhesion (Ahuja et al., 1997; Smart, 2005). These six theories are discussed briefly in the following part.

The electronic theory proposes that electron transfer occurs upon contact of adhering surfaces due to differences in their electronic structure while *the wetting theory* is primarily concerned with liquid systems and takes surface and interfacial energies into account. It involves the capability of a liquid to spread spontaneously onto a surface as a prerequisite for the development of adhesion. *The adsorption theory* explains the attachment of adhesives on the basis of van der Waals' forces and hydrogen bonding. It has been suggested that these forces are the primary contributors to the adhesive interaction. On the other hand, *the diffusion theory* describes interdiffusion of polymer chains across an adhesive interface. This process is driven by concentration gradients and is affected by the available molecular chain lengths and their mobilities. *The mechanical theory* presumes that adhesion is caused by an interlocking of a liquid adhesive into irregularities on a rough surface. Finally, *the fracture theory* is a bit different from the other five in that, it relates the adhesive strength to the forces that are required for the detachment of the two involved surfaces after adhesion.

The mechanism of mucoadhesion is generally divided in two stages: first contact stage and then consolidation stage (Lohani and Chaudhary, 2012). The contact stage is mainly characterized by the contact between the mucoadhesive and the mucous membrane, with swelling and spreading of the formulation, leading to its deep contact with the mucus layer (Hagerstrom et al., 2003). In the next step (i.e., consolidation), the mucoadhesive materials are activated by the presence of moisture. Moisture plasticizes the system, which in turn allows the mucoadhesive molecules to break free and to link up by weak hydrogen bonds and van der Waals forces (Smart, 2005).

Mucoadhesive DDS are based on the adhesion of a drug or carrier to the mucous membrane (Mahajan et al., 2013). To enhance this adherence, a suitable carrier is required. Ideal mucoadhesive polymers must have some specific characteristics, some of them presented as follows.

1. *High molecular weight*: An ideal mucoadhesive polymer has a high molecular weight up to 100.00 kDa or more. This high molecular

weight is required to increase the adhesiveness between the polymer and mucus (Huang et al., 2000).

2. *Long-chain polymers*: Chain length of the polymer must be long enough to increase the interpenetration and it should not be too long to hinder the diffusion (Sudhakar et al., 2006).

3. *Degree of cross-linking*: Chain mobility and resistance to dissolution are directly influenced by the degree of cross-linking. Highly cross-linked polymers swell in presence of water and retain their original structure. Swelling favors controlled release of the drug and promotes the mucus/polymer interpenetration (Sudhakar et al., 2006).

4. *Flexibility of polymer chain*: A flexible polymer chain promotes the interpenetration of the polymer within the mucus network (Imam et al., 2003).

5. *Concentration of the polymer*: An optimum concentration of polymer is required to promote the mucoadhesive strength (Ugwoke et al., 2005). However, this factor mainly depends on the dosage form.

6. *Optimum hydration*: Excessive hydration has some adverse effects and leads to decreased mucoadhesive strength due to the formation of a slippery mucilage (Mortazavi and Smart, 1993).

7. *Optimum pH*: Mucoadhesion is optimum at low pH conditions but at higher pH values a change in the conformation leads to a rod-like structure which is more available for interpenetration and interdiffusion. At highly elevated pH values, positively charged polymers create polyelectrolyte complexes with mucus and exhibit strong mucoadhesive forces (Peppas and Huang, 2004).

In addition, ideal mucoadhesive polymers should be of spatial conformation and high viscosity. They should also be nontoxic, economic, biocompatible, and preferably biodegradable (Tangri and Madhav, 2011).

Various mucoadhesive polymers can broadly be categorized into two groups: first-generation mucoadhesive polymers and novel second-generation mucoadhesive polymers (Andrews et al., 2009).

First-generation mucoadhesive polymers are usually divided into three subcategories: anionic, cationic, and nonionic polymers. Among these, anionic and cationic polymers have been proved to exhibit the highest mucoadhesive strength (Ludwig, 2005). Due to their high mucoadhesive functionality and low toxicity, anionic polymers are known to be the most widely used mucoadhesive polymers for pharmaceutical purposes (Andrews et al., 2009). Typical examples of anionic polymers include polyacrylic acid (PAA) and sodium carboxymethylcellulose (NaCMC). Of the cationic polymer systems, chitosan seems to be the most extensively investigated one in the extant scientific literature. A separate section has been allocated to chitosan in the following parts.

Novel second-generation mucoadhesive polymers, unlike first-genera-tion nonspecific ones, are less vulnerable to mucus turnover rates and some species of these polymers bind directly to mucosal surfaces (Andrews et al., 2009). The most widely investigated systems in this group are lectins. Lectins belong to a group of structurally diverse proteins and glycoproteins that are able to bind reversibly to some specific carbohydrate residues (Clark et al., 2000). After initial mucosal cell-binding, lectins can either remain on the cell surface or in the case of receptor-mediated adhesion, possibly become internalized through an endocytosis process (Lehr, 2000).

Thiolated polymers are also an important group of the novel second-generation mucoadhesive polymers (Andrews et al., 2009). The presence of free thiol groups in the polymeric skeleton helps to the formation of disul-phide bonds with that of the cysteine-rich subdomains present in mucin, which promotes the mucoadhesive properties of the polymers (Albrecht et al., 2006).

10.2.3   MECHANISMS OF GELATION

The terms gels and hydrogels are used interchangeably by biomaterials scientists to refer to polymeric cross-linked network structures (Gulrez et al., 2011). Gels are defined as a considerably dilute cross-linked system, and are principally categorized as weak or strong depending on their flow behavior in steady-state (Ferry, 1980). Edible gels are widely used in the food industry and principally refer to gelling polysaccharides (i.e., hydrocolloids) (Taggart and Mitchel, 2009). The term *hydrogel* refers to three-dimensional network structures which are obtained from a class of synthetic and/or natural poly-mers being able to retain and absorb significant amount of water (Rosiak and Yoshii, 1999).

Gelation refers to the linking of macromolecular chains together which leads to progressively larger branched yet soluble polymers depending on the conformation and the structure of the starting material (Gulrez et al., 2011). The mixture of these polydisperse soluble branched polymer is referred to as "sol." Continuation of the linking process leads to increasing the size of the branched polymer with more solubility. This "infinite polymer" is referred to as "gel" or "network" and is permeated with finite branched polymers

(Gulrez et al., 2011). The transition from a system with a certain number of branched polymer to infinite molecules is known as sol–gel transition or gelation and the critical point where gel appears first is known as the gel point (Rubinstein and Colby, 2003). Gelation can take place in two ways: by physical linking (i.e., physical gelation) or by chemical linking (i.e., chemical gelation) (Gulrez et al., 2011). Physical gels are generally subcategorized as strong physical gels and weak gels. Strong physical gels have strong physical bonds between polymer chains and are permanently effective at a given set of experimental conditions. Therefore, strong physical gels are analogous to chemical gels. Lamellar microcrystals, glassy nodules, or double and triple helices are examples of strong physical bonds. Weak physical gels have reversible links formed by temporary associations between chains (Gulrez et al., 2011). These associations have finite lifetimes and reform and break continuously. Hydrogen bonds, block copolymer micelles, and ionic associations are examples of weak physical bonds. On the other hand, chemical gelation includes the formation of covalent bonds and on every occasion results in a strong gel. The three main chemical gelation processes involve vulcanization, condensation, and addition polymerization (Gulrez et al., 2011).

10.3 NANOGELS

Intense research studies on various carriers have been conducted to improve delivery of drugs in the body, to minimize side effects, and attain a desired local action in the target sites. These carriers involve molecules such as lipoproteins, natural agents such as erythrocytes, and artificial elements such as liposomes (Eckmann et al., 2014). Few of these carriers appeared in practical medical use; more are in the clinical trials and many more are in the preliminary stages and in laboratory design and development. It is very unlikely that one type of the carrier can suite all medical goals. Yet, some properties including biocompatibility, multifunctionality, adequate pharmacokinetics, and responsiveness to the microenvironment represent highly attractive attributes for many drug carriers. Hydrogel NPs or nanogels, are an important group of these carriers and have been the subject of attention and research due to their unique and attractive features.

Nanogels typically consist of highly hydrated, cross-linked hydrophilic polymers (Ferrer et al., 2014; Kabanov and Vinogradov, 2009). More generally, a gel can be defined as a soft material which combines the properties of fluids and solids. Nanogels are defined as submicron sized three-dimensionally cross-linked polymer networks (Zhang et al., 2016). Nanogels are formed by hydrogel particulate entities with a nanometer size and because of this, they have the features of NPs and hydrogels at the same time.

Based on the materials, NPs generally consist of lipid NPs, inorganic NPs and polymer NPs. Nanogels belong to the latter type. Nanogels are mainly prepared from polymer precursors or via heterogeneous polymerization of monomers (Zhang et al., 2016). Unique properties of nanogels make them a promising candidate for a lot of applications. Numerous research studies demonstrate that nanogels are ideal drug delivery carriers due to their responsiveness to a wide variety of environmental stimuli, flexibility and versatility, ability for site-specific and time-controlled delivery of bioactive agents, excellent drug loading capacity, and extended circulation times (Oh et al., 2007).

Nanogels can be formulated to respond to various external stimuli, which can lead to changes in various attributes, such as viscoelasticity, permeability, and hydrophobicity (Eckmann et al., 2014). The external stimuli that can elicit such responses involve changes in pH, photosensitivity and light exposure, temperature, and ionic strength as well as exposure to chemicals, magnetic fields, and biological agents (Bawa et al., 2009). These characteristics can be incorporated into the design of nanogels, helping nanogels to emerge in various fields such as biosensing and imaging (Peng et al., 2010), tissue engineering (Hayashi et al., 2009), and medical diagnostics (Oishi et al., 2007).

In addition, flexibility of nanogels offers various opportunities for their use as targeted drug delivery vehicles (Eckmann et al., 2014). Due to their unique physicochemical structure, nanogels can also be formulated to have a number of particular features, such as (1) higher stability via their cross-linked structure to extend their circulation time in the bloodstream; (2) deformability to promote binding within the targeting tissue, and (3) modular drug loading and release profiles, which can notably enhance drug loading efficiency as well as bioavailability and thereby reduce side effects and drug toxicity. Besides, nanogels are able to absorb high amounts of water or biological fluids while retaining their structure, which is attributed to the presence of hydrophilic groups such as $-CONH-$, $-OH$, $-SO_3H$, and $-CONH_2-$ in the polymer (Hamidi et al., 2008).

Furthermore, nanogels have attracted considerable attention as one of the most versatile DDS especially for site-specific and time-controlled drug delivery, owing to their combined features of NPs and hydrogels (Zhang et al., 2016). Nanogels that are physically synthesized can provide a platform which is able to encapsulate different types of bioactive compounds, particularly biomacromolecules and hydrophobic drugs but their mechanical stability has been shown to be poor. On the other hand, chemically cross-linked nanogels have larger flexibility and a wider application.

Besides, nanogels have been suggested to be suitable nanomedicine carriers as compared to other NPs notably in terms of drug loading. Nanogels can be synthesized or prepared even in the absence of the drug to be loaded since drug loading in nanogels can be effectively achieved later on when the nanogels are swollen and equilibrated in biological fluids or in water (Soni and Yadav, 2016). Due to the existence of cross-links in nanogels, they show swelling property instead of being dissolved. Also, drug loading occurs spontaneously in nanogels and compared to other conventional NPs, nanogels allow much higher drug loading. Additionally, the methods of preparation of nanogels are simpler and do not require the use of organic solvents or mechanical energy (Vinogradov et al., 2004). Consequently, the loaded drug is not exposed to any vigorous conditions during preparation. After administration, nanogels safely carry the drug, move within the cells, and release the contents in the target site *in vivo* (Soni and Yadav, 2016).

Finally, since nanogels are generally too large to be quickly removed by renal clearance following intravenous administration, they often exhibit extended circulation time that allows for the targeted drug delivery to tissues that express specific disease markers (Eckmann et al., 2014). This is accomplished by attaching antibodies, ligands, or other molecules having molecular recognition specificity to the surface of nanogels.

Nanogels meet many, if not all, of the key basic requirements of a flexible nanocarrier delivery vehicle. Among nanogels, polysaccharide-based as well as reduction-sensitive nano/microgels have drawn attention as a vesicle of different pharmaceutical agents due to their unique properties.

10.3.1 REDUCTION-SENSITIVE NANO/MICROGELS

In the past decade, reduction-sensitive biodegradable conjugates and polymers have emerged as an interesting class of biomedical materials that can be applied for the development of sophisticated delivery systems for both biotherapeutics as well as "classical" low molecular weight drugs (Meng et al., 2009). These materials usually have disulfide linkage or linkages in the main chain, at the side chain, or in the cross-linker. The disulfide bonds, nevertheless sufficiently stable in the circulation and in the extracellular environment, may be susceptible to rapid cleavage under a reductive environment through thiol–disulfide exchange reactions (Raina and Missiakas, 1997). This quick-response chemical degradation behavior is different from common hydrolytically degradable polymers such as polycarbonates and aliphatic polyesters in which carbonate bonds and the ester usually display gradual degradation kinetics inside body with longer degradation time (Ikada and Tsuji, 2000).

Thiol–disulfide exchange reactions, being rapid and readily reversible, play a significant role in sustaining proper biological functions of living cells, including stabilization of enzymatic activity, redox cycles, and protein structures (Go and Jones, 2008). Glutathione tripeptide (g-glutamyl-cysteinyl-glycine; GSH) is known to be the most abundant low molecular weight biological thiol and GSH/glutathione disulfide (GSSG) is the primary redox couple in animal cells (Wu et al., 2004). In body fluids and on the cell surface and in extracellular matrices, the proteins are rich in stabilizing disulfides due to a relatively high redox potential as a result of a low concentration of GSH (nearly 2–20 mM) (Meng et al., 2009). On the other hand, the concentration of GSH inside cells (0.5–10 mM), that is kept reduced by NADPH and glutathione reductase, preserves a highly reducing environment inside the cells (Wu et al., 2004). The large difference in reducing potential between the intracellular and extracellular environments may be utilized for triggered intracellular delivery of various bioactive molecules including DNA, antisense oligonucleotide, small interfering ribonucleic acid (siRNA), proteins, and drugs with low molecular weight (Saito et al., 2003). Moreover, of particular interest is the fact that tumor tissues are highly reducing and hypoxic compared with normal tissues (Kuppusamy et al., 1998), with at least fourfold higher concentrations of GSH in the tumor tissues over normal tissues (Kuppusamy et al., 2002), rendering the reducible bioconjugates applicable and valuable for tumor-specific drug delivery.

There are various reduction-sensitive delivery systems including liposomes, polymeric micelles, polymersomes, DNA containing NPs, nanotubes, multilayered thin films, and polyion complex micelles. Reduction-sensitive nano- and microgel delivery systems also form an important group here and we would like to highlight them in this chapter. Reduction-sensitive nano- and microgels form a highly promising functional biomaterials platforms that have enormous potential for the design and development of sophisticated drug and gene delivery systems.

Reduction-sensitive nano- and microgels have been the subject of different research studies. For example, Kommareddy and Amiji (2005) synthesized thiolated gelatin by covalent modification of the primary amino groups of Type B gelatin using 2-iminothiolane. The thiolated gelatins developed with 20 and 40 mg of 2-iminothiolane (SHGel-20 and SHGel-40) per gram of gelatin demonstrated a comparable cell viability profile to that of the unmodified gelatin. *In vitro* release studies of fluorescein isothiocyanate-labeled dextran (molecular weight 70 kDa) demonstrated enhanced release by about 40% of the loading in the presence of GSH. The thiolated gelatin NPs (SHGel-20) carrying plasmid DNA encoding for enhanced green fluorescent protein (EGFP-N1) were well effective in transfecting NIH-3T3 murine fibroblast cells and showed sustained expression of GFP for up to 96 h. In a subsequent study, polyethylene glycol (PEG)-modified thiolated gelatin (PEG–SHGel) NPs were developed as a long-circulating passively targeted delivery system that quickly responded to intracellular GSH to enhance DNA release and transfection (Kommareddy and Amiji, 2007).

Plunkett et al. (2005) synthesized chymotrypsin and reduction responsive polyacrylamide microscopic hydrogels using a tetrapeptide sequence, Ac–Cys–Tyr–Lys–Cys–NH$_2$ (CYKC), as a cross-linker. These microscopic hydrogels dissolved when exposed to a solution of R-chymotrypsin, while control hydrogels cross-linked with the tetrapeptide, Ac–Cys–Ser–Lys–Cys–NH$_2$ (CSKC), were not influenced by R-chymotrypsin. Both the CYKC and CSKC cross-linked hydrogels were eroded in the presence of the disulfide reducing agent tris(2-carboxyethyl) phosphine (TCEP). Andac et al. (2008) prepared biodegradable macroporous poly(2-hydroxyethyl methacrylate) (PHEMA) cryogels by blending two cross-linkers, PEG diacrylate and a disulfide-containing water soluble cross-linker, N,N′-bis(methacryloyl)-L-cystine (MAS-S) (2008). The cryogels disintegrated into small pieces when subjected to reductive agents like dithiothreitol (DTT). The degradation time was shown to be controlled by both the content of S–S bonds and DTT concentration.

Robust biodegradable nanogels cross-linked with disulfide linkages have been developed by inverse miniemulsion atom transfer radical polymerization (ATRP) of oligo(ethylene oxide) monomethylether methacrylate (OEOMA) and disulfide-functionalized dimethacrylate cross-linker (Fig. 10.1) (Bae et al., 2008). These nanogels were proven to be nontoxic to cells in the concentration range tested. Degradation profiles demonstrated that over 90% of the nanogels were degraded within 3 h in the presence of 20% glutathione in water. The biodegradation of nanogels could be employed to trigger release of encapsulated molecules, such as water-soluble rhodamine 6G (R6G) and doxorubicin (Dox) (Bae et al., 2008) and rhodamine

B isothiocyanate–dextran (RITC–Dx) (Meng et al., 2009). It was further demonstrated that hydroxyl-functionalized nanogels could be prepared by introducing 2-hydroxyethyl acrylate (HEA) during inverse miniemulsion ATRP, which could subsequently be derivatized with biotin to form bioconjugates with avidin (Bae et al., 2008).

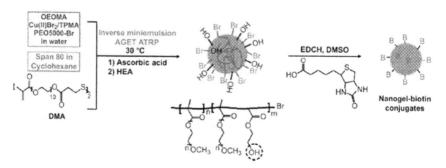

FIGURE 10.1 Biodegradable nanogels cross-linked with disulfide linkages. (Reprinted with permission from Meng, F.; Hennink, W. E.; Zhong, Z. Reduction-sensitive Polymers and Bioconjugates for Biomedical Applications. *Biomaterials* **2009**, *30* (12), 2180–2198. © 2009 Elsevier)

Schmitz et al. (2007) prepared chitosan-thiobutylamidine by treating chitosan with 2-iminothiolane, which formed nanosized coacervates with pDNA. These thiolated chitosan–DNA particles obtained after oxidation were more stable, but were vulnerable to dissociation in a reducing environment as found intracellularly, releasing about 50% of pDNA within 3 h. More transfection studies with Caco2 cells demonstrated a promoted expression of transgene as compared to chitosan-based systems.

10.3.2 ADVANTAGES OF NANOGELS

Nanogels, like other nanosize drug carriers, exhibit numerous advantages for drug delivery when compared to other delivery systems. Some of their advantages are as follows.

10.3.2.1 SIZE CONTROL

Nanogel size and surface properties can be chemically controlled to limit the rate of clearance by phagocytic cells and to enable either active or passive cell targeting (Eckmann et al., 2014). Due to their tiny volume, nanogels have the ability to reach the smallest capillary vessels and to penetrate the tissues either through transcellular or paracellular pathways (Goncalves et al., 2010). In addition, the nanoscale dimension of nanogels enables them to respond rapidly to environmental changes such as temperature and pH (Soni and Yadav, 2016). Furthermore, the larger surface area that nanogel dispersions have, make them suitable for *in vivo* applications (Mbuya et al., 2016).

10.3.2.2   HIGH ENCAPSULATION STABILITY

In order to provide minimum side effects or toxicity and maximum therapeutic effects, drug molecules loaded into the nanogel must be retained and not be transported out while circulating. Cross-linking of polymer components within the nanogel can be utilized to control drug release and drug encapsulation (Eckmann et al., 2014). Nanogels have low buoyant density, high dispersion stability in aqueous media and substantial drug loading capacity. In addition, nanogels can encapsulate compounds with low or high molecular weights and can notably extend their activity in biological environments (Mbuya et al., 2016).

10.3.2.3   RESPONSE TO STIMULI

Nanogels can be designed to respond to specific stimuli and can retain high drug encapsulation stability while circulating. Then the drug is released readily in response to the appropriate stimulus when the target site is reached.

10.3.2.4   EASE OF SYNTHESIS

One of the attractive features of nanogels is their easy synthesis. In addition, the scalability of laboratory-based nanogel development to industrial-scale production for clinical applications and the use of "green" approaches to nanogel manufacturing are among the key considerations for cost and environmental impact (Li et al., 2008).

10.3.2.5   TARGETING

Site-specific delivery of nanogel carriers is possible by either coupling to their surface affinity ligands binding to target determinants or by passive targeting approaches including extravasation in the pathological sites and retention in the microvasculature (Eckmann et al., 2014). Upon intravenous injection, nanogels can reach the sites that cannot be easily accessed by hydrogels. Nanogels are also ideal for intracellular delivery and can be safely delivered into the cytoplasm of the cell (Toita et al., 2011).

10.3.2.6   LOW TOXICITY

The nanogels themselves are known to be highly biocompatible and free from toxicity, and are considered biodegradable with nontoxic degradation products that are easily cleared from the body (Eckmann et al., 2014).

10.3.2.7   CONTROLLED AND SUSTAINED DRUG RELEASE

Drug transport needs to occur at the target site and needs to provide both therapeutic efficacy and reduced side effects. Drug loading should be high enough to achieve therapeutic goals. Cross-linking of different components within nanogels makes them efficient in controlled and sustained drug release.

10.3.2.8   EFFICIENCY AND VERSATILITY

Nanogels increase the efficacy of therapeutic nucleoside analogs (Vinogradov et al., 2005). In addition, weakly cross-linked polyelectrolyte nanogels can embrace biomacromolecules of the opposite charge, whereas biomacromolecules cannot be usually accommodated in hydrogels due to the effects of cross-linking density and excluded volume (Kabanov and Vinogradov, 2009). And last but not least, nanogels can be used for effective delivery of biopharmaceuticals in cells and also for increasing drug delivery across cellular barriers (Park et al., 2010).

10.3.3   CLASSIFICATION OF NANOGELS

Nanogels are generally classified in two main ways. The first classification is based on their responsive behavior. In this way, nanogels can be either stimuli-responsive or nonresponsive (Sultana et al., 2013). Nonresponsive nanogels swell as a result of absorbing water, whereas stimuli-responsive microgels swell or deswell based on environmental factors such as pH, temperature, ionic strength, and magnetic field.

The second classification is based on the type of linkages present in the network chains of gel structure (Sultana et al., 2013). In this way, nanogels are subdivided into two main categories, physically cross-linked gels and chemically cross-linked gels.

10.3.3.1   PHYSICALLY CROSS-LINKED GELS

Physically cross-linked gels or pseudogels are known to be formed by weaker linkages through either (1) hydrogen bonding, (2) hydrophobic, electrostatic interactions, or (3) van der Waals forces. A few simple methods are available to derive physical gels. These gels are sensitive and this sensitivity is deter-mined by temperature, polymer composition, concentrations of the polymer and cross-linking agent, and ionic strength of the medium (Sultana et al., 2013). The association of amphiphilic block copolymers and complexation of oppositely-charged polymeric chains lead to the formation of nanogels in a few minutes (Harada and Kataoka, 1999). Physical gels can also be formed by the aggregation and/or self-assembly of polymeric chains (Yallapu et al., 2007). Liposome-modified nanogels (Sakaguchi et al., 2008), micellar nano-gels (Li et al., 2006), and hybrid nanogels (Akiyoshi et al., 2000) are some of the nanogels in this group.

10.3.3.2   CHEMICALLY CROSS-LINKED GELS

Chemically cross-linked gels consist of permanent chemical linkages (i.e., covalent bonds) through the gel networks. The properties of these gel systems are influenced by the chemical linkages and the existing functional groups. So far, various nanogels have been synthesized using different approaches for chemical linking of polymeric chains. Hydrophilic polymers and hydrophilic–hydrophobic copolymers are usually formulated by the polymerization of vinyl monomers in the presence of multifunctional cross-linkers. These cross-linking points make it possible to modify physicochemical properties of the gel systems. Some versatile cross-linking agents are already known and explored (Yallapu et al., 2007).

Figure 10.2 illustrates some nanogel formulation strategies.

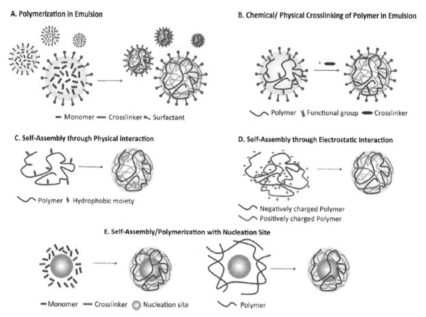

FIGURE 10.2   Nanogel formulation strategies. (A) Hydrophilic monomers and cross-linkers in a water-in-oil emulsion, stabilized by surfactants. Upon the addition of a catalyst, polymerization occurs within the emulsion droplets, forming nanogels. (B) Hydrophilic polymer modified with functional groups that allow physical/chemical cross-linking to form nanogels. (C) Polymer modified with hydrophobic moieties for self-assembly into nanogels. (D) Positively and negatively charged polymer self-assembly through electrostatic interaction. (E) Polymerization of monomers and cross-linkers shell or self-assembly of polymer modified with hydrophobic moieties in presence of nucleation sites. (Reprinted with permission from Chana, M.; Almutairi, A. Nanogels as Imaging Agents for Modalities Spanning the Electromagnetic Spectrum. *Mater. Horiz.* **2016,** *3* (21), 21–40. © 2015 The Royal Society of Chemistry 2015)

10.4   POLYSACCHARIDE-BASED NANOGEL DRUG DELIVERY SYSTEMS

Polysaccharides are considered as a group of carbohydrates with a large polymeric oligosaccharides formed through glycosidic linkages among multiple monosaccharide repeats (Zhang et al., 2013). The main natural sources of polysaccharides are animals (e.g., chitosan, chitin, and glycosaminoglycan), plants (e.g., pectin, cellulose, and starch), algal sources (e.g., agar, alginate, and carrageenan), and microbial sources (e.g., dextran, pullulan, xanthan gum, and gellan gum) (Aspinall, 1983). Depending on the composition of monosaccharide units, polysaccharides can be classified as homopolymers (i.e., formed from the same monosaccharide repeats such as glycogen, cellulose, starch, levan, pullulan, pectin, etc.) or heteropolymers (i.e., formed from different monosaccharide units such as chitosan, hyaluronic acid (HA), heparin, chondroitin sulfate (CS), dermatan sulfate, keratan sulfate, and heparan sulfate) (Debele et al., 2016; Posocco et al., 2015).

Polysaccharides are known as the most promising and attractive biomaterials in nanomedicine due to their low cost processing, abundance in nature, biodegradability, biocompatibility, nontoxicity, bioactivity, and water solubility (Rodrigues et al., 2015). Furthermore, polysaccharides contain numerous reactive functional groups (such as amino, hydroxyl, and carboxylic acid groups) on their backbone, which contributes to their functional and structural diversity (Saravanakumar et al., 2012). Due to these multifunctional groups, polysaccharides can be easily amended biochemically or chemically to various kinds of polysaccharide derivatives (Hafrén et al., 2006). The main approaches for polysaccharide modifications are (1) chemical oxidation of primary alcohols to aldehyde and carboxylic acids, (2) ester and ether formation using sugar hydroxyl group as nucleophiles, (3) formation of amide bonds between saccharide carboxyl group and heteroatomic nucleophiles using coupling agents, (4) enzymatic oxidation of primary alcohols to uronic acid, and (5) nucleophilic reactions of the amines of some polysaccharides (Debele et al., 2016). More information about molecular modification of polysaccharide and resulting bioactivities can be found in the review article by Li et al. (2016). In short, due to biodegradability of polysaccharides and nontoxic end products, currently numerous studies have been conducted on polysaccharides and their derivatives for their potential application as NPs in DDS (Debele et al., 2016).

In this section, different kinds of polysaccharide-based nanogels DDS, including those derived from chitosan, heparin, HA, alginate, pullulan, pectin, levan, CS, cellulose, starch, cyclodextrins (CDs), and dextran will be discussed briefly. Most of these polysaccharides are illustrated in Figure 10.3.

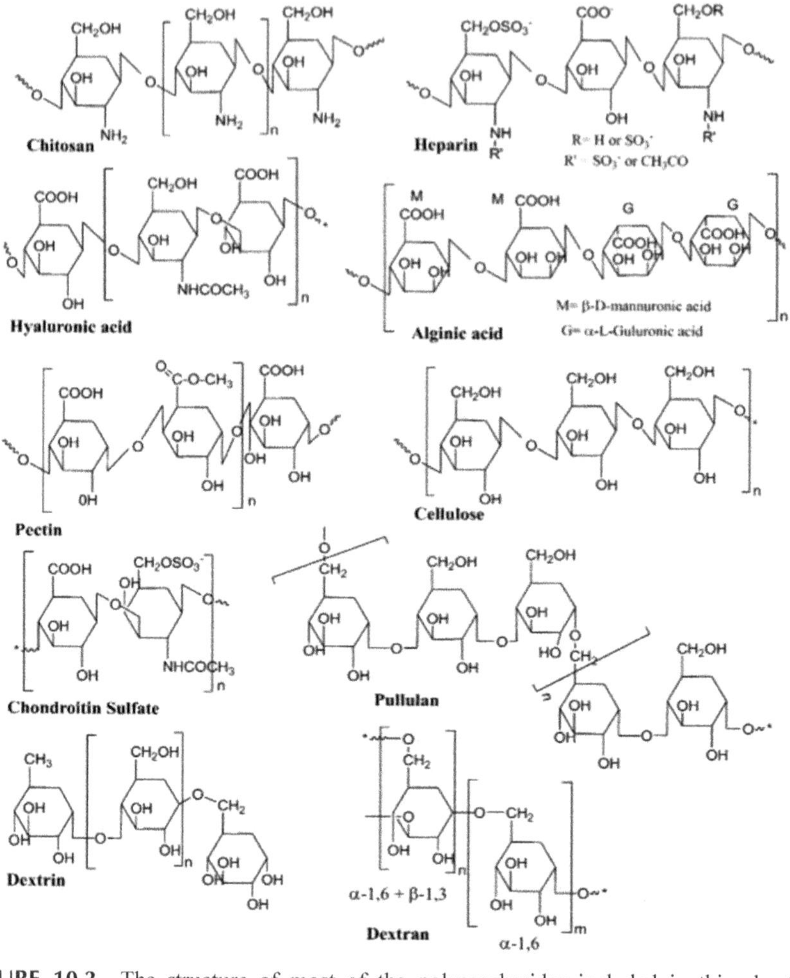

FIGURE 10.3 The structure of most of the polysaccharides included in this chapter. (Reprinted with permission from Debele, T. A.; Mekuria, S. L.; Tsai, H. C. Polysaccharide Based Nanogels in the Drug Delivery System: Application as the Carrier of Pharmaceutical Agents. *Mater. Sci. Eng.* C **2016**, *68*, 964–981. © 2016 Elsevier)

10.4.1 CHITOSAN

Chitosan is a deacetylated chitin (poly-N-acetyl glucosamine) derivative, containing β-1,4-linked glucosamine (2-amino-2-deoxy-β-D-glucose) and small amounts of N-acetyl glucosamine (Nagpal et al., 2010). Chitosan is the second most abundant polysaccharide next to cellulose and is used quite widely for pharmaceutical applications due to its biocompatibility, reactive functional groups, gel forming capability, biodegradability, high charge density, and nontoxicity (Sinha et al., 2004). Chitosan consists of multifunctional groups which can lead to hydrogen bonding and the linear molecule exhibits an adequate chain flexibility (Grenha, 2012). In addition, the cationic polyelectrolyte nature of chitosan offers a strong electrostatic interaction with mucus, which leads unique features like mucoadhesion which enhances cellular uptake into cancer cells and promotes targeted drug release (Debele et al., 2016; Sogias et al., 2008). In order to synthesize chitosan-based nanogels, a chemical cross-linker (e.g., glutaraldehyde) or an ionic cross-linker [e.g., tripolyphosphate (TPP)] might be required to react with amino groups on chitosan (Wang et al., 2005). TPP is a multivalent anion and nontoxic ionic cross-linker, which is able to form a gel by ionic interaction between negatively-charged counterions of TPP and positively-charged amino groups of chitosan (Liu et al., 2008). Calvo et al. (1997) synthesized and reported TPP-cross-linked chitosan NPs for the first time. Currently a number of chitosan and chitosan-based nanogels have been synthesized via different protocols and are used as carriers in DDS (Debele et al., 2016).

10.4.2 HEPARIN

Heparin is a highly sulfated, biocompatible, water-soluble, biodegradable, and naturally-derived anionic polysaccharide composed of repeating units of 2-deoxy-2-6-O-sulfo-α-D-glucose, 2-O-sulfo-L-iduronic acid, 2-acetamido-2-deoxy-α-D-glucose, β-D-glucuronic acid, and α-L-iduronic acid (Li et al., 2011). Anticoagulation and inhibition of both angiogenesis and tumor growth are among the biological functions of heparin (Mousa and Petersen, 2009). Several *in vivo* studies proved that heparin has a tendency to interact with multiple molecules such as heparinase, growth factors, P- and L-selectins, which plays an important role in cancer metastasis (Vlodavsky et al., 2007). Some research studies demonstrated that heparinase, the sole heparan sulfate degrading endoglycosidase, is involved in cancer progression and for this reason is a valid target for cancer treatments (Debele et al., 2016). The overexpressed heparinase in cancer cells enhances the degradation of extracellular matrix (ECM), hence aiding the extravasation of cancer cells, which is crucial in the metastatic cascade (Ilan et al., 2006). Consequently, since heparin interacts with heparinase, it has the tendency to hinder the activity of heparinase, which in turn reduces cancer metastasis (Debele et al., 2016). Likewise, heparin is also able to interact with P- and L-selectins, which have been shown to enhance cancer metastasis by mediating interactions between cells (Stevenson et al., 2007). Moreover, heparin can restrain angiogenesis by binding with growth factors such as vascular endothelial growth factors (VEGF) and fibroblast growth factor (FGF), which can foster the development of endothelial cells (Yagi et al., 2005). Additionally, heparin has various functional groups such as hydroxyl, amino, and carboxylic groups which provide ample opportunities for derivatization via chemical methods (Xu et al., 2011). More recent research has proposed that heparin-based nanocarriers, such as nanogels and micelles, can be used as platforms for delivery of small pharmaceutical agents and macromolecules such as peptides, protein, and nucleic acids (Debele et al., 2016). Heparin can be used to construct both physical and chemical cross-linked nanogels (Yang et al., 2015).

10.4.3 HYALURONIC ACID

HA is a nonsulfated glycosaminoglycan (GAG) that is available naturally in high concentrations in different soft connective tissues such as vitreous humor, synovial fluid, the umbilical cord as well as skin (Mero and Campisi, 2014). HA is a linear heteropolysaccharide that consists of repeating monosaccharide units of D-glucuronic acid and N-acetyl-D-glucosamine which are linked together via alternating β-1,3 and β-1,4 glycosidic bonds (Prestwich, 2011). HA is a critical part of the ECM and its structural and biological properties mediate its activity in shock absorbing and structure stabilizing, in the lubrication of joints, in water homeostasis of tissue, and in the physiological functions via interaction with cell surface receptors (Liao et al., 2005).

For the last three decades, HA and its derivatives have witnessed wide medical applications. Major therapeutic applications of HA involve surgery and wound healing, ophthalmology as well as rheumatology and orthopedic surgery (Debele et al., 2016). HA can be chemically redesigned due to its multifunctional groups (such as primary and secondary hydroxyl groups, carboxylic acid of glucuronic acid, and the N-acetyl groups), which may modify the properties of HA derivatives, including its biological activity and hydrophobicity (Schanté et al., 2011). Most common chemical modification approaches of HA have been reviewed in detail by Schanté et al. (2011), which involves amidation and esterification of carboxyl groups as well as etherification and esterification of hydroxyl groups of HA (Debele et al., 2016). In addition, deacetylation of the N-acetyl group of HA recovers an amino group that can subsequently react with carboxyl-containing molecules using the same amidation tactics. Primarily these HA derivatives can be classified into two categories, namely, as living and monolithic (Glenn and Jing-wen, 2008). Living derivatives of HA can shape new covalent bonds in the presence of cells, therapeutic agents, and tissues. In contrast, monolithic HA derivatives are terminally modified forms of HA which cannot initiate new chemical bonds in the presence of tissues or cells (Burdick and Prestwich, 2011). There are various studies that have investigated the application of HA-based nanogels as the vesicle of pharmaceutical agents. Many cancer stem cells and drug-resistant tumors express high levels of CD44 receptor, which is a cellular glycoprotein-binding HA (Rachid et al., 2008). Consequently, drug encapsulated into HA nanogels or conjugated to HA will be internalized by cells via CD44 receptors and this enables intracellular delivery of drugs (Tripodo et al., 2015).

10.4.4 ALGINATE

Alginate is a linear anion polyelectrolyte polysaccharide containing β-D-mannuronic acid (M units) and α-L-guluronic acid (G units) (d'Ayala, 2008). The monosaccharide repeats of alginate can be located in blocks of repeating G residues (G–G blocks), blocks of repeating M residues (M–M blocks), or blocks of mixed M and G residues (M–G blocks) (Debele et al., 2016). Alginates with a more content of G blocks give gels of substantially higher strength compared to alginates rich in M blocks and this is due to the fact that G residues demonstrate a more potent affinity for divalent ions than the M residues (George and Abraham, 2006). Hence, the positioning of monosaccharide repeats and M/G ratio changes the physical and chemical characteristics of alginate (Pawar and Edgar, 2012). Like other polyelectrolyte polysaccharides, alginate consists of multifunctional groups, such as polyhydroxyl and carboxyl groups, that are distributed along the backbone. These groups are highly reactive and offer great opportunity for chemical modifications such as amidation, oxidation, sulfation, esterification, and grafting (Vallée et al., 2009). As a result, biological and physicochemical properties such as hydrophobicity and solubility may be modified and the type of modification depends on the techniques used and the desired applications (Sarika et al., 2015a, 2015b). Alginate-based nanogels are generally synthesized via ionic cross-linking using divalent cations such as Ca^{2+}, Zn^{2+}, Cu^{2+}, or Mn^{2+} that cooperatively bind between the G blocks of neighboring alginate chains (Oh et al., 2008). However, ionically cross-linked alginates lose their mechanical characteristics over time *in vitro* due to loss or an outward flux of cross-linking ions into the nearby medium (LeRoux et al.,

1999). This constraint can be overcome by using multifunctional cross-linker to form stable covalent cross-links that control the swelling and mechanical properties of the gels (De Santis et al., 2014).

10.4.5 PULLULAN

Pullulan is a nonionic, water-soluble, nontoxic, unbranched, and biodegradable polysaccharide which is produced from starch by the fungus-like yeast (Gaur et al., 2015). Pullulan is a homopolysaccharide and structurally consists of maltotriose units with three monosaccharides in the repeating units of α-1,4-linked glucose molecules, that are polymerized linearly via α-1,6-linkages (Miyahara et al., 2012). Pullulan is composed of only polyhydroxyl functional groups in which its derivatizations are meant to introduce reactive or charged groups for functionality or alter its water solubility (Kobayashi et al., 2009). A number of tactics have been explored to convert pullulan to a reactive derivative to use it for different applications including DDS (Coviello et al., 2007). A suitable chemical derivatization of pullulan polysaccharide is needed to synthesize pullulan-based nanogels since pullulan polysaccharide is not a natural gelling polysaccharide (Debele et al., 2016). Pullulan derivatization has been reviewed in detail by Singh et al. (2015). Different research studies have synthesized self-assembled pullulan-based nanogels by dispersing pullulan, a water-soluble polysaccharide, carrying hydrophobic substituents such as polylactic acid, cholesterol, fat-soluble vitamins, and bile acids (Fujioka-Kobayashi et al., 2012).

10.4.6 PECTIN

Pectin is a significant linear heteropolysaccharide that forms about one third of higher plants' cell wall (Voragen et al., 2009). The major components of natural pectin are, D-galacturonic acid units, which are joined through α-(1→4)-glycosidic linkages (Yapo et al., 2007). Nowadays, the interest in using pectin and pectin derivatives for drug delivery has been increased due to its ready availability and low production cost (Debele et al., 2016). Pectin consists of multicarboxyl and hydroxyl groups that are distributed on the backbone as well as a certain amount of neutral sugars available as side chains, all of which help to the diversity of its derivatives. Chen et al. (2015) reviewed the different techniques that are used for pectin modification. These techniques include chain elongation, substitution, and depolymerization.

D-polygalacturonic acid residues that are available in pectin polysaccharide might be used as the targeting group without addition of other targeted moieties for the asialoglycoprotein receptor expressed cells (such as liver cells) (Debele et al., 2016). Consequently, the physical encapsulation of anticancer drug into pectin-based NPs makes it possible to design DDS targeted to hepatocellular carcinoma cells such as HepG2 (Debele et al., 2016).

10.4.7 LEVAN

Levan is a homopolysaccharide that is composed of β-D-fructofuranose with β-(2→6) linkages between fructose rings. Levan is formed by the action of levansucrase enzyme by many fungi bacteria and actinomycets, through conversion of sucrose (Toksoy Oner et al., 2016). This naturally occurring polymer is strongly biocompatible, adhesive, and soluble in water and oil (Kang et al., 2009). The unique properties that distinguish levan from most other polysaccharides has long attracted attention. The applications of levan in hair care products and whiteners as well as its medical applications in healing wounds and burned tissue, anti-irritant, antioxidant and anti-inflammatory activities, weight loss, and cholesterol control are well documented (Toksoy Oner et al., 2016). Levan-based DDS has been demonstrated to form NPs through self-assembly (Renuart and Viney, 2000). Levan has been the subject of many research studies. For instance, Kazak Sarilmiser and Toksoy Oner (2014) explored anticancer activity of oxidized levan against various human cancer cell lines and reached the conclusion that increasing oxidation degree and dose enhances anticancer activity. Sezer et al. (2011) explored different biodegradable levan-based NP systems with different particular charge, size, and release profiles. The study showed that levan NPs can be used effectively as drug carriers for proteins and peptide. In addition, biological, morphological, and structural properties of ternary blend films of levan with polyethylene oxide and chitosan have been investigated (Bostan et al., 2014). These films were proved to be of interest in biomedical and tissue engineering and can be used in wound healing bandages or surgical sealants.

10.4.8 CHONDROITIN SULFATE

CS is a linear sulfated GAGs heteropolysaccharide, which is widely available in animals and in some bacteria. CS is considered as one of the major components of the ECM of many connective tissues (such as bone, skin, and tendons) and has a range of bioactivity, such as anticoagulation, anti-inflammation, antioxidants, and antiapoptotic (Egea et al., 2010). CS has gained many interests as a potential therapeutic agent for Osteoarthritis (OA) (Uebelhart, 2008). As a result, CS has been recommended by the Osteoarthritis Research Society International (OARSI) for the management of knee OA and by the European League against Rheumatism (EULAR) for the management of knee and hip OA (Zhang et al., 2010). In addition, as a member of GAGs family, CS has multifunctional groups (such as carboxyl and hydroxyl groups) which can be modified to obtain additional unique properties (Debele et al., 2016). Due to its water solubility, CS cannot be incorporated into a self-organizing nanogel in aqueous solution. Hence, different hydrophobic moieties have been conjugated to the CS backbone to develop the nanogels (Huang et al., 2009).

10.4.9   CELLULOSE

Cellulose is the most abundant natural polymer and is widely distributed over a variety of sources, including marine animals, plants, bacterial sources, and even amoeba (Abeer et al., 2014). Structurally, cellulose is a linear homopolysaccharide, composed of anhydro-D-glucopyranose units joined by β-(1→4)-glycosidic linkage (Debele et al., 2016). Cellulose contains hydroxyl functional groups, hence has a tendency to self-assemble and develops extended networks via intra/intermolecular hydrogen bonds. Due to their low cost, excellent mechanical properties, abundant availability, biodegradability, biocompatibility, and minimum cytotoxicity, cellulose and its derivatives have been the subject of several studies for potential use in medicine (Wu et al., 2013). Cellulose and its derivatives are well-known in pharmaceutical industries as "excipients," which are used to achieve the right drug concentration and to control the rate of drug release (Debele et al., 2016). Accordingly, several studies have been conducted on cellulose-based nanogels as the carrier of anticancer drugs by considering the stability of the nanogels structure and the biological properties of the tumor environments (Li et al., 2015).

10.4.10   STARCH

Starch is an abundant, well-known, and inexpensive natural polysaccharide in higher plants and a part of human and animal diets (Santander-Ortega et al., 2010). Starch is mainly found in the leaves of all green plants and in the fruits, seeds, stems, tubers, and roots of most plants (Mischnick and Momcilovic, 2010). Structurally, starch is a homopolysaccharide made up of amylose and amylopectin composed of D-glucose units that are joined together by glycosidic linkages (Beneke et al., 2009). Amylose is a linear polysaccharide with α-(1→4)-glycosidic linkage, while amylopectin is an immensely branched polysaccharide with both α-(1→6) and α-(1→4)-glycosidic linkages (Song and Jane, 2000). Due to its biodegradability, biocompatibility, and relatively easy isolation process from plant sources, starch is currently used in nanomedicine including DDS (Marques et al., 2002). In addition, most of the D-glucopyranose residues of starch consist of several hydroxyl groups which can be chemically modified to introduce unique properties such as hydrophobic character, hydrophilic character, and positive or negative charges (Robyt, 2008). So far, several starch and starch derivative-based NPs have been developed and used as the carriers of pharmaceutical agents (Santander-Ortega et al., 2010). Furthermore, the partial hydrolysis, either chemically or enzymatically, products of starch or glycogen play a variety of applications in the field of biomedicine (Debele et al., 2016).

10.4.11 CYCLODEXTRINS

CDs, also known as cycloamyloses, cyclomaltoses, or Schardinger dextrins, are biocompatible cyclic oligosaccharides that are composed of α-(1,4)-linked glucopyranose subunits (Mischnick and Momcilovic, 2010). The most common CDs are α, β, and γ, containing six, seven, and eight gluco-pyranose units, respectively (Wintgens et al., 2012). In addition to naturally available CDs, many CD derivatives have been developed via etherification and esterification reaction of primary and secondary hydroxyl groups of CDs (Boger et al., 1978). Based on the nature and type of substituents, CD derivatives have various physiochemical properties such as stability, solubility, and affinity to accommodate molecules apart from original CDs (Debele et al., 2016). This structure makes it possible for CDs to encapsulate numerous hydrophobic pharmaceutical agents into their cavity and CDs have been identified as auspicious vesicle in the pharmaceutical fields (Layre et al., 2009). Due to their low toxicity, nonimmunogenicity, and biocompatibility, CDs-based nanogels have been synthesized as the carrier of several pharmaceutical agents (Blanco-Fernandez et al., 2011). Moya-Ortega et al. (2012)

provided a detailed review regarding the characterization, advantages, and preparation of CD-based nanogels.

10.4.12   DEXTRAN

Dextran is a polysaccharide synthesized from sucrose by various strain of bacteria such as *Streptococcus, Acetobacter*, and *Leuconostoc* species (Uretimi, 2005). Dextran consists of various lengths of glucose units that are joined through glycosidic linkage of α-$(1{\rightarrow}6)$ and α-$(1{\rightarrow}3)$ (Debele et al., 2016). Dextran-based products are highly hydrophilic and biocompatible, exhibit low protein adsorption, and are clinically of interest due to their anti-thrombotic effects and their usage in intravenous fluids. In addition, dextran has numerous industrial applications in food, chemical, and pharmaceutical industries as stabilizer, adjuvant, carrier, and emulsifier (Bhavani and Nisha, 2010). Dextran, like other polysaccharides, has polyhydroxyl groups which enables a variety of opportunities for derivatization with other elements to form dextran-based NPs (Xu et al., 2015). Various research studies have described the synthesis of polylactide-grafted dextran copolymers and their possibility as biomedical materials (Nagahama et al., 2007).

10.5   SYNTHESIS OF NANOGELS

Depending on the structural featuring of polysaccharides, NPs synthesized from polysaccharides are generally classified as self-assembly, polyelectro-lyte complexation, covalent cross-linking, and ionic cross-linking (Mizrahy and Peer, 2012).

10.5.1   SELF-ASSEMBLY

Self-assembly is generally seen in nature and it is a pervasive process in biology where it plays numerous significant roles and initiates the formation of a range of biological structures (Rodell et al., 2015). In self-assembling systems, each individual constituent (i.e., molecules, atoms, polymers, and colloids) interacts in predefined ways which leads to the spontaneous self-organization of higher order structures (Ozin et al., 2009). When hydrophilic polysaccharide chains such as heparin, chitosan, or dextran are grafted with hydrophobic segments, amphiphilic copolymers are developed (Myrick et al., 2014). Amphiphilic polysaccharides tend to form NPs upon contact with an aqueous environment via inter- or intramolecular associations between hydrophobic moieties (Debele et al., 2016). Due to its amphiphilic nature, levan has also been reported to form NPs by self-assembly in water (Renuart and Viney, 2000). These polysaccharide derived NPs have unique properties, such as small particle size with core–shell structure, thermodynamic stability, and mechanical strength, all of which are determined by hydrophobic/hydrophilic components of copolymers (Wen and Oh, 2014). Particularly, polysaccharide derivatizations with hydrophobic moieties are known as promising drug carriers, since their core parts, which are surrounded by a hydrophilic outer shell, make it easier to encapsulate hydrophobic drugs (Mendes et al., 2013). In addition, hydrophilic DDS lead to less drug toxicity since they are less absorbed by normal tissues and can accumulate in tumor tissues through the enhanced permeability and retention effect (Myrick et al., 2014).

10.5.2   POLYELECTROLYTE COMPLEXATION

Polyelectrolyte complexations (PECs) are fundamentally formed through well-built electrostatic interactions between charged groups of at least two oppositely charged partners (Sarika et al., 2015a, 2015b). The first and foremost driven force for PEC formation is an increase in entropy caused by the release of the counter ions in the neighboring solution (Rolland et al., 2012). Both cationic and anionic polysaccharides such as alginate, chitosan, HA, heparin, CS, CDs, and dextran sulfate can form PEC with oppositely charged polymers by intermolecular electrostatic interaction (Berger et al., 2004). The stability and formation of PEC is essentially determined by factors such as the molecular weight of polyelectrolyte, the structure of polyelectrolyte, the duration of the interaction, surface charge density of polyelectrolyte, mixing ratio of the polyelectrolyte, temperature, pH, nature of the solvents, and ionic strength (Maciel et al., 2015). The main advantage of the PEC-based NPs is their simplicity of synthesis and cost-effectiveness (Debele et al., 2016).

10.5.3   COVALENT CROSS-LINKING

The other method to synthesize polysaccharide NPs is via covalent cross-linking, which is done through the introduction of covalent bonds between polysaccharide chains (Nitta and Numata, 2013). Covalent cross-linking techniques generally include Schiff base reaction, radical polymeriza-tions, click chemistries, photoreactions, and thiol–disulfide exchange reaction (Zhang et al., 2015). Furthermore, various forms of cross-linker, such as glutaraldehyde, have been used to cross-link polysaccharide (e.g., chitosan)-based NPs (Schiffman and Schauer, 2007). Synthesis of NPs though covalent cross-linking leads to relatively robust NPs, but possible unwanted side reactions with the active ingredient and the toxicity of glutar-aldehyde on cell viability limits the usefulness of this synthesis method in the field of drug delivery (Migneault et al., 2004). However, various forms of biocompatible cross-linkers have been identified recently and have been used to develop polysaccharide-based NPs via covalent cross-linking (Park et al., 2002). With the help of the water soluble condensation agents, such as carbodiimide, several natural di- and tricarboxylic acids (such as tartaric acid, malic acid, tricarboxylic acid, and succinic acid) have been utilized to formulate intermolecular cross-linking of polysaccharide NPs (Bodnar et al., 2005). The covalent cross-linking method makes it possible to form polycation, polyampholyte, and polyanion NPs which are known as stable in aqueous media at low pH, and at mild alkaline and neutral conditions (Debele et al., 2016). Generally, covalently cross-linked polysaccharide-based NPs can be synthesized via several methods, including emulsion cross-linking, solvent evaporation, reverse micellar, thermal cross-linking, or spray drying (Kumari et al., 2010). The main importance of nano-gels prepared via covalent bonds through chemical cross-linking is their colloidal stability under *in vivo* environments, which is critical for leakage of drug release caused by unwanted dissociation of the gel network (Matai and Gopinath, 2016).

10.5.4 IONIC CROSS-LINKING

Compared to the covalent cross-linkers (such as formaldehyde and glutar-aldehyde), ionic cross-linkers (such as sodium sulfate, sodium tripolyphos-phate, and magnesium sulfate) are more advantageous owing to simple experimental procedures and their mild preparation conditions (Jatariu et al., 2011). Ionic cross-linker tends to form a gel via ionic interaction between polyelectrolyte polysaccharides (both anions and cations) through their counterions (Kuo and Ma, 2001). Furthermore, low-molecular weight polycations and polyanions could be used as ionic cross-linkers for polyanionic and polycationic polysaccharides, respectively (Alvarez-Lorenzo et al., 2013). Factors that may affect synthesis of NPs via ionic cross-linking involve the ionic strength and pH of the solvent, the molecular weight and type of polysaccharide, the ratio of ionic cross-linker to polysaccharide, and polysaccharide concentration (Debele et al., 2016). NPs formed via ionic cross-linkers are known as less robust ones than covalently cross-linked ones, which may be because of the differences in bond strength (Liu et al., 2008).

10.6 APPLICATIONS OF NANOGELS

10.6.1 CANCER THERAPY

Cancer treatment involves targeted delivery of drugs with expected low toxicities to surrounding tissues and high therapeutic efficacy (Soni and Yadav, 2016). Due to their easy tailoring properties and their ability to efficiently encapsulate various therapeutics through simple mechanisms, nanogels are being investigated as drug delivery agents for targeting cancer. Nanogels are efficiently internalized by the target cells and are not accumulated in nontarget tissues, which decreases the therapeutic dosage and minimizes side effects. However, translation of the nanogel concept into a viable therapeutic application for the treatment of cancer needs more research and exploration.

Nanogel expresses excellent merit for antitumor drug delivery. First, the porosity of nanogel network provides an ideal storage cavern for loading drugs, avoids early release, and acts as a shield to environmental hazards and degradation (Yallapu et al., 2011). For instance, nanogels can significantly sustain the efficiency of decitabine owing to their enhanced stability and their ability of bypassing the nucleoside transporters (Vijayaraghavalu and Labhasetwar, 2013). Second, nanogels allow a high flexibility in carrier shape, from a sphere to ellipse (Na et al., 2000). This property is critical in increasing the duration of blood circulation through microcapillaries.

In addition, nanogels can easily be active and passive targets to tumor tissue (Zhang et al., 2016). The nanosize nanogels endow themselves an improved enhanced permeability and retention effect, while active targeting can be efficiently achieved through introducing various functional groups into nanogels (Yallapu et al., 2011). Furthermore, prolonged circulation time could be obtained because of the small size and hydrophilicity of the nanogels, which minimizes the uptake by macrophages (Zhang et al., 2016). In addition to the targeting, other properties of nanogels have been investigated in cancer therapy, such as chargeability and sensitivity. Nanogels have excellent pH-responsive ability and tumor is a special tissue with slightly acidic extracellular microenvironment, thus pH-sensitive nanogels can increase the efficacy of the loaded drug (Manchun et al., 2012). In addition to effective delivery of the hydrophobic photosensitizer, nanogels are also suitable delivery systems for encapsulating the high atomic number molecules (such as platinum and gold) which can promote the biological effect of X-irradiation (Zhang et al., 2016).

10.6.2 TREATMENT OF DIABETICS

Glucose molecules are known to be able to enter and diffuse easily through gels. Therefore, when levels are high, large amount of glucose passes through the gel. This increases acidity, which leads to the release of the insulin. This line of enquiry is at its beginning stages. Paddock (2013) reported the preparation of an injectable nanonetwork which responds to glucose and releases insulin. It contains a mixture of oppositely charged NPs which attract each other. This keeps the gel together and does not allow the NPs to drift away in the body. Dextran was used to make the nanogel respond to increased acidity.

10.6.3 TREATMENT OF AUTOIMMUNE DISEASE

Nanogels are attractive for researchers because of their intrinsic ability to enable greater systemic accumulations of their cargo and to bind more immune cells *in vivo* than free fluorescent tracer, which is believed to permit high, localized concentrations of mycophenolic acid (Sultana, et al., 2013). Nanogel drug delivery system increases the longevity of the patient and delays the onset of kidney damage, which is a common complication of lupus (Look et al., 2013).

10.6.4 OPHTHALMIC APPLICATIONS

Nanogels have also had some ophthalmic applications. For example, pH-sensitive polyvinylpyrrolidone-poly(acrylic acid) (PVP/PAAc) nanogels prepared by γ-radiation-induced polymerization of acrylic acid in an aqueous solution of polyvinylpyrrolidone as a template polymer were utilized to encapsulate pilocarpine to maintain a sufficient concentration of the pilocarpine at the target site for prolonged period of time (Abd El-Rehim et al., 2013).

10.6.5 TREATMENT OF NEURODEGENERATIVE DISEASE

Nanogel is considered as a promising system for delivery of oligonucleotides (ODN) to the brain. Nanogels bound or encapsulated with spontaneously negatively-charged ODN leads to the formation of stable aqueous dispersion of polyelectrolyte complex with particle sizes less than 100 nm which can be effectively transported across the blood–brain barrier (BBB). The transport efficacy is further increased when the surface of the nanogel is modified with transferrin or insulin (Sultana, et al., 2013). Vinogradov et al. (2004) developed a novel system for ODN delivery to the brain based on nanoscale network of cross-linked PEG and polyethylenimine for the treatment of neuro-degenerative diseases.

10.6.6 ANTI-INFLAMMATORY ACTION

Nanogels have also been used for anti-inflammatory purposes. Shah et al. (2012) used poly(lactide-co-glycolic acid) and chitosan to prepare bi-layered NPs. Hydroxypropyl methyl cellulose (HPMC) and Carbopol were used in appropriate viscosity to prepare the nanogels. Two anti-inflammatory drugs, spantide II and ketoprofen, which are known as effective against allergic contact dermatitis and psoriatic plaque were applied topically along with nanogel. The result showed that nanogel increases potential for the percutaneous delivery of spantide II and ketoprofen to the deeper skin layers for treatment of various skin inflammatory disorders (Shah et al., 2012).

10.6.7 IN STOPPING BLEEDING

Protein-based nanogels have been shown to stop bleeding through protein matrix self-assembly after application at the site of injury (Burgess, 2012).

10.7 FUTURE PERSPECTIVES

Unique properties of nanogels make them promising candidates for a lot of applications. Numerous research studies demonstrate that nanogels are ideal drug delivery carriers due to their responsiveness to a wide variety of environmental stimuli, flexibility and versatility, ability for site-specific and time-controlled delivery of bioactive agents, excellent drug loading capacity, and extended circulation times (Oh et al., 2007).

Nanogels meet many of the basic requirements of a flexible nanocarrier delivery vehicle. Among nanogels, polysaccharide-based as well as reduction-sensitive nano/microgels have received attention as a vesicle of different pharmaceutical agents. Natural inherent features such as biocompatibility and biodegradability make polysaccharide-based nanocarriers a high potential platform for developing DDS.

The future impact of nanogel-based drug delivery includes nontoxic or minimally toxic methods which enhances site-specific delivery and reduces side effects. Although nanogels can be very stable, highly biocompatible, and stimulus responsive, their practical clinical application is still limited at present.

Several important factors must be taken into consideration in the future development of nanogels. These include their shape, size, and surface modification, all of which will influence their circulation duration and their molecular recognition; biodegradability, which will influence both their nanotoxicology and their drug delivery and; their responsiveness to different stimuli such as pH changes or temperature and biomolecule levels, all of which can affect targeted drug delivery and drug release. It is also needed to develop and understand a variety of important physical and biological mechanisms that can be used to transport a sufficient amount of drug efficiently and effectively across any number of physiological barriers to accumulate in specific therapeutic sites. Nanogels are probably one of the best DDS to provide sustained or controlled release of the drug. However, designing perfect nanogel-based delivery systems needs further research and studies.

KEYWORDS

- nanogel
- hydrogel
- drug delivery system
- polysaccharide
- nanoparticle
- bioadhesive

REFERENCES

Abd El-Rehim, H. A.; Swilem, A. E.; Klingner, A.; Hegazy, E. S. A.; Hamed, A. A. Developing the Potential Ophthalmic Applications of Pilocarpine Entrapped Into Polyvinylpyrrolidone–Poly(Acrylic Acid) Nanogel Dispersions Prepared By γ Radiation. *Biomacromolecules* **2013,** *14* (3), 688–698.

Abeer, M. M.; Amin, M.; Iqbal, M. C.; Martin, C. A Review of Bacterial Cellulose-based Drug Delivery Systems: Their Biochemistry, Current Approaches and Future Prospects. *J. Pharm. Pharmacol.* **2014,** *66* (8), 1047–1061.

Ahuja, A.; Khar, R. K.; Ali, J. Mucoadhesive Drug Delivery Systems. *Drug Dev. Ind. Pharm.* **1997,** *23* (5), 489–515.

Akiyoshi, K.; Kang, E. C.; Kurumada, S.; Sunamoto, J.; Principi, T.; Winnik, F. M. Controlled Association of Amphiphilic Polymers in Water: Thermosensitive Nanoparticles Formed by Self-assembly of Hydrophobically Modified Pullulans and Poly(N-isopropylacrylamides). *Macromolecules* **2000,** *33* (9), 3244–3249.

Albrecht, K.; Greindl, M.; Kremser, C.; Wolf, C.; Debbage, P.; Bernkop-Schnürch, A. Comparative *In vivo* Mucoadhesion Studies of Thiomer Formulations Using Magnetic Resonance Imaging and Fluorescence Detection. *J. Control. Release* **2006,** *115* (1), 78–84.

Alvarez-Lorenzo, C.; Blanco-Fernandez, B.; Puga, A. M.; Concheiro, A. Crosslinked Ionic Polysaccharides for Stimuli-sensitive Drug Delivery. *Adv. Drug Deliv. Rev.* **2013,** *65* (9), 1148–1171.

Andac, M.; Plieva, F. M.; Denizli, A.; Galaev, I. Y.; Mattiasson, B. Poly(Hydroxyethyl Methacrylate)-based Macroporous Hydrogels with Disulfide Cross-linker. *Macromol. Chem. Phys.* **2008,** *209* (6), 577–584.

Andrews, G. P.; Laverty, T. P.; Jones, D. S. Mucoadhesive Polymeric Platforms for Controlled Drug Delivery. *Eur. J. Pharm. Biopharm.* **2009,** *71* (3), 505–518.

Aspinall, G. O. Classification of Polysaccharides. In *The Polysaccharides*; Aspinall, G. O., Ed.; Academic Press: New York, 1983; pp 1–10.

Bae, K. H.; Mok, H.; Park, T. G. Synthesis, Characterization, and Intracellular Delivery of Reducible Heparin Nanogels for Apoptotic Cell Death. *Biomaterials* **2008,** *29* (23), 3376–3383.

Bajpai, A. K.; Shukla, S. K.; Bhanu, S.; Kankane, S. Responsive Polymers in Controlled Drug Delivery. *Prog. Polym. Sci.* **2008,** *33* (11), 1088–1118.

Bawa, P.; Pillay, V.; Choonara, Y. E.; Du Toit, L. C. Stimuli-responsive Polymers and Their Applications in Drug Delivery. *Biomed. Mater.* **2009,** *4* (2), 022001.

Beneke, C. E.; Viljoen, A. M.; Hamman, J. H. Polymeric Plant-derived Excipients in Drug Delivery. *Molecules* **2009,** *14* (7), 2602–2620.

Berger, J.; Reist, M.; Mayer, J. M.; Felt, O.; Peppas, N. A.; Gurny, R. Structure and Interactions in Covalently and Ionically Crosslinked Chitosan Hydrogels for Biomedical Applications. *Eur. J. Pharm. Biopharm.* **2004,** *57* (1), 19–34.

Bhavani, A. L.; Nisha, J. Dextran—The Polysaccharide with Versatile Uses. *Int. J. Pharm. Biol. Sci.* **2010,** *1* (4), 569–573.

Blanco-Fernandez, B.; Lopez-Viota, M.; Concheiro, A.; Alvarez-Lorenzo, C. Synergistic Performance of Cyclodextrin–Agar Hydrogels for Ciprofloxacin Delivery and Antimicrobial Effect. *Carbohydr. Polym.* **2011,** *85* (4), 765–774.

Bodnar, M.; Hartmann, J. F.; Borbely, J. Preparation and Characterization of Chitosan-based Nanoparticles. *Biomacromolecules* **2005,** *6* (5), 2521–2527.

Boger, J.; Corcoran, R. J.; Lehn, J. M. Cyclodextrin Chemistry. Selective Modification of All Primary Hydroxyl Groups of α-and β-Cyclodextrins. *Helv. Chim. Acta* **1978,** *61* (6), 2190–2218.

Bostan, M. S.; Mutlu, E. C.; Kazak, H.; Keskin, S. S.; Toksoy Oner, E.; Eroglu, M. S. Comprehensive Characterization of Chitosan/PEO/Levan Ternary Blend Films. *Carbohydr. Polym.* **2014,** *102*, 993–1000.

Burdick, J. A.; Prestwich, G. D. Hyaluronic Acid Hydrogels for Biomedical Applications. *Adv. Mater.* **2011,** *23* (12), H41–H56.

Burgess, R. *Understanding Nanomedicine: An Introductory Textbook*; CRC Press: Boca Raton, Florida, 2012; pp 139–142.

Calvo, P.; Remuñan-López, C.; Vila-Jato, J. L.; Alonso, M. J. Chitosan and Chitosan/Ethylene Oxide-propylene Oxide Block Copolymer Nanoparticles as Novel Carriers for Proteins and Vaccines. *Pharm. Res.* **1997,** *14* (10), 1431–1436.

Chana, M.; Almutairi, A. Nanogels as Imaging Agents for Modalities Spanning the Electromagnetic Spectrum. *Mater. Horiz.* **2016,** *3* (21), 21–40.

Chen, J.; Liu, W.; Liu, C. M.; Li, T.; Liang, R. H.; Luo, S. J. Pectin Modifications: A Review. *Crit. Rev. Food Sci. Nutr.* **2015,** *55* (12), 1684–1698.

Clark, M. A.; Hirst, B. H.; Jepson, M. A. Lectin-mediated Mucosal Delivery of Drugs and Microparticles. *Adv. Drug Deliv. Rev.* **2000,** *43* (2), 207–223.

Coelho, J. F.; Ferreira, P. C.; Alves, P.; Cordeiro, R.; Fonseca, A. C.; Góis, J. R.; Gil, M. H. Drug Delivery Systems: Advanced Technologies Potentially Applicable in Personalized Treatments. *EPMA J.* **2010,** *1* (1), 164–209.

Coviello, T.; Matricardi, P.; Marianecci, C.; Alhaique, F. Polysaccharide Hydrogels for Modified Release Formulations. *J. Control. Release* **2007,** *119*, 5–24.

d'Ayala, G. G.; Malinconico, M.; Laurienzo, P. Marine Derived Polysaccharides for Biomedical Applications: Chemical Modification Approaches. *Molecules* **2008,** *13* (9), 2069–2106.

Dames, P.; Gleich, B.; Flemmer, A.; Hajek, K.; Seidl, N.; Wiekhorst, F.; Eberbeck, D.; Bittmann, I.; Bergemann, C.; Weyh, T.; Trahms, L.; Rosenecker, J.; Rudolph, C. Targeted Delivery of Magnetic Aerosol Droplets to the Lung. *Nat. Nanotechnol.* **2007,** *2*, 495–499.

De Santis, S.; Diociaiuti, M.; Cametti C.; Masci, G. Hyaluronic Acid and Alginate Covalent Nanogels by Template Cross-linking in Polyion Complex Micelle Nanoreactors. *Carbohydr. Polym.* **2014**, *101*, 96–103.

Debele, T. A.; Mekuria, S. L.; Tsai, H. C. Polysaccharide Based Nanogels in the Drug Delivery System: Application as the Carrier of Pharmaceutical Agents. *Mater. Sci. Eng.* C **2016**, *68*, 964–981.

Devi, V. K.; Jain, N.; Valli, K. S. Importance of Novel Drug Delivery Systems in Herbal Medicines. *Pharmacogn. Rev.* **2010**, *4* (7), 27–31.

Dhar, P.; Fischer, T. M.; Wang, Y.; Mallouk, T. E.; Paxton, W. F.; Sen, A. Autonomously Moving Nanorods at a Viscous Interface. *Nano Lett.* **2006**, *6* (1), 66–72.

Eckmann, D. M.; Composto, R. J.; Tsourkas, A.; Muzykantov, V. R. Nanogel Carrier Design for Targeted Drug Delivery. *J. Mater. Chem. B* **2014**, *2* (46), 8085–8097.

Egea, J.; García, A. G.; Verges, J.; Montell, E.; López, M. G. Antioxidant, Antiinflammatory and Neuroprotective Actions of Chondroitin Sulfate and Proteoglycans. *Osteoarthr. Cartil.* **2010**, *18*, S24–S27.

Ferrer, M. C.; Dastgheyb, S.; Hickok, N. J.; Eckmann, D. M.; Composto, R. J. Designing Nanogel Carriers for Antibacterial Applications. *Acta Biomater.* **2014**, *10*, 2105–2111.

Ferry, J. D. *Viscoelastic Properties of Polymers*, 3rd ed.; John Wiley & Sons: Canada, 1980; pp 486–544.

Fujioka-Kobayashi, M.; Ota, M. S.; Shimoda, A.; Nakahama, K. I.; Akiyoshi, K.; Miyamoto, Y.; Iseki, S. Cholesteryl Group- and Acryloyl Group-bearing Pullulan Nanogel to Deliver BMP2 and FGF18 for Bone Tissue Engineering. *Biomaterials* **2012**, *33*, 7613–7620.

Gandhi, S. D.; Pandya, P. R.; Umbarkar, R.; Tambawala, T.; Shah, M. A. Mucoadhesive Drug Delivery Systems an Unusual Maneuver for Site Specific Drug Delivery System. *Int. J. Pharma. Sci.* **2011**, *2* (3), 132–152.

Gaur, R.; Singh, R.; Gupta, M.; Gaur, M. K. *Aureobasidium pullulans*, an Economically Important Polymorphic Yeast with Special Reference to Pullulan. *Afr. J. Biotechnol.* **2015**, *9*, 7989–7997.

George, M.; Abraham, T. E. Polyionic Hydrocolloids for the Intestinal Delivery of Protein Drugs: Alginate and Chitosan—A Review. *J. Control. Release* **2006**, *114* (1), 1–14.

Glenn, D. P.; Jing-wen, K. Chemically-modified HA for Therapy and Regenerative Medicine. *Curr. Pharm. Biotechnol.* **2008**, *9*, 242–245.

Go, Y. M.; Jones, D. P. Redox Compartmentalization in Eukaryotic Cells. *Biochim. Biophys. Acta.* **2008**, *1780* (11), 1273–1290.

Golan, D. E.; Tashjian, A. H.; Armstrong, E. J. *Principles of Pharmacology: The Pathophysiologic Basis of Drug Therapy*, 2nd ed.; Lippincott Williams & Wilkins: Philadelphia, 2011; pp 49–62.

Goncalves, C.; Pereira, P.; Gama, M. Self-assembled Hydrogel Nanoparticles for Drug Delivery Applications. *Materials* **2010**, *3*, 1420–1460.

Grenha, A. Chitosan Nanoparticles: A Survey of Preparation Methods. *J. Drug Target.* **2012**, *20* (4), 291–300.

Gulrez, S. K.; Phillips, G. O.; Al-Assaf, S. Hydrogels: Methods of Preparation, Characterisation and Applications. In *Progress in Molecular and Environmental Bioengineering—From Analysis and Modeling to Technology Applications*; Carpi, A., Ed.; INTECH: Croatia, 2011; pp 117–151.

Hafrén, J.; Zou, W.; Córdova, A. Heterogeneous 'Organoclick' Derivatization of Polysaccharides. *Macromol. Rapid Commun.* **2006**, *27* (16), 1362–1366.

Hagerstrom, H.; Edsman, K.; Strømme, M. Low-frequency Dielectric Spectroscopy as a Tool for Studying the Compatibility Between Pharmaceutical Gels and Mucous Tissue. *J. Pharm. Sci.* **2003,** *92* (9), 1869–1881.

Hamidi, M.; Azadi, A.; Rafiei, P. Hydrogel Nanoparticles in Drug Delivery. *Adv. Drug Deliv. Rev.* **2008,** *60* (15), 1638–1649.

Harada, A.; Kataoka, K. Chain Length Recognition: Core-shell Supramolecular Assembly from Oppositely Charged Block Copolymers. *Science* **1999,** *283* (5398), 65–67.

Hayashi, C.; Hasegawa, U.; Saita, Y.; Hemmi, H.; Hayata, T.; Nakashima, K.; Noda, M. Osteoblastic Bone Formation is Induced by Using Nanogel-crosslinking Hydrogel as Novel Scaffold for Bone Growth Factor. *J. Cell. Physiol.* **2009,** *220* (1), 1–7.

Huang, Y.; Leobandung, W.; Foss, A.; Peppas, N. A. Molecular Aspects of Muco-and bioadhesion: Tethered Structures and Site-specific Surfaces. *J. Control. Release* **2000,** *65* (1), 63–71.

Huang, S. J.; Sun, S. L.; Feng, T. H.; Sung, K. H.; Lui, W. L.; Wang, L. F. Folate-mediated Chondroitin Sulfate-Pluronic® 127 Nanogels as a Drug Carrier. *Eur. J. Pharm. Sci.* **2009,** *38* (1), 64–73.

Ikada, Y.; Tsuji, H. Biodegradable Polyesters for Medical and Ecological Applications. *Macromol. Rapid Commun.* **2000,** *21* (3), 117–132.

Ilan, N.; Elkin, M.; Vlodavsky, I. Regulation, Function and Clinical Significance of Heparanase in Cancer Metastasis and Angiogenesis. *Int. J. Biochem. Cell Biol.* **2006,** *38* (12), 2018–2039.

Imam, M. E.; Hornof, M.; Valenta, C.; Reznicek, G.; Bernkop-Schnürch, A. Evidence for the Interpenetration of Mucoadhesive Polymers into the Mucous Gel Layer. *STP Pharma. Sci.* **2003,** *13* (3), 171–176.

Jain K. K. Current Status and Future Prospects of Drug Delivery Systems. In *Drug Delivery Systems*; Jain, K. K., Ed.; Humana: New York, 2008; pp 1–56.

Jatariu, A. N.; Popa, M.; Curteanu, S.; Peptu, C. A. Covalent and Ionic Co-cross-linking-An Original Way to Prepare Chitosan-Gelatin Hydrogels for Biomedical Applications. *J. Biomed. Mater. Res. Part A.* **2011,** *98* (3), 342–350.

Joung, Y. K.; Jang, J. Y.; Choi, J. H.; Han, D. K.; Park, K. D. Heparin-conjugated Pluronic Nanogels as Multi-drug Nanocarriers for Combination Chemotherapy. *Mol. Pharm.* **2012,** *10* (2), 685–693.

Kabanov, A. V.; Vinogradov, S. V. Nanogels as Pharmaceutical Carriers: Finite Networks of Infinite Capabilities. *Angew. Chem. Int. Ed.* **2009,** *48* (30), 5418–5429.

Kang, S. A.; Jang, K. H.; Seo, J. W.; Kim, K. H.; Kim, Y. H.; Rairakhwada, D.; Rhee, S. K. Levan: Applications and Perspectives. In *Microbial Production of Biopolymers and Polymer Precursors: Applications and Perspectives*; Rehm, B., Ed.; Caister Academic Press: Norfolk, 2009; pp 145–162.

Kazak Sarilmiser, H.; Toksoy Oner, E. Investigation of Anti-cancer Activity of Linear and Aldehyde-activated Levan from *Halomonas smyrnensis* AAD6 T. *Biochem. Eng. J.* **2014,** *92*, 28–34.

Khanvilkar, K.; Donovan, M. D.; Flanagan, D. R. Drug Transfer Through Mucus. *Adv. Drug Deliv. Rev.* **2001,** *48* (2), 173–193.

Kobayashi, H.; Katakura, O.; Morimoto, N.; Akiyoshi, K.; Kasugai, S. Effects of Cholesterol-bearing Pullulan (CHP)-nanogels in Combination with Prostaglandin E1 on Wound Healing. *J. Biomed. Mater. Res. B Appl. Biomater.* **2009,** *9* (1), 55–60.

Kommareddy, S.; Amiji, M. Preparation and Evaluation of Thiol-modified Gelatin Nanoparticles for Intracellular DNA Delivery in Response to Glutathione. *Bioconjugate Chem.* **2005,** *16* (6), 1423–1432.

Kommareddy, S.; Amiji, M. Poly(ethylene glycol)–modified Thiolated Gelatin Nanoparticles for Glutathione-responsive Intracellular DNA Delivery. *Nanomed. Nanotechnol. Biol. Med.* **2007,** *3* (1), 32–42.

Kozlowska, D.; Foran, P.; MacMahon, P.; Shelly, M. J.; Eustace, S.; O'Kennedy, R. Molecular and Magnetic Resonance Imaging: The Value of Immunoliposomes. *Adv. Drug Deliv. Rev.* **2009,** *61* (15), 1402–1411.

Kuhn, S. J.; Hallahan, D. E.; Giorgio, T. D. Characterization of Superparamagnetic Nanoparticle Interactions with Extracellular Matrix in an *In vitro* System. *Ann. Biomed. Eng.* **2006,** *34* (1), 51–58.

Kumari, A.; Yadav, S. K.; Yadav, S. C. Biodegradable Polymeric Nanoparticles Based Drug Delivery Systems. *Colloids Surf. B* **2010,** *75* (1), 1–18.

Kuo, C. K.; Ma, P. X. Ionically Crosslinked Alginate Hydrogels as Scaffolds for Tissue Engineering: Part 1. Structure, Gelation Rate and Mechanical Properties. *Biomaterials* **2001,** *22* (6), 511–521.

Kuppusamy, P.; Afeworki, M.; Shankar, R. A.; Coffin, D.; Krishna, M. C.; Hahn, S. M.; Zweier, J. L. *In vivo* Electron Paramagnetic Resonance Imaging of Tumor Heterogeneity and Oxygenation in a Murine Model. *Cancer Res.* **1998,** *58* (7), 1562–1568.

Kuppusamy, P.; Li, H.; Ilangovan, G.; Cardounel, A. J.; Zweier, J. L.; Yamada, K.; Mitchell, J. B. Noninvasive Imaging of Tumor Redox Status and its Modification by Tissue Glutathione Levels. *Cancer Res.* **2002,** *62* (1), 307–312.

Lai, S. K.; Wang, Y. Y.; Hanes, J. Mucus-penetrating Nanoparticles for Drug and Gene Delivery to Mucosal Tissues. *Adv. Drug Deliv. Rev.* **2009,** *61* (2), 158–171.

Layre, A. M.; Volet, G.; Wintgens, V.; Amiel, C. Associative Network Based on Cyclodextrin Polymer: A Model System for Drug Delivery. *Biomacromolecules* **2009,** *10* (12), 3283–3289.

Lehr, C. M. Lectin-mediated Drug Delivery: The Second Generation of Bioadhesives. *J. Control. Release* **2000,** *65* (1), 19–29.

LeRoux, M. A.; Guilak, F.; Setton, L. A. Compressive and Shear Properties of Alginate Gel: Effects of Sodium Ions and Alginate Concentration. *J. Biomed. Mater. Res.* **1999,** *47,* 46–53.

Li, J.; Yu, S. Y.; Yao, P.; Jiang, M. Lysozyme-dextran Core-shell Nanogels Prepared via a Green Process. *Langmuir* **2008,** *24,* 3486–3492.

Li, L.; Moon, H. T.; Park, J. Y.; Heo, Y. J.; Choi, Y.; Tran, T. H.; Huh, K. M. Heparin-based Self-assembled Nanoparticles for Photodynamic Therapy. *Macromol. Res.* **2011,** *19* (5), 487–494.

Li, S.; Xiong, Q.; Lai, X.; Li, X.; Wan, M.; Zhang, J.; Zhang, D. Molecular Modification of Polysaccharides and Resulting Bioactivities. *Compr. Rev. Food Sci. Food Saf.* **2016,** *15* (2), 237–250.

Li, Y. Y.; Zhang, X. Z.; Kim, G. C.; Cheng, H.; Cheng, S. X.; Zhuo, R. X. Thermosensitive Y-shaped Micelles of Poly(oleic acid-Y-N-isopropylacrylamide) for Drug Delivery. *Small* **2006,** *2* (7), 917–923.

Li, Z.; Xu, W.; Zhang, C.; Chen, Y.; Li, B. Self-assembled Lysozyme/Carboxymethylcellulose Nanogels for Delivery of Methotrexate. *Int. J. Biol. Macromol.* **2015,** *75,* 166–172.

Liao, Y. H.; Jones, S. A.; Forbes, B.; Martin, G. P.; Brown, M. B. Hyaluronan: Pharmaceutical Characterization and Drug Delivery. *Drug Delivery* **2005,** *12* (6), 327–342.

Lieleg, O.; Ribbeck, K. Biological Hydrogels as Selective Diffusion Barriers. *Trends Cell Biol.* **2001,** *21* (9), 543–551.

Liu, Z.; Jiao, Y.; Wang, Y.; Zhou, C.; Zhang, Z. Polysaccharides-based Nanoparticles as Drug Delivery Systems. *Adv. Drug Deliv. Rev.* **2008,** *60* (15), 1650–1662.

Lohani, A.; Chaudhary, G. P. Mucoadhesive Microspheres: A Novel Approach to Increase Gastroretention. *Chron Young Sci.* **2012,** *3* (2), 121–128.

Look, M.; Stern, E.; Wang, Q. A.; DiPlacido, L. D.; Kashgarian, M.; Craft, J.; Fahmy, T. M. Nanogel-based Delivery of Mycophenolic Acid Ameliorates Systemic Lupus Erythematosus in Mice. *J. Clin. Invest.* **2013,** *123* (4), 1741–1749.

Ludwig, A. The Use of Mucoadhesive Polymers in Ocular Drug Delivery. *Adv. Drug Deliv. Rev.* **2005,** *57* (11), 1595–1639.

Luo, C.; Sun, J.; Sun, B.; He, Z. Prodrug-based Nanoparticulate Drug Delivery Strategies for Cancer Therapy. *Trends Pharmacol. Sci.* **2014,** *35* (11), 556–566.

Maciel, V. B. V.; Yoshida, C. M.; Franco, T. T. Chitosan/Pectin Polyelectrolyte Complex as a pH Indicator. *Carbohydr. Polym.* **2015,** *132*, 537–545.

Mahajan, P.; Kaur, A.; Aggarwal, G.; Harikumar, S. L. Mucoadhesive Drug Delivery System: A Review. *Int. J. Drug Dev. Res.* **2013,** *5* (1), 11–20.

Manchun, S.; Dass, C. R.; Sriamornsak, P. Targeted Therapy for Cancer Using pH-Responsive Nanocarrier Systems. *Life Sci.* **2012,** *90* (11), 381–387.

Marques, A. P.; Reis, R. L.; Hunt, J. A. The Biocompatibility of Novel Starch-based Polymers and Composites: *In vitro* Studies. *Biomaterials* **2002,** *23* (6), 1471–1478.

Matai, I.; Gopinath, P. Chemically Crosslinked Hybrid Nanogels of Alginate and PAMAM Dendrimers as Efficient Anticancer Drug Delivery Vehicles. *ACS Biomater. Sci. Eng.* **2016,** *2* (2), 213–223.

Mbuya, V.; Vishal, G.; Tenzin, T. Application of Nanogels in Reduction of Drug Resistance in Cancer Chemotherapy. *J. Chem. Pharm. Res.* **2016,** *8* (2), 556–561.

Mendes, A. C.; Baran, E. T.; Reis, R. L.; Azevedo, H. S. Self-assembly in Nature: Using the Principles of Nature to Create Complex Nanobiomaterials. *Wiley Interdiscip. Rev. Nanomed. Nanobiotechnol.* **2013,** *5* (6), 582–612.

Meng, F.; Hennink, W. E.; Zhong, Z. Reduction-sensitive Polymers and Bioconjugates for Biomedical Applications. *Biomaterials* **2009,** *30* (12), 2180–2198.

Mero, A.; Campisi, M. Hyaluronic Acid Bioconjugates for the Delivery of Bioactive Molecules. *Polymers* **2014,** *6* (2), 346–369.

Migneault, I.; Dartiguenave, C.; Bertrand, M. J.; Waldron, K. C. Glutaraldehyde: Behavior in Aqueous Solution, Reaction with Proteins, and Application to Enzyme Crosslinking. *Biotechniques* **2004,** *37* (5), 790–806.

Mischnick, P.; Momcilovic, D. Chemical Structure Analysis of Starch and Cellulose Derivatives. *Adv. Carbohydr. Chem. Biochem.* **2010,** *64*, 117–210.

Miyahara, T.; Nyan, M.; Shimoda, A.; Yamamoto, Y.; Kuroda, S.; Shiota, M.; Akiyoshi, K.; Kasugai, S. Exploitation of a Novel Polysaccharide Nanogel Cross-linking Membrane for Guided Bone Regeneration (GBR). *J. Tissue Eng. Regen. Med.* **2012,** *6*, 666–672.

Mizrahy, S.; Peer, D. Polysaccharides as Building Blocks for Nanotherapeutics. *Chem. Soc. Rev.* **2012,** *41* (7), 2623–2640.

Mortazavi, S. A.; Smart, J. D. An Investigation into the Role of Water Movement and Mucus Gel Dehydration in Mucoadhesion. *J. Control. Release* **1993,** *25* (3), 197–203.

Mousa, S. A.; Petersen, L. J. Anti-cancer Properties of Low-molecular-weight Heparin: Preclinical Evidence. *Thromb Haemost.* **2009**, *102* (2), 258–267.

Moya-Ortega, M. D.; Alvarez-Lorenzo, C.; Sigurdsson, H. H.; Concheiro, A.; Loftsson, T. Crosslinked Hydroxypropyl-β-cyclodextrin and γ-Cyclodextrin Nanogels for Drug Delivery: Physicochemical and Loading/Release Properties. *Carbohydr. Polym.* **2012**, *87* (3), 2344–2351.

Myrick, J. M.; Vendra, V. K.; Krishnan, S. Self-assembled Polysaccharide Nanostructures for Controlled-release Applications. *Nanotechnology Rev.* **2014**, *3* (4), 319–346.

Na, K.; Park, K. H.; Kim, S. W.; Bae, Y. H. Self-assembled Hydrogel Nanoparticles from Curdlan Derivatives: Characterization, Anti-cancer Drug Release and Interaction with a Hepatoma Cell Line (HepG2). *[J. Control. Release* **2000**, *69* (2), 225–236.

Nagahama, K.; Mori, Y.; Ohya, Y.; Ouchi, T. Biodegradable Nanogel Formation of Poly-lactide-grafted Dextran Copolymer in Dilute Aqueous Solution and Enhancement of its Stability by Stereocomplexation. *Biomacromolecules* **2007**, *8* (7), 2135–2141.

Nagpal, K.; Singh, S. K.; Mishra, D. N. Chitosan Nanoparticles: A Promising System in Novel Drug Delivery. *Chem. Pharm. Bull.* **2010**, *58* (11), 1423–1430.

Nitta, S. K.; Numata, K. Biopolymer-based Nanoparticles for Drug/Gene Delivery and Tissue Engineering. *Int. J. Mol. Sci.* **2013**, *14* (1), 1629–1654.

Oh, J. K.; Siegwart, D. J.; Matyjaszewski, K. Synthesis and Biodegradation of Nanogels as Delivery Carriers for Carbohydrate Drugs. *Biomacromolecules* **2007**, *8* (11), 3326–3331.

Oh, J. K.; Drumright, R.; Siegwart, D. J.; Matyjaszewski, K. The Development of Microgels/Nanogels for Drug Delivery Applications. *Prog. Polym. Sci.* **2008**, *33* (4), 448–477.

Oishi, M.; Sumitani, S.; Nagasaki, Y. On-off Regulation of 19F Magnetic Resonance Signals Based on pH-sensitive PEGylated Nanogels for Potential Tumor-specific Smart 19F MRI Probes. *Bioconjugate Chem.* **2007**, *18* (5), 1379–1382.

Olmsted, S. S.; Padgett, J. L.; Yudin, A. I.; Whaley, K. J.; Moench, T. R.; Cone, R. A. Diffusion of Macromolecules and Virus-like Particles in Human Cervical Mucus. *Biophys. J.* **2001**, *81* (4), 1930–1937.

Ozin, G. A.; Hou, K.; Lotsch, B. V.; Cademartiri, L.; Puzzo, D. P.; Scotognella, F.; Thomson, J. Nanofabrication by Self-assembly. *Mater. Today* **2009**, *12* (5), 12–23.

Paddock, C. Nanogel to Manage Type 1 Diabetes. Medical News Today [Online] May 17, 2013, http://www.medicalnewstoday.com/articles/260664.php (accessed Nov 29, 2016).

Park, S. N.; Park, J. C.; Kim, H. O.; Song, M. J.; Suh, H. Characterization of Porous Collagen/Hyaluronic Acid Scaffold Modified by 1-ethyl-3-(3-dimethylaminopropyl) Carbodiimide Cross-linking. *Biomaterials* **2002**, *23* (4), 1205–1212.

Park, W.; Park, S. J.; Na, K. Potential of Self-organizing Nanogel with Acetylated Chondroitin Sulfate as an Anti-cancer Drug Carrier. *Colloids Surf. B* **2010**, *79* (2), 501–508.

Pawar, S. N.; Edgar, K. J. Alginate Derivatization: A Review of Chemistry, Properties and Applications. *Biomaterials* **2012**, *33* (11), 3279–3305.

Peng, H. S.; Stolwijk, J. A.; Sun, L. N.; Wegener, J.; Wolfbeis, O. S. A Nanogel for Ratiometric Fluorescent Sensing of Intracellular pH Values. *Angew. Chem. Int. Ed. Engl.* **2010**, *49*, 4246–4249.

Peppas, N. A.; Huang, Y. Nanoscale Technology of Mucoadhesive Interactions. *Adv. Drug Deliv. Rev.* **2004**, *56* (11), 1675–1687.

Plunkett, K. N.; Berkowski, K. L.; Moore, J. S. Chymotrypsin Responsive Hydrogel: Application of a Disulfide Exchange Protocol for the Preparation of Methacrylamide Containing Peptides. *Biomacromolecules* **2005**, *6* (2), 632–637.

Posocco, B.; Dreussi, E.; de Santa, J.; Toffoli, G.; Abrami, M.; Musiani, F.; Dapas, B. Polysaccharides for the Delivery of Antitumor Drugs. *Materials* **2015**, *8* (5), 2569–2615.

Prestwich, G. D. Hyaluronic Acid-based Clinical Biomaterials Derived for Cell and Molecule Delivery in Regenerative Medicine. *J. Control. Release* **2011**, *155* (2), 193–199.

Rachid, M.; Pamela, K.; Tobias, N.; Irina, N.; Markus, W. B.; Margot, Z. CD44 and EpCAM: Cancer-initiating Cell Markers. *Curr. Mol. Med.* **2008**, *8*, 784–804.

Raina, S.; Missiakas, D. Making and Breaking Disulfide Bonds. *Annu. Rev. Microbiol.* **1997**, *51* (1), 179–202.

Renuart, E.; Viney, C. Biological Fibrous Materials: Self-assembled Structures and Optimised Properties. In *Pergamon Materials Series*; Elices, M., Ed.; Pergamon/Elsevier Science: Oxford, 2000; pp 221–267.

Robinson, D. H.; Mauger, J. W. Drug Delivery Systems. *Am. J. Health Syst. Pharm.* **1991**, *48* (10), S14–S23.

Robyt, J. F. Starch: Structure, Properties, Chemistry, and Enzymology. *Glycoscience* **2008**, 1437–1472.

Rodell, C. B.; Mealy, J. E.; Burdick, J. A. Supramolecular Guest–Host Interactions for the Preparation of Biomedical Materials. *Bioconjugate Chem.* **2015**, *26* (12), 2279–2289.

Rodrigues, S.; Cardoso, L.; da Costa, A. M. R.; Grenha, A. Biocompatibility and Stability of Polysaccharide Polyelectrolyte Complexes Aimed at Respiratory Delivery. *Materials* **2015**, *8* (9), 5647–5670.

Rolland, J.; Guillet, P.; Schumers, J. M.; Duhem, N.; Préat, V.; Gohy, J. F. Polyelectrolyte Complex Nanoparticles from Chitosan and Poly(Acrylic Acid) and Polystyrene-block-poly(Acrylic Acid). *J. Polym. Sci. A Polym. Chem.* **2012**, *50* (21), 4484–4493.

Rosiak, J. M.; Yoshii, F. Hydrogels and Their Medical Applications. *Nucl. Instr. Meth. Phys. Res. B* **1999**, *151* (1), 56–64.

Round, A. N.; Berry, M.; McMaster, T. J.; Stoll, S.; Gowers, D.; Corfield, A. P.; Miles, M. J. Heterogeneity and Persistence Length in Human Ocular Mucins. *Biophys. J.* **2002**, *83* (3), 1661–1670.

Rubinstein, M.; Colby, R. H. *Polymer Physics*; Oxford University Press: Oxford, 2003; pp 199–252.

Saito, G.; Swanson, J. A.; Lee, K. D. Drug Delivery Strategy Utilizing Conjugation via Reversible Disulfide Linkages: Role and Site of Cellular Reducing Activities. *Adv. Drug Deliv. Rev.* **2003**, *55* (2), 199–215.

Sakaguchi, N.; Kojima, C.; Harada, A.; Kono, K. Preparation of pH-sensitive Poly(Glycidol) Derivatives with Varying Hydrophobicities: Their Ability to Sensitize Stable Liposomes to pH. *Bioconjugate Chem.* **2008**, *19* (5), 1040–1048.

Sanders, N.; Rudolph, C.; Braeckmans, K.; De Smedt, S. C.; Demeester, J. Extracellular Barriers in Respiratory Gene Therapy. *Adv. Drug Deliv. Rev.* **2009**, *61*, 115–127.

Santander-Ortega, M. J.; Stauner, T.; Loretz, B.; Ortega-Vinuesa, J. L.; Bastos-González, D.; Wenz, G.; Lehr, C. M. Nanoparticles Made from Novel Starch Derivatives for Transdermal Drug Delivery. *J. Control. Release* **2010**, *141* (1), 85–92.

Saravanakumar, G.; Jo, D. G.; Park, J. H. Polysaccharide-based Nanoparticles: A Versatile Platform for Drug Delivery and Biomedical Imaging. *Curr. Med. Chem.* **2012**, *19* (19), 3212–3229.

Sarika, P. R.; Kumar, P. A.; Raj, D. K.; James, N. R. Nanogels Based on Alginic Aldehyde and Gelatin by Inverse Miniemulsion Technique: Synthesis and Characterization. *Carbohydr. Polym.* **2015a**, *119*, 118–125.

Sarika, P. R.; Pavithran, A.; James, N. R. Cationized Gelatin/Gum Arabic Polyelectrolyte Complex: Study of Electrostatic Interactions. *Food Hydrocoll.* **2015b,** *49,* 176–182.

Schanté, C. E.; Zuber, G.; Herlin, C.; Vandamme, T. F. Chemical Modifications of Hyaluronic Acid for the Synthesis of Derivatives for a Broad Range of Biomedical Applications. *Carbohydr. Polym.* **2011,** *85* (3), 469–489.

Schiffman, J. D.; Schauer, C. L. Cross-linking Chitosan Nanofibers. *Biomacromolecules* **2007,** *8* (2), 594–601.

Schmitz, T.; Bravo-Osuna, I.; Vauthier, C.; Ponchel, G.; Loretz, B.; Bernkop-Schnürch, A. Development and *In vitro* Evaluation of a Thiomer-based Nanoparticulate Gene Delivery System. *Biomaterials* **2007,** *28* (3), 524–531.

Sezer, A. D.; Kazak, H.; Toksoy Oner, E.; Akbuğa, J. Levan-based Nanocarrier System for Peptide and Protein Drug Delivery: Optimization and Influence of Experimental Parameters on the Nanoparticle Characteristics. *Carbohydr. Polym.* **2011,** *84* (1), 358–363.

Shah, P. P.; Desai, P. R.; Patel, A. R.; Singh, M. S. Skin Permeating Nanogel for the Cutaneous Co-delivery of Two Anti-inflammatory Drugs. *Biomaterial* **2012,** *33* (5), 1607–1617.

Sigurdsson, H. H.; Kirch, J.; Lehr, C. M. Mucus as a Barrier to Lipophilic Drugs. *Int. J. Pharm.* **2013,** *453* (1), 56–64.

Sims, D. E.; Horne, M. M. Heterogeneity of the Composition and Thickness of Tracheal Mucus in Rats. *Am. J. Physiol. Lung Cell Mol. Physiol.* **1997,** *273* (5), L1036–L1041.

Singh, R. S.; Kaur, N.; Kennedy, J. F. Pullulan and Pullulan Derivatives as Promising Biomolecules for Drug and Gene Targeting. *Carbohydr. Polym.* **2015,** *123,* 190–207.

Sinha, V. R.; Singla, A. K.; Wadhawan, S.; Kaushik, R.; Kumria, R.; Bansal, K.; Dhawan, S. Chitosan Microspheres as a Potential Carrier for Drugs. *Int. J. Pharm.* **2004,** *274* (1), 1–33.

Smart, J. D. The Basics and Underlying Mechanisms of Mucoadhesion. *Adv. Drug Deliv. Rev.* **2005,** *57* (11), 1556–1568.

Sogias, I. A.; Williams, A. C.; Khutoryanskiy, V. V. Why is Chitosan Mucoadhesive? *Biomacromolecules* **2008,** *9* (7), 1837–1842.

Song, Y.; Jane, J. Characterization of Barley Starches of Waxy, Normal, and High Amylose Varieties. *Carbohydr. Polym.* **2000,** *41* (4), 365–377.

Soni, G.; Yadav, K. S. Nanogels as Potential Nanomedicine Carrier for Treatment of Cancer: A Mini Review of the State of the Art. *Saudi Pharm. J.* **2016,** *24* (2), 133–139.

Stevenson, J. L.; Varki, A.; Borsig, L. Heparin Attenuates Metastasis Mainly due to Inhibition of P-and L-selectin, but Non-anticoagulant Heparins Can Have Additional Effects. *Thromb. Res.* **2007,** *120,* S107–S111.

Sudhakar, Y.; Kuotsu, K.; Bandyopadhyay, A. K. Buccal Bioadhesive Drug Delivery—A Promising Option for Orally Less Efficient Drugs. *J. Control. Release* **2006,** *114* (1), 15–40.

Sultana, F.; Imran-Ul-Haque, M.; Arafat, M.; Sharmin, S. An Overview of Nanogel Drug Delivery System. *J. App. Pharm. Sci.* **2013,** *3,* S95–S105.

Taggart, P.; Mitchel, J. R. Starch. In *Handbook of Hydrocolloids,* 2nd ed.; Phillips, G. O., Williams, P. A., Eds.; Woodhead Publishing Limited: Cambridge, England, 2009; pp 108–140.

Tangri, P.; Madhav, N. S. Oral Mucoadhesive Drug Delivery Systems: A Review. *Int. J. Biopharm.* **2011,** *2* (1), 36–46.

Tiwari, G.; Tiwari, R.; Sriwastawa, B.; Bhati, L.; Pandey, S.; Pandey, P.; Bannerjee, S. K. Drug Delivery Systems: An Updated Review. *Int. J. Pharm. Investig.* **2012,** *2,* 2–11.

Toita, S.; Sawada, S. I.; Kiyoshi, K. Polysaccharide Nanogel Gene Delivery System with Endosome-escaping Function: Co-delivery of Plasmid DNA and Phospholipase A 2. *J. Control. Release* **2011,** *155* (1), 54–59.

Toksoy Oner, E.; Hernández, L.; Combie, J. Review of Levan Polysaccharide: From a Century of Past Experiences to Future Prospects. *Biotechnol. Adv.* **2016,** *34* (5), 827–844.

Tripodo, G.; Trapani, A.; Torre, M. L.; Giammona, G.; Trapani, G.; Mandracchia, D. Hyaluronic Acid and Its Derivatives in Drug Delivery and Imaging: Recent Advances and Challenges. *Eur. J. Pharm. Biopharm.* **2015,** *97,* 400–416.

Uebelhart, D. Clinical Review of Chondroitin Sulfate in Osteoarthritis. *Osteoarthr. Cartil.* **2008,** *16,* S19–S21.

Ugwoke, M. I.; Agu, R. U.; Verbeke, N.; Kinget, R. Nasal Mucoadhesive Drug Delivery: Background, Applications, Trends and Future Perspectives. *Adv. Drug Deliv. Rev.* **2005,** *57* (11), 1640–1665.

Uretimi, D. Production of Dextran by Newly Isolated Strains of *Leuconostoc mesenteroides* PCSIR-4 and PCSIR-9. *Türk Biyokimya Dergisi Turk J Biochem].* **2005,** *31* (1), 21–26.

Vallée, F.; Müller, C.; Durand, A.; Schimchowitsch, S.; Dellacherie, E.; Kelche, C.; Leonard, M. Synthesis and Rheological Properties of Hydrogels Based on Amphiphilic Alginateamide Derivatives. *Carbohydr. Res.* **2009,** *344* (2), 223–228.

Vijayaraghavalu, S.; Labhasetwar, V. Efficacy of Decitabine-loaded Nanogels in Overcoming Cancer Drug Resistance Is Mediated via Sustained DNA Methyltransferase 1 (DNMT1) Depletion. *Cancer Lett.* **2013,** *331* (1), 122–129.

Vinogradov, S. V.; Batrakova, E. V.; Kabanov, A. V. Nanogels for Oligonucleotide Delivery to the Brain. *Bioconjugate Chem.* **2004,** *15* (1), 50–60.

Vinogradov, S. V.; Zeman, A. D.; Batrakova, E. V.; Kabanov, A.V. Polyplex Nanogel Formulations for Drug Delivery of Cytotoxic Nucleoside Analogs. *J. Control. Release* **2005,** *107* (1), 143–157.

Vlodavsky, I.; Ilan, N.; Nadir, Y.; Brenner, B.; Katz, B. Z.; Naggi, A.; Sasisekharan, R. Heparanase, Heparin and the Coagulation System in Cancer Progression. *Thromb. Res.* **2007,** *120,* S112–S120.

Voragen, A. G. J.; Coenen, G. J.; Verhoef, R. P.; Schols, H. A. Pectin, a Versatile Polysaccharide Present in Plant Cell Walls. *Struct. Chem.* **2009,** *20,* 263–275.

Wang, L. Y.; Ma, G. H.; Su, Z. G. Preparation of Uniform Sized Chitosan Microspheres by Membrane Emulsification Technique and Application as a Carrier of Protein Drug. *J. Control. Release* **2005,** *106* (1), 62–75.

Wen, Y.; Oh, J. K. Recent Strategies to Develop Polysaccharide-based Nanomaterials for Biomedical Applications. *Macromol. Rapid Commun.* **2014,** *35* (21), 1819–1832.

Wintgens, V.; Layre, A. M.; Hourdet, D.; Amiel, C. Cyclodextrin Polymer Nanoassemblies: Strategies for Stability Improvement. *Biomacromolecules* **2012,** *13* (2), 528–534.

Wu, G.; Fang, Y. Z.; Yang, S.; Lupton, J. R.; Turner, N. D. Glutathione Metabolism and its Implications for Health. *J. Nutr.* **2004,** *134* (3), 489–492.

Wu, L.; Zhou, H.; Sun, H. J.; Zhao, Y.; Yang, X.; Cheng, S. Z.; Yang, G. Thermoresponsive Bacterial Cellulose Whisker/Poly(NIPAM-co-BMA) Nanogel Complexes: Synthesis, Characterization, and Biological Evaluation. *Biomacromolecules* **2013,** *14* (4), 1078–1084.

Xu, Z.; Shen, G.; Xia, X.; Zhao, X.; Zhang, P.; Wu, H.; Liang, S. Comparisons of Three Polyethyleneimine-derived Nanoparticles as a Gene Therapy Delivery System for Renal Cell Carcinoma. [*J. Transl. Med.* **2011,** *9,* 46.

Xu, W.; Ding, J.; Xiao, C.; Li, L.; Zhuang, X.; Chen, X. Versatile Preparation of Intracellular-acidity-sensitive Oxime-linked Polysaccharide-Doxorubicin Conjugate for Malignancy Therapeutic. *Biomaterials* **2015,** *54,* 72–86.

Yagi, H.; Miyamoto, S.; Tanaka, Y.; Sonoda, K.; Kobayashi, H.; Kishikawa, T.; Nakano, H. Clinical Significance of Heparin-binding Epidermal Growth Factor-like Growth Factor in Peritoneal Fluid of Ovarian Cancer. *Br. J. Cancer* **2005,** *92* (9), 1737–1745.

Yallapu, M. M.; Reddy, M. K.; Labhasetwar, V. Nanogels: Chemistry to Drug Delivery. *J. Biomed. Nanotech.* **2007,** *3*, 131–171.

Yallapu, M. M.; Jaggi, M.; Chauhan, S. C. Design and Engineering of Nanogels for Cancer Treatment. *Drug Discov. Today* **2011,** *16*, 457–463.

Yang, X.; Du, H.; Liu, J.; Zhai, G. Advanced Nanocarriers Based on Heparin and its Derivatives for Cancer Management. *Biomacromolecules* **2015,** *16* (2), 423–436.

Yapo, B. M.; Lerouge, P.; Thibault, J. F.; Ralet, M. C. Pectins from Citrus Peel Cell Walls Contain Homogalacturonans Homogenous with Respect to Molar Mass, Rhamnogalacturonan I and Rhamnogalacturonan II. *Carbohydr. Polym.* **2007,** *69*, 426–435.

Zempsky, W. T.; Cote, C. J.; Berlin, C. Alternative Routes of Drug Administration—Advantages and disadvantages (Subject Review). *Peds* **1998,** *101* (4), 730–731.

Zhang, W.; Nuki, G.; Moskowitz, R. W.; Abramson, S.; Altman, R. D.; Arden, N. K.; Dougados, M. OARSI Recommendations for the Management of Hip and Knee Osteoarthritis: Part III: Changes in Evidence Following Systematic Cumulative Update of Research Published Through January 2009. *Osteoarthr. Cartil.* **2010,** *18* (4), 476–499.

Zhang, N.; Wardwell, P. R.; Bader, R. A. Polysaccharide-based Micelles for Drug Delivery. *Pharmaceutics* **2013,** *5* (2), 329–352.

Zhang, X.; Malhotra, S.; Molina, M.; Haag, R. Micro-and Nanogels with Labile Crosslinks–from Synthesis to Biomedical Applications. *Chem. Soc. Rev.* **2015,** *44* (7), 1948–1973.

Zhang, H.; Zhai, Y.; Wang, J.; Zhai, G. New Progress and Prospects: The Application of Nanogel in Drug Delivery. *Mater. Sci. Eng. C* **2016,** *60*, 560–568.

CHAPTER 11

NANOCAPSULES: RECENT APPROACH IN THE FIELD OF TARGETED DRUG DELIVERY SYSTEMS

BHUSHAN RAJENDRA RANE[1*], ASHISH S. JAIN[1], NAYAN A. GUJARATHI[2], and RAJ K. KESERVANI[3]

[1]*Department of Pharmaceutics, Shri D. D. Vispute College of Pharmacy & Research Center, Mumbai University, New Panvel, India*

[2]*Sandip Institute of Pharmaceutical Sciences, Pune University, Nashik, India*

[3]*Faculty of B. Pharmacy, CSM Group of Institutions, Allahabad, India*

Corresponding author. E-mail: rane7dec@gmail.com

ABSTRACT

Nanocapsule is an emerging nanocarrier drug delivery system which is based on science of nanotechnology and medicine. Various nanocarriers are developed with an aim to target and deliver various therapeutically active molecules such as anticancer agents, nucleic acids, proteins, and peptides for their therapeutic response by nonphagocytic routes. Nanocapsule is one of the most promising carriers for the development of efficient targeted drug delivery in cancer treatment. It overcomes various cellular barriers with an aim to improve the delivery of active moieties including the potential therapeutic biomolecules (i.e., proteins and nucleic acids). Dispersed polymer nanocapsules can be used as nanosized drug carriers to get controlled release as well as efficient drug targeting. Drug-loaded polymeric nanocapsules have showed possible applications in the field of drug delivery systems. Enormous research efforts have been performed in order to develop modern nanoparticulate drug delivery systems. However, newly developed drug molecules with moderate biopharmaceutical profile are still missing. The entrapment of drug molecules can protect them from the biological environment and facilitate their transport through biological barriers. Therefore, nanocarriers especially nanocapsules can give promising therapeutic benefits in the field of drug delivery system.

11.1   INTRODUCTION

In the recent era, the attention of the scientists from all over the world is toward the development of materials and carrier for targeted drug delivery system which can be achieved by designing the system at nano level. The nanotechnology plays a huge role in the development of such materials with desired properties; it also shows significant applications in all branches of science and technology as well as in medicine and pharmacy. At present, modern pharmaceutical industries are engaged in the formation of drug with almost all ideal characteristics as the drug can be targeted to specific site and shows no damage to healthy organs and tissues, and avoid toxic effects of drug on organisms in general. All the researchers worldwide oriented their research to develop a system with targeted and improved drug delivery properties. Various controlled release systems have been developed in last decades, among them are the colloidal drug delivery systems such as nano-capsules. It gained a special interest in the development of targeted drug delivery due to considerable results of this system as compared to the free drugs (Damge et al., 1990).

The development of nanoparticles or nanocapsule colloidal drug delivery system is based on some aspects stated as: (1) sufficient drug encapsula-tion, (2) target the drug to specific organ or tissue, and (3) controlled release behavior must be shown by the formulated system.

Nanoparticles are solid colloidal particles that are also called as nano-spheres and nanocapsules. One of the fundamental characteristics of nanoparticles is their size that is generally considered around 5–10 nm with an upper size limit of ~1000 nm, although the range generally obtained is 100–500 nm.

Definition: Nanocapsules are vesicular systems in which the drug is confined to a cavity consisting of an inner liquid core surrounded by a poly-meric membrane. In this case, the active substances are usually dissolved in the inner core but may also be adsorbed to the capsule surface (Huguette et al., 2002).

The nanocapsules can be analogus to vesicular systems in which a drug is confined in a cavity consisting of an inner liquid core surrounded by a polymeric membrane (Quintanar et al., 1998). However, seen from a general level, they can be defined as nano-vesicular systems that exhibit a typical core–shell structure in which the drug is confined to a reservoir or within a cavity surrounded by a polymer membrane or coating (Letchford et al., 2007; Anton et al., 2008). The cavity can contain the active substance in liquid or solid form or as a molecular dispersion (Fessi et al., 1989; Devissaguet et al., 1991; Radtchenko et al., 2002). Likewise, this reservoir can be lipophilic or hydrophobic according to the preparation method and raw materials used. Also, considering the operative limitations of preparation methods, nanocapsules can also carry the active substance on their surfaces or imbibed in the polymeric membrane (Khoee et al., 2008).

Nanocapsules are also called as polymeric nanoparticles (Jager et al., 2007) if they contain a nonionic surfactants polymeric wall and also macromolecules, phospholipids (Beduneau et al., 2006; Mohanraj and Chen, 2006), and sometimes an oil core (Adriana et al., 2008). Polymeric nanoparticles can be prepared by two techniques: (1) the interfacial polymerization and (2) interfacial nano-deposition. Nanocapsules have an ability to carry the biomedical substances and are used during controlled release and targeting of therapeutic agents against the protection of proteins, enzymes, foreign cells, etc. (Diaspro et al., 2002).

Nanocapsules, are a typical class of nanoparticles, which are made up of core materials and a shell (protective matrix) (Benita, 1998) and various drugs can be incorporated inside the polymeric shell. The outer coat of nanoparticles is pyrophoric and can be easily oxidized; thus, nanoparticles are of great interest in targeted drug delivery system. In past five decades, nanoparticles have been used as drug carriers and many of such nanoparticulate systems in the nanosize have been in development. Various researchers investigated the use of nanoparticles in the treatment of cancer therapy and diagnosis, known as nanomedicines. Therapeutically, such nanosystems have been more advantageous than free drugs in treatment of various diseases (Shen et al., 2010). The most important advantages of such systems are (1) release of drug in sustain manner, (2) efficiency, (3) improved bioavailability, and (4) decreased drug toxicity. Nanocapsules of submicron size reach to the target site after administration through IV route and release the drug.

11.1.1   BENEFITS OF NANOCAPSULE (ARENAS ET AL., 2006; BEDUNEAU ET AL., 2006)

1. Nanocapsules have been developed as drug delivery systems for several drugs by different routes of administrations such as oral and parenteral.
2. Nanocapsules reduce the toxicity of drugs. Improve the stability of the drug either in biological fluid or in the formulation.
3. Nanocapsules are promising active vectors due to their capacity to release drugs; their subcellular size allows relatively higher intracellular uptake than other particulate systems. They can improve the stability of active substances.
4. Nano-encapsulated systems as an active substance carriers include high drug encapsulation efficiency due to optimized drug solubility in the core, low polymer content compared to other nanoparticulated systems such as nanospheres, drug polymeric shell protection against degradation factors such as pH and light, and the reduction of tissue irritation due to the polymeric shell.
5. Some newly developed powerful drug molecules is strongly limited by their inadequate biopharmaceutical profiles. Moreover, it is sometimes difficult to synthesize new drugs with adequate stability and permeability properties. In this condition, the drug is incapable to reach adequate biological compartment. Therefore, development of appropriate delivery system for these drugs would be a step forward for their clinical exploitation.
6. Not only the entrapment of drugs in nanocapsules protects them from the biological environment, but they also make their transport possible through biological barriers.

11.1.2   ADVANTAGES (BOAL ET AL., 2000)

1. Increases the stability of any volatile pharmaceutical agent, easily and cheaply fabricated in large quantities by a multitude of methods.
2. They offer a significant improvement over traditional oral and intravenous methods of administration in terms of efficiency and effectiveness.
3. Delivers a higher concentration of pharmaceutical agent to a desired location.

4. The choice of polymer and the ability to modify drug release from polymeric nanoparticles have made them ideal candidates for cancer therapy and delivery of vaccines, contraceptives, and targeted antibiotics.
5. Polymeric nanoparticles can be easily incorporated into other activities related to drug delivery, such as tissue engineering.

11.1.3 MECHANISMS OF DRUG RELEASE (DOUGLAS AND ZENO, 1999)

The polymeric drug carriers deliver the drug at the tissue site by any one of the general physicochemical mechanisms:

1. By the swelling of the polymer nanoparticles because of hydration followed by release through diffusion.
2. By an enzymatic reaction resulting in rupture or cleavage or degradation of the polymer at the site of delivery, thereby releasing the drug from the entrapped inner core.
3. Dissociation of the drug from the polymer and its de-adsorption/release from the swelled nanoparticles.
4. Slow release—the capsule releases drug molecules slowly over a longer period of time (e.g., for slow delivery of a substance in the body).
5. Quick release—the capsule shell breaks upon contact with a surface.
6. Specific release—the shell is designed to break open when a molecular receptor binds to a specific chemical (e.g., upon encountering a tumor or protein in the body).
7. Moisture release—the shell breaks and releases drug in the presence of water (e.g., in soil).
8. Heat release—the shell releases drug only when the environment warms above a certain temperature.
9. pH release—nanocapsule breaks only in specific acid or alkaline environment (e.g., in the stomach or inside a cell).
10. Ultrasound release—the capsule is ruptured by an external ultrasound frequency.
11. Magnetic release—a magnetic particle in the capsule ruptures the shell when exposed to a magnetic field.

12. DNA nanocapsule—the capsule smuggles a short strand of foreign DNA into a living cell which, once released, and hijacks cell machinery to express a specific protein (used for DNA vaccines).
13. The thickness of the capsule wall can be precisely tuned in the range of a few nanometers by choosing coating materials and number of layers.
14. The pore size on shell wall membrane can be controlled through different polyelectrolyte pairs and assembly conditions.
15. Wide varieties of shell materials, includes charged polymers, lipids, proteins, and magnetic nanoparticles and can be used during shell assembly. Shells can be switched between "open" and "closed" states for trigger release (Zhang, 2004).

11.1.4 RATIONALE BEHIND NANOCAPSULE FORMULATION

Macromolecules and drug compounds showed some disadvantages of stability which lead to the decreased bioavailability, less therapeutic response, toxicity, etc. There is a need to develop a stable formulation which will be able to protect the drug from the environment to overcome above-mentioned problems. The interest in research on nanocapsules has increased considerably because of their advantages over conventional dosage forms. Due to submicron size of the system, nanocapsules have become more advantageous for extensive use of applications. The production of nanocapsules is based on their applications. These materials may present different magnetic behaviors from their corresponding counterparts. Researchers in China have succeeded in synthesizing a new type of intermetallic nanocapsule that can be applied in cryogenic magnetic refrigerator devices (Berger, 2006). Various drugs have difficulty in marketing due to their objectionable side effects; but if they encapsulated in polymeric shell, system will be able to deliver the drug to targeted site in reproducible manner (Radhika et al., 2011). The targeting of drug by nanocapsule system has various challenges in order to achieve greater therapeutic effect, but it is an opportunity for new researchers to develop stable and more efficient novel nanocapsule for improved effects.

11.2 MATERIALS OF COMPOSITION

Nanocapsule means a core which is oily or aqueous in nature, and is surrounded by a thin polymer membrane (Dongwoo et al., 2010).

11.2.1 ACTIVE DRUG MOLECULE

Variety of macromolecules, therapeutic agents, diagnostic agents, proteins, peptides, nutritional compounds, immunological compounds, etc. can be incorporated in nanocapsule shell. Nanocapsule approach is applicable to both the type of drug molecules of hydrophilic and lipophilic class.

11.2.2 POLYMERS

Natural and synthetic polymers are used during formulation of nanocapsules. Both hydrophilic and lipophilic surfactants can be used to stabilize the nanocapsules (Adriana et al., 2007; Xinfei et al., 2010). Generally, the lipophilic surfactant is a natural lecithin of relatively low phosphatidylcholine content, whereas hydrophilic one is synthetic: anionic (lauryl sulfate), cationic (quaternary ammonium), or more commonly nonionic poly(oxyethylene)–poly(propylene) glycol.

Nanocapsules can be prepared without surfactants but sometimes aggregation occurs on storage. Various polymers used during nanocapsule development such as natural, synthetic, and semisynthetic; sometimes a broad range of vegetable and mineral oils also used for the preparation of nanocapsules, in some cases pure compounds are also useful such as ethyl oleate and benzyl benzoate (Bruce, 1993; Wan et al., 1992).

11.2.3 POLYMERS USED IN PREPARATION OF NANOPARTICLES

The different applications of polymers are presented in Table 11.1. (Watnasirichaikul et al., 2000; Wan et al., 1992; Vonarbourg et al., 2009; Redhead et al., 2001; Raffin et al., 2002; Qiang et al., 2001; Puglisi et al. 1995; Lambert et al., 2000).

The polymers should possess following properties:

1. Be compatible with the body in the terms of adaptability
2. Nontoxic
3. Nonantigenic
4. Biodegradable
5. Biocompatible with other excipients

Natural Polymers

The most commonly used natural polymers in preparation of polymeric nanocapsules are (Kompella et al., 2003; Kumar et al., 2004; Li et al., 2001):

- Gum arabica
- Chitosan
- Gelatin
- Sodium alginate
- Lecithin
- Albumin

Semisynthetic Polymers

- Diacyl beta cyclodextrin
- Hydroxyl propyl methyl cellulose
- Hydroxyl propyl methyl cellulose phthalate

Synthetic Polymers Used in Nanocapsule Preparation as Follows:

- Polylactides (PLA)
- Polyglycolides (PGA)
- Poly(lactide co-glycolides) (PLGA)
- Polyanhydrides
- Polyorthoesters
- Polycyanoacrylates
- Polycaprolactone
- Polyglutamic acid
- Polymalic acid
- Poly(N-vinyl pyrrolidone)
- Poly(methyl methacrylate)
- Poly(vinyl alcohol)
- Poly(acrylic acid)
- Polyacrylamide
- Poly(ethylene glycol)
- Poly(methacrylic acid)

TABLE 11.1 Materials Used for Encapsulation of Nanocapsules.

Encapsulation material	Application
Aqueous monomers (ethylene diamine, hexa-methylene diamine, and 1,4-diaminobutane)	Profound effects on the drug (curcuminoid) loading capacity
Chitosan, gelatin, and alginate	Hydrophilicity of ART crystals was improved after encapsulation
Isocyanates have been successfully encapsulated into polystyrene, and hydroxyl and amine functionalized nanospheres using a commercially available blocked isocyanate	The thermally dissociated isocyanate can be utilized as an active functional group in coatings and adhesive applications
Poly(epsilon-caprolactone)	Reduces the percutaneous drug absorption through stripped skin
PIBCA	Targeting action
Spherical hybrid assemblies based on cationic surfactants and anionic porous polyoxometalate nanocapsules $[\{(Mo)Mo_5O_{21}(H_2O)_6\}_{12}\{Mo_2O_4(SO_4)\}_{30}]^{72}$ (Mo_{132} for short) are fabricated by the method combining an electrostatic encapsulation process	Not only presents a new route to assemble Mo_{132} nanocapsules but also demonstrates a new concept of using the microenvironment of supramolecular assemblies to adjust the ion-trapping properties of Mo_{132}
Xerogels	To diminish the burst release of drugs from xerogel mesopores instead to Aerosil 200

ART, artemisinin; PIBCA, poly(isobutylcyanoacrylate).

11.3 FORMULATION OF NANOCAPSULES

11.3.1 PREPARATION TECHNIQUE

The preparation technique involves infiltration of polymer–drug conjugates into mesoporous shells, followed by cross-linking with infiltrated polymer chains and lastly removal of silica, leads to formation of nanocapsule.

Abovementioned approach is more advantageous than the conventional techniques for the preparation of nanocapsules.

Capsules can be prepared in stepwise manner as follows:

In Step 1: the positively or negatively surface charged polymer is added.

In Step 2: ultrathin polymer film can be formed by self-assembly mechanism; resulted capsule wall will have thickness of 8–50 nm due to approximately 4–20 layers of polymer coat. Additionally, nanocapsules can be modified to dock antibodies on the surfaces while in some cases core can be removed to incorporate another material inside the shell. Apart from this interfacial polymerization, interfacial precipitation and interfacial nano-deposition methods can be used for the preparation of nanocapsules. In organic phase solvent, polymer, oil, and drug can be added, while in aqueous phase water and surfactant can be incorporated.

11.3.2   METHODS OF PREPARATION

11.3.2.1   POLYMERIZATION METHOD

The nanoparticles can be formed by polymerizing monomers in aqueous solution and the drug loading can be achieved by placing drug in the polymer solution or by adsorption mechanism. For the purification of nanoparticles ultracentrifugation method can be used to remove the traces of chemicals used during polymerization process, later on nanoparticles are resuspended in isotonic surfactant-free medium, for example, polybutylcyanoacrylate or polyalkylcyanoacrylate nanoparticles (Qiang et al., 2001; Boudad et al., 2001). The yield and particle size of nanoparticles depend on the ratio of surfactant and stabilizers (Puglisi et al., 1995).

11.3.2.2   INTERFACIAL POLYMERIZATION

Alternative method to bulk polymerization is an interfacial polymeriza-tion (Lambert et al., 2000; Morgan, 1987; Jang et al., 2006) as previous one requires high-temperature condition for the condensation of polymers. In this technique, different monomers are present in different immiscible solvents reacting at the interface to form capsule shell.

11.3.2.3   ARC-DISCHARGE METHOD

Arc-discharge method (Song et al., 2006) is used very rarely in the prepara-tion of self-assembled aggregates. Some modifications made in this method such as change in hydrogen pressure, using an anode of the material gado-linium–aluminum alloy ($GdAl_2$), or by changing the elements on the basis of their evaporation pressures to synthesize the aggregates, by using inter-metallic compounds $GdAl_2$ as the core material and amorphous Al_2O_3 to form shell (Yosida et al., 1994; Ziyi et al., 2000). Regularly aligned three-dimensional self-assembled nanocapsules can be formed by arc-discharge method (Zhang, 2004).

11.3.2.4 EMULSION POLYMERIZATION

The pre-emulsion can be prepared by (Yang et al., 2008) magnetic blending of Part A and Part B separately for a period of 10 min, subsequently Part B can be added to Part A by continue mechanical agitation at 1800 rpm for 30 min. The resulted pre-emulsion cooled to below refrigeration temperature and shifted to sonication but the achieved particle size still remained below 250 nm. Abovementioned emulsion was transferred to a three-neck round bottom flask, which was equipped with a mechanical stirrer, reflux condenser, and a nitrogen inlet, and degassed for 30 min (Jackson and Lee, 1991). Lastly, temperature was increased to 70°C and preserved for 8 h to complete the process of polymerization.

Part A consists: styrene (40 g) + divinyl benzene (0.8 g) + 0.82 g AIBN (2,2'-azobisisobutyronitrile) + 40 g Desmodur BL3175A while,

Part B consists: 1.71 g SDS (sodium dodecyl sulfate) + 1.63 g Igepal CO-887 + 220 g of water.

Other preparation methods for nanocapsules include laser vaporization–condensation (Samy et al., 1996), solution–liquid–solid method (Boal et al., 2000), catalytic vapor–liquid–solid growth (Zhu et al., 2001), organic reagent-assisted method (Qingyi et al., 2002), charge transferring (Kensuke et al., 2003), chemical vapor deposition (Kimberly et al., 2004), and electron irradiation deposition (Sung et al., 2007).

11.3.3 ENCAPSULATION OF NANOCAPSULES

The encapsulation technology has been used to prepare the various particulate systems of different size with different applications in the field of food (Stenekes et al., 2001), medicine, and biology (Sarah et al., 2009). In many of the techniques for encapsulation, the use of isocyanates is preferred to build the shell or matrix for the functional materials encapsulation (Matkan and Treleaven, 1987) and is also used during encapsulation of materials which is to be released at predetermined rate, or making pressure on sensitive copying paper (Irii and Shiozaki, 1987).

Xerogels and Aerosil 200 used in encapsulation process leads to delayed release of drug from the formulation (Arenas et al., 2006). The Aerosil 200 has shown the bursting of nanocapsules; to diminish such effect, different approaches have been proposed (Slowing et al., 2007). Sometimes polymeric nanocapsules have also been used as coasting material for drug-loaded xerogel. The technique of fabrication is used to encapsulate DNA in an oily core for blood injection (Vonarbourg et al., 2009).

11.4   CHARACTERIZATION OF NANOCAPSULES

Nanocapsules were characterized mainly for their structural and morphological properties, particle size distribution, and drug-loading capacity.

11.4.1   PARTICLE SIZE

Particle size is a basic evaluation parameter, which plays an important role in nanocapsule formulation; as it directly reflects the bioavailability, toxicity, *in vivo* distribution and targeting specificity. Sometimes it also influences the loading capacity, release, and time for therapeutic activity. Smaller size particles have greater surface area, thus release of drug is very fast; while in case on larger size particles diffusion rate is slow (Redhead et al., 2001). In some cases, such as the rate of poly(D, L-lactide-co-glycolide) (PLGA) polymer degradation revealed an enhancement with an increase in particle size *in vitro* (Dunne et al., 2000). Determination of the particle size is by zeta sizer, photon correlation spectroscopy, or dynamic light scattering (Repka, 2002).

11.4.2   SURFACE PROPERTIES OF NANOCAPSULES

The surface properties of nanocapsule such as charge can be characterized by zeta potential (Couvreur et al., 2002). Surface coating is also evaluated by checking their circulation within the body after administration (Jang et al., 2006).

11.4.3   FLUORESCENCE QUENCHING

To confirm the localization of nanocapsules, which contains the oligonucleotides in aqueous core is by quenching of fluorescence (Lambert et al., 2000; Bingyun and Daniel, 2003; Daniel et al., 2010).

11.4.4 X-RAY DIFFRACTION STUDIES

After X-ray diffraction study the XRD pattern reveals the phase composition of formulated nanocapsule (Aiyer et al., 1995). It can be performed by powder XRD on a Rigaku D/max-2000 diffractometer with graphite monochromatized CuKα; at a voltage of 50 kV and a current of 250 mA.

11.4.5 SCANNING ELECTRON MICROSCOPY (SEM)

To confirm the branching aggregates, small clusters, big clusters (clusters are nothing but flocculent structure formed due combining of small particles) (Sung et al., 2007), and branches of self-similar attributes of the structure (Watnasirichaikul et al., 2000), which is characterized by a Philips XL-30 scanning electron microscope (SEM).

11.4.6 DIFFERENTIAL SCANNING CALORIMETRY

Differential scanning calorimetry (DSC) is conducted for the samples (in open and closed pan) to check and verify the changes observed when keeping the sample at different temperature conditions. All the changes were observed and recorded and finally concluded the thermal behavior of sample (Douglas and Zeno, 1999).

11.4.7 TRANSMISSION ELECTRON MICROSCOPY

The prepared nanoparticles characterized by transmission emission microscopy (TEM) to examine shell and core structure. The transport of insulin across the epithelium mucosa from the biodegradable nanocapsules can be checked by TEM as well as, transport of particularly insulin-loaded nanocapsules in experimental rats can be observed (Kepczynski et al., 2009; Huguette et al., 2003).

11.4.8 HIGH-RESOLUTION TRANSMISSION ELECTRON MICROSCOPY

The complete morphological study of nanoparticles can be examined by High-resolution Transmission Electron Microscopy (HRTEM); through which researcher is able to know the core and shell structure (Song et al., 2006; Zhang et al., 2001).

11.4.9 X-RAY PHOTOELECTRON SPECTROSCOPY

X-ray photoelectron spectroscopy (XPS) is highly specific to the solid surface due to the narrow range of photoelectrons which gets excited and gives information about structure of the system by ESCALAB-250 with a monochromatic X-ray source (an aluminum Kα line of 1486.6 eV energy and 150 W). To describe the valency of surface aluminum atoms present on the nanocapsules, ESCALAB-250 is used at a depth of 1.6 nm. To find the composition of surface material, binding energies and peak areas (with equivalent sensitivity factor) can be used as both are characteristic properties of the element. It also provides the chemical bonding information for the surface material (Pohlmann et al., 2008).

11.4.10 SUPERCONDUCTING QUANTUM INTERFERENCE DEVICE

Quantum design MPMS-7s or MPMS-5s superconducting quantum interference device can be used to determine the magnetic properties of nanocapsules. The slight changes in the magnetic flux can be checked by superconducting quantum interference device (SQUID) due to its sensitivity (Liu et al., 2009).

11.4.11 MULTIANGLE LASER LIGHT SCATTERING

To control the entrapment and release of encapsulated materials these studies reveals lot of information regarding the same. In nanocapsules, the vault particle has an incredible potential for the compound encapsulation, release, and its protection (Kedersha, 1991); vaults means a capsule-like structure with a very thin shell of size approximately 2 nm which is surrounded by a large internal cavity. In environmental and medical detoxification the vaults with binding sites for toxic metals have great importance (Sangwoo et al., 2010). The presence of vault in solution is confirmed by using the multiangle laser light scattering (MALLS) (Leonard et al., 2003; Stephen et al., 2001).

11.4.12 FOURIER-TRANSFORM INFRARED ANALYSIS

Fourier-transform infrared (FTIR) spectrophotometer is useful in the confirmation of characteristic functional groups in the nanocapsule and its material of composition (core/shell) (Bouchemal et al., 2004, Benvenutti et al., 2002).

11.4.13 DETERMINATION OF THE pH OF NANOCAPSULE

Nanocapsules formulation pH was measured using a digital pH meter at room temperature. Nanocapsules dispersion pH values fall within a range of 3.0–7.5.

11.4.14 MEAN NANOCAPSULES

The mean particle size of nanocapsules prepared from performed polymers is in general between 250 and 500 nm. Double emulsification method has concluded that particle size depends on the internal and external surfactants that determine droplet size, the interaction at the interface, and the structural conformation of the nanocapsules wall.

11.4.15 DETERMINATION OF DRUG CONTENT

Drug content was determined by dissolving 1 mL of prepared nanocapsules in 20 mL of acetonitrile. Appropriate quantity of sample was then subjected to the UV spectrophotometer. The absorbance for each sample was measured and compared with the standard.

11.4.16 PARTICLE SIZE DISTRIBUTION AND PARTICLE CHARGE/ZETA POTENTIAL

Particle size distribution is an important aspect during the formulations of nano systems. Nanocapsules were characterized for their particle size distribution and zeta potential using Malvern zeta sizer.

11.4.17 IN VITRO DRUG RELEASE

In vitro dissolution studies were carried out using USP type 11 dissolution apparatus. The study was carried out in 100 mL of buffer (pH 3.0). The nanocapsules suspension was placed in dialysis membrane and dipped in dissolution medium which was kept inert thermostatically at $37 \pm 0.5°C$. The stirring rate was maintained at 100 rpm. At predetermined time intervals 5 mL of sample were withdrawn and assessed for drug release spectrophoto-metrically. After each withdrawal 5 mL of fresh dissolution medium was added to dissolution jar.

11.5 APPLICATIONS OF NANOCAPSULES

Nanocapsules have a wide variety of applications in the field of drug delivery, protein and peptide, hormones, cell, DNA; also in treatment of cancer, breast cancer, melanomas, and many more. Nanocapsule shows wide variety of advantages over conventional dosage forms due to the micronized size and high reproducibility. It has some applications in the field of life sciences, agrochemicals, genetic engineering, cosmetics products, wastewater treatments, cleaning products, etc.

11.5.1 IN DELIVERY OF DRUG

In comparison to microparticles the nanocapsules have shown faster onset of action due to small particle size, high surface area which results into faster dissolution, absorption, and finally increased bioavailability. On the surface of nanocapsule shell, various therapeutic agents can be adsorbed for medicinal applications, for example, antibodies, drug particles, and tiny gold particles.

11.5.2 ORAL DELIVERY OF PROTEINS AND PEPTIDES

Nanocapsules   can   be   used   as   carriers   for   biodegradable poly(isobutylcyanoacrylate) (Puglisi et al., 1995; Hildebrand and Tack, 2000) to administer peptides and proteins via oral route.

Many more systems and drugs can be degraded in the GIT by digestive enzymes, by using this technique bioactive molecules can be encapsulated and protected from enzymatic and hydrolytic degradation, for example, insulin nanoparticles (Damge et al., 1990).

11.5.3   IN HORMONE-DEPENDENT BREAST CANCER

Specific siRNAs encapsulated in nanocapsules can be used to target estrogen receptor alpha (ERα) Jack et al. (2008); after administration via IV route the estradiol stimulated MCF-7 cell xenografts result into decrease in tumor growth and finally there is decrease in ERα expression in tumor cells. It indicates that a new approach, based on ERα–siRNA delivery, could be used in treatment of breast cancers (hormone dependent).

11.5.4   MAGNETIC RESONANCE IMAGING-GUIDED NANOROBOTIC SYSTEMS FOR THERAPEUTIC AND DIAGNOSTIC APPLICATIONS

The nanorobotic systems can be used in diagnosis and in some therapeutic applications by the MRI-guided approach (Panagiotis et al., 2011). It is specifically based on a Magnetic resonance imaging (MRI) technique, which forces the system at targeted site by stimulating magnetic nanocapsules.

11.5.5   FOR IN VIVO DELIVERY OF PLASMIDS LIVER CELL-TYPE

The literature reported that the cost effective and efficient delivery can be achieved by viral vectors (Betsy et al., 2006). Due to insufficient *in vivo* delivery systems, the use of nonviral vectors for gene therapy has been held up.

11.5.6   IN TREATMENT OF CANCER BY USING NUCLEAR NANOCAPSULES WITH RADIOACTIVE MATERIALS

Various radioactive compounds such as radium, uranium, and astatine are able to emit α-particles of high velocity due to the radioactive decay which is used in treatment of cancer. After use of very low and optimized penetrating power of very large alpha particle, the targeting can be achieved at single cellular level (Deutsch et al., 1986).

11.5.7   AS A SELF-HEALING MATERIAL

The polymeric nanocapsules have adequate strength, longer shelf life, and possess excellent binding ability to the host material, it contains healing material and thus can be used for self-healing. The nanocapsules will have novel therapeutic applications in the research of medicine and technology due to functionalized surface areas and their walls with the possibility of forming and using nanosized objects.

11.5.8   IN FOOD SCIENCE AND AGRICULTURE (LIPOSOMAL NANOCAPSULES)

Self-assembled bilayered lipoidal barrier system (liposome) has an ability to enclose the hydrophilic and lipophilic moieties. Liposome systems can be enclosed into nanocapsule for the effective drug delivery and targeting for most sensitive drugs which result into the stabilization of active drug moieties from enzymatic degradation, environmental conditions, and various other factors such as pH, temperature, and ionic strength.

11.5.   IN VIVO HAIR GROWTH PROMOTION EFFECTS OF COSMETIC PREPARATIONS-CONTAINING HINOKITIOL-LOADED POLY(EPSILON-CAPROLACTONE) NANOCAPSULES

Nanocapsules system which contains active moiety hinokitiol (HKL), which is prepared by an emulsion diffusion method (Hwang and Kim, 2008) promotes the hair growth and shows some positive changes in the hair follicles.

11.5.10   SUN SCREEN COSMETICS COMPRISING TIO_2 NANOCAPSULES

To protect the skin from harmful UV rays nanocapsules of TiO_2 are prepared by dispersing TiO_2 with surfactant. Some surface-treating agents such as aluminum stearate and isostearic are also used for oleophilic surface treatment.

11.5.11 MOLECULAR DESIGN OF PROTEIN-BASED NANOCAPSULES FOR STIMULUS-RESPONSIVE CHARACTERISTICS

The heat shock protein (sHSP), that is, Hsp16.5 is very small unit (Sao et al., 2009) which is able to form a homogeneous complex. It consists of 24 subunits having approximate molecular weight of 400 kDA; it possessess very high thermal stability. From the various research studies, it is revealed that Mut6 acts as a stimulus-responsive nanocapsule.

11.5.12 IN MELANOMAS TREATMENT

Melanomas (cancer) are nothing but the violent tumors associated with dismal prognosis especially when they have metastasized.

In cancer, the best possible treatment is the use of dacarbazine approved by FDA (Curie, 2011) but effectiveness is not more than 20%. In targeting of specific cancer cell, two approaches are used most of the times; first is passive enhanced permeability retention phenomenon due to the composition, size, and stealth properties of the nanocapsules and second is active targeting by coupling with various antibodies.

In treatment of cancer, the strategic magnetic drug delivery system is most suitable by using active anticancer moiety and magneto hyperthermia (Falqueiro et al., 2011).

11.5.13 SELF-ASSEMBLED DNA NANOCAPSULES FOR DRUG DELIVERY

DNA is basic and most useful material for all the researchers' those engaged in nanotechnology-related work (Berger, 2009). Tetrahedron and octahedron of cubical shape formed by the process of nanofabrication and encapsulate within DNA by the molecular self-assembly. The targeted drug delivery of these molecules in the body parts and its tissues or cells is then studied (Table 11.2).

TABLE 11.2   Applications of Nanocapsules.

Application	Drug	Mode of preparation
Agrochemicals	Abamectin nanocapsules	Emulsion polymerization
	Cypermethrin nanocapsules	Microemulsion polymerization
	Pyrethrum nanocapsules	Microemulsion polymerization
Anti-inflammatory drugs	Diclofenac sodium	Sol–gel method
	Indomethacin-loaded nanocapsules	Interfacial polymerization
Antiseptics	Monodisperse polymer nanocapsule	Precipitation
Cosmetics	Hinokitiol-loaded poly(epsilon-caprolactone) nanocapsules	Emulsion–diffusion method
Diabetes	Insulin-loaded biodegradable poly(isobutylcyanoacrylate) nanocapsules	Interfacial polymerization
Nanocapsules for cancer	Artemisinin	Nanoencapsulation method
	CPT and doxorubicin	Sol–gel method
	Cisplatin	Repeated freezing and thawing of a concentrated solution of cisplatin in the presence of negatively charged phospholipids
	Indomethacin-loaded polyisobutyl-cyanoacrylate nanocapsules	Interfacial polymerization
	Lipid nanocapsules loaded with rhenium-188 (LNC188Re-SSS)	Phase-inversion process
Nanocapsule for topical use	Chlorhexidine	Interfacial polymerization method

CPT, camptothecin.

Throughout the world, the researchers recently published their conclusions in *Nano Letters* (Wang et al., 2008) and explained the wide use of this approach by preparing nanocapsules using various therapeutically active drugs for cancer treatment, including synthetic polymers. They also investigated the better applications of thick-walled shell polymerized nanocapsules in treatment of tumors, which is nonimmunogenic, biodegradable, and biocompatible.

11.6 FUTURE PROSPECTIVE

In future, nanocapsules will be advantageous in all the fields of technology and medicine. It was reported that conventional dressings must be removed if skin layer was affected (Radhika et al., 2011) while the dressings of nanocapsule material did not want to be removed; it works automatically to discharge antibiotics when the wound becomes infected. Such dressings are also used in the ulcers and in military people during war. Nanocapsule dressings release antiobiotics after activation due to presence of pathogenic microorganisms and thus cure the wound before it becomes severe.

11.7 CONCLUSION

Nanocapsules are one of the nanoparticulate systems prepared by the interfacial polymerization and interfacial nano-deposition. They possess biochemical, electrical, optical, and magnetic properties. Nanocapsules identified on the basis of various characterization parameters by investigating the outer shell and core. In the field of genetic engineering, agrochemicals, cosmetics, cleaning products, components with adhesive property, nanocapsules have greater applications. Various biochemical's (such as enzymes, proteins) oils, catalyst, micro and nano inorganic systems, adhesives, latex systems, and biological cells can be encapsulated. Most importantly various active pharmaceutical ingredients can be encapsulated to achieve desired therapeutic response after administration to target at the specific site.

KEYWORDS

- nanocapsule
- particle size
- targeted release
- nanocarrier
- cancer treatment

REFERENCES

Adriana, R. P.; Leticia, C.; Graziela, M.; Leonardo, U. S.; Nadya, P. D.; Silvia, S. G. Structural Model of Polymeric Nanospheres Containing Indomethacin Ethyl Ester and *In vivo* Antiedematogenic Activity. *Int. J. Nanotechnol.* **2007,** *4* (5), 454–467.

Adriana, R. P.; Leticia, S. F.; Rodrigo, P. S.; Alberto, M. D.; Edilson, V. B.; Tania, M. H. C., et al. Nanocapsule@Xerogel Microparticles Containing Sodium Diclofenac: A New Strategy to Control the Release of Drugs. *Int. J. Pharm.* **2008,** *358* (1–2), 292–295.

Aiyer, H. N.; Seshadri, R.; Raina, G.; Sen, R.; Rahul, R. Study of Carbon Nanocapsules (Onions) and Spherulitic Graphite by Stm and Other Techniques. *Fullerene Sci. Tech.* **1995,** *3* (6), 765–777.

Arenas, L. T.; Dias, S. L. P.; Moro, C. C.; Costa, T. M. H.; Benvenutti, E. V.; Lucho, A. M. S., et al. Structure and Property Studies of Hybrid Xerogels Containing Bridged Positively Charged 1 4-diazoniabi Cycle 2.2.2 Octane Dichloride. *J. Colloidal Interface Sci.* **2006,** *297* (1), 244–250.

Beduneau, A.; Saulnier, P.; Anton, N.; Hindre, F.; Passirani, C.; Rajerison, H., et al. Pegylated Nanocapsules Produced by an Organic Solvent Free Method: Evaluation of Their Stealth Properties. *Pharm. Res.* **2006,** *23* (9), 2190–2199.

Benita, S. Microparticulate Drug Delivery Systems: Release Kinetic Models. In *Microspheres, Microcapsules and Liposomes*; Arshady, R., Ed.; The MML Series; Citrus Books: London, 1998; pp 255–278.

Benvenutti, E. V.; Pavan, F. A.; Gobbi, S. A.; Costa, T. M. H. FTIR Thermal Analysis on Anilinepropylsilica Xerogel. *J. Therm. Anal. Calorym.* **2002,** *68* (1), 199–206.

Berger, M. Rare-earth Nanocapsules as a New Type of Nanomaterial for Cryogenic Magnetic Refrigerators. Nanowerk Spotlight, 2006. http://www.nanowerk.com/spotlight/spotid=991. php (accessed April 04, 2012).

Berger, M. Integrating Nanotube-based NEMS into Large Scale MEMS. *Nanowerk LLC* **2009,** *5,* 316–322. http://www.nanowerk.com/spotlight/spotid=11804.php. (accessed August 14, 2016).

Betsy, T. K.; Gretchen, M. U.; Mark, T. R.; Clifford, J. S. Targeted Nanocapsules for Liver Cell-type Delivery of Plasmids *In vivo. Mol. Ther.* **2006,** *13,* S415.

Bingyun, S.; Daniel, T. C. Spatially and Temporally Resolved Delivery of Stimuli to Single Cells. *J. Am. Chem. Soc.* **2003,** *125* (13), 3702–3703.

Boal, A. K.; Ilhan, F.; Derouchey, J. E.; Thurn, A. T.; Russell, T. P.; Rotello, V. M. Self-assembly of Nanoparticles into Structured Spherical and Network Aggregates. *Nature* **2000,** *404* (6779), 746–748.

Bouchemal, K.; Briançon, S.; Perrier, E.; Fessi, H.; Bonnet, I.; Zydowicz, N. Synthesis and Characterization of Polyurethane and Poly(Ether Urethane) Nanocapsules Using a New Technique of Interfacial Polycondensation Combined to Spontaneous Emulsification. *Int. J. Pharm.* **2004,** *269* (1), 89–100.

Boudad, H.; Legrand, P.; Lebas, G.; Cheron, M.; Duchene, D.; Ponchel, G., et al. Combined Hydroxypropyl-beta-cyclodextrin and Poly(Alkylcyanoacrylate) Nanoparticles Intended for Oral Administration of Saquinavir. *Int. J. Pharm.* **2001,** *218* (1–2), 113–124.

Bruce, M. N. Hybrid Nanocomposite Materials Between Inorganic Glasses and Organic Polymers. *Adv. Mater.* **1993,** *5* (6), 422–433.

Couvreur, P.; Barratt, G.; Fattal, E.; Legrand, P.; Vauthier, C. Nanocapsule Technology—A Review. *Crit. Rev. Ther. Drug Carrier Syst.* **2002,** *19* (2), 99–134.

Daniel, T. C.; Polina, B. S.; Kimberly, A. D. Investigating Lyophilization of Lipid Nanocapsules with Fluorescence Correlation Spectroscopy. *Langmuir* **2010,** *26* (12), 10218–10222. (PMC Free Article)

Damge, C.; Michel, C.; Aprahamian, M.; Couvreur, P.; Devissaguet, J. P. Nanocapsules as Carriers for Oral Peptide Delivery. *J. Control. Release* **1990,** *13* (2–3), 233–239.

Deutsch, E.; Libson, K.; Vanderheyden, J. L.; Ketring, A. R.; Maxon, H. R. The Chemistry of Rhenium and Technetium as Related to the Use of Isotopes of These Elements in Therapeutic and Diagnostic Nuclear Medicine. *Int. J. Rad. Appl. Instrum. B* **1986,** *13* (4), 465–477.

Diaspro, A.; Krol, S.; Cavalleri, O.; Silvano, D.; Gliozzi, A. Microscopical Characterization of Nanocapsules Templated on Ionic Crystals and Biological Cells Toward Biomedical Applications. *IEEE Trans. Nanobiosci.* **2002,** *1* (3), 110–115.

Dongwoo, K.; Eunju, K.; Jiyeong, L.; Soonsang, H.; Wokyung, S.; Namseok, L., et al. Direct Synthesis of Polymer Nanocapsules: Self-assembly of Polymer Hollow Spheres Through Irreversible Covalent Bond Formation. *J. Am. Chem. Soc.* **2010,** *132* (28), 9908–9919.

Douglas, A. W.; Zeno, W. W. Blocked Isocyanates III: Part A. Mechanisms and Chemistry. *Prog. Org. Coat.* **1999,** *36* (3), 148–172.

Dunne, M.; Corrigan, O. I.; Ramtoola, Z. Influence of Particle Size and Dissolution Conditions on the Degradation Properties of Polylactide-co-glycolide Particles. *Biomaterials* **2000,** *21* (16), 1659–1668.

Falqueiro, M.; Primo, F. L.; Morais, P. C.; Mosiniewicz, S.; Suchocki, P.; Tedesco, A. C. Selol-loaded Magnetic Nanocapsules: A New Approach for Hyperthermia Cancer Therapy. *J. Appl. Phys.* **2011,** *109*, 07B306.

Haolong, L.; Yang, Y.; Yizhan, W.; Chunyu, W.; Wen, L.; Lixin, W. Self-assembly and Ion-trapping Properties of Inorganic Nanocapsule-surfactant Hybrid Spheres. *Soft Matter* **2011,** *7* (6), 2668–2673.

Hildebrand, G. E.; Tack, J. W. Microencapsulation of Peptides and Proteins. *Int. J. Pharm.* **2000,** *196* (2), 173–176.

Huguette, P. A.; Malam, A.; Danielle, J.; Patrick, C.; Christine, V. Lipid Nanocapsules Loaded with Rhenium-188 Reduce Tumor Progression in a Rat Hepatocellular Carcinoma Model. *IEEE Trans. Nanobiosci.* **2002,** *1* (3), 110–115.

Huguette, P. A.; Malam, A.; Danielle, J.; Patrick, C.; Christine, V. Visualization of Insulin-loaded Nanocapsules: *In vitro* and *In vivo* Studies After Oral Administration to Rats. *Pharm. Res.* **2003,** *20* (7), 1071–1084.

Hwang, S. L.; Kim, J. C. *In vivo* Hair Growth Promotion Effects of Cosmetic Preparations Containing Hinokitiol-loaded Poly(ε-caprolacton) Nanocapsules. *J. Microencapsul.* **2008,** *25* (5), 351–356.

Irii, S.; Shiozaki, T. Manufacture of Microcapsules. JP Patent 62,193,641, Japan, 1987.

Jack, M. R.; Celine, B.; Laurence, M.; Herv, H.; Vronique, M.; Elisabeth, C., et al. Physico-chemical Characteristics and Preliminary *In vivo* Biological Evaluation of Nanocapsules Loaded with siRNA Targeting Estrogen Receptor Alpha. *Biomacromolecules* **2008,** *9* (10), 2881–2890.

Jackson, L. S.; Lee, K. Microencapsulation and the Food Industry. *Lebensm. Wiss. Technol.* **1991,** *1* (1), 289–297.

Jager, A.; Stefani, V.; Guterres, S. S.; Pohlmann, A. R. Physico-chemical Characterization of Nanocapsule Polymeric Wall Using Fluorescent Benzazole Probes. *Int. J. Pharm.*, **2007,** *338* (1–2), 297–305.

Jang, J.; Bae, J.; Park, E. Selective Fabrication of Poly(3, 4-ethylenedioxythiophene) Nano-capsules and Mesocellular Foams Using Surfactant-mediated Interfacial Polymerization. *Adv. Mater.* **2006,** *18* (3), 354–358.

Kedersha, N. L. Vault Ribonucleoprotein Particles Open into Flower-like Structures with Octagonal Symmetry. *J. Cell Biol.* **1991,** *112* (2), 225–235. (PMC Free Article)

Kensuke, N.; Hideaki, I.; Yoshiki, C. Temperature-dependent Reversible Self-assembly of Gold Nanoparticles into Spherical Aggregates by Molecular Recognition Between Pyrenyl and Dinitrophenyl Units. *Langmuir* **2003,** *19* (13), 5496–5501.

Kepczynski, M.; Bednar, J.; Lewandowska, J.; Staszewska, M.; Nowakowska, M. Hybrid Silicasilicone Nanocapsules Obtained in Catanionic Vesicles. Cryo-TEM Studies. *J. Nanosci. Nanotechnol.* **2009,** *9* (5), 3138–3143.

Kimberly, A. D.; Knut, D.; Magnus, W. L.; Thomas, M.; Werner, S.; Reine, W., et al. Synthesis of Branched 'Nanotrees' by Controlled Seeding of Multiple Branching Events. *Nat. Mater.* **2004,** *3* (6), 380–384.

Kompella, U. B.; Bandi, N.; Ayalasomayajula, S. P. Subconjunctival Nano- and Micaroparti-cles Sustain Retinal Delivery of Budesonide, a Corticosteroid Capable of Inhibiting VEGF Expression. *Invest. Ophthalmol. Vis. Sci.* **2003,** *44* (3), 1192–1201.

Kumar, M. N. V. R.; Bakowsky, U.; Lehr, C. M. Preparation and Characterization of Cationic PLGA Nanospheres as DNA Carriers. *Biomaterials* **2004,** *25* (10), 1771–1777.

Lambert, G.; Fattal, E.; Pinto-Alphandary, H.; Gulik, A.; Couvreur, P. Polyisobutylcyano-acrylate Nanocapsules Containing an Aqueous Core as a Novel Colloidal Carrier for the Delivery of Oligonucleotides. *Pharm. Res.* **2000,** *17* (6), 707–714.

Leonard, H. R.; Hal, M.; Bruce, D.; Jeffrey, Z.; James, H. In *The Development of Vault Nano Capsules,* NSF Nanoscale Science and Engineering Grantees Conference, December, 16–18, 2003, National Science Foundation, Arlington, Virginia,1–3.

Li, Y. P.; Pei, Y. Y.; Zhou, Z. H.; Zhang, X. Y.; Gu, Z. H.; Ding, J.; Zhou, J. J.; Gao, X. J., et al. PEGylated Polycyanoacrylate Nanoparticles as Tumor necrosis Factor-alpha Carriers. *J. Control. Release* **2001,** *71* (3), 287–296.

Liu, X. G.; Li, B.; Geng, D. Y.; Cui, W. B.; Yang, F.; Xie, Z. G., et al. (Fe, Ni)/C Nanocapsules for Electromagnetic-wave-absorber in the Whole Ku-band. *Carbon* **2009,** *47* (2), 470–474.

Curie, M. Novel Lipid Nanocapsules of Anti Alpha 1 Sodium Pump Subunit SiRNA to Specifically Target Metastatic Melanomas. *European Commission Euraxess,* **2011,** *11,* 37.

Matkan, J.; Treleaven, R. J. Particles Containing Releasable Fill Material and Method of Making Same. U.S. Patent 4681806A, July 21, 1987.

Mohanraj, V. J.; Chen, Y. Nanoparticles: A Review. *Trop. J. Pharm. Res.* **2006,** *5* (1), 561–573.

Morgan, P. W. Interfacial Polymerisation. In *Encyclopaedia of Polymer Science,* 2nd ed.; Wiley: New York, 1987; pp 231–237.

Panagiotis, V.; Matthieu, F.; Antoine, F.; Constantinos, M. MRI-guided Nanorobotic Systems for Therapeutic and Diagnostic Applications. *Annu. Rev. Biomed. Eng.* **2011,** *13* (1), 157–184.

Pohlmann, R.; Beck, R. C. R.; Lionzo, M. I. Z.; Coasta, T. M. H.; Benvenutti, E. V.; Re, M. I., et al. Surface Morphology of Spray-dried Nanoparticle-coated Microparticles Designed as an Oral Drug Delivery System. *Braz. J. Chem. Eng.* **2008,** *25* (2), 389–398.

Puglisi, G.; Fresta, M.; Giammona, G.; Ventura, C. A. Influence of the Preparation Condi-tions on Poly(Ethylcyanoacrylate) Nanocapsule Formation. *Int. J. Pharm.* **1995,** *125* (2), 283–287.

Qiang, Z.; Zancong, S.; Tsuneji, N. Prolonged Hypoglycemic Effect of Insulin-loaded Poly-butylcyanoacrylate Nanoparticles After Pulmonary Administration to Normal Rats. *Int. J. Pharm.* **2001,** *218* (1–2), 75–80.

Qingyi, L.; Feng, G.; Dongyuan, Z. The Assembly of Semiconductor Sulfide Nanocrystallites with Organic Reagents as Templates. *Nanotechnology* **2002,** *13* (6), 741–745.

Radhika, P. R.; Sasikanth S.; Sivakumar, T. Nanocapsules: A New Approach for Drug Delivery. *Int. J. Pharma. Sci. Res.* **2011,** *2* (6), 1426–1429.

Raffin, P. A.; Weiss, V.; Mertins, O.; Pesce, S. N.; Stanisçuaski, G. S. Spray-dried Indometh-acin-loaded Polyester Nanocapsules and Nanospheres: Development, Stability Evaluation and Nanostructure Models. *Eur. J. Pharm. Sci.* **2002,** *16* (4–5), 305–312.

Redhead, H. M.; Davis, S. S.; Illum, L. Drug Delivery in Poly(lactide-co-glycolide) Nanopar-ticles Surface Modified with Poloxamer 407 and Poloxamine 908: *In vitro* Characterisation and *In vivo* Evaluation. *J. Control. Release* **2001,** *70* (3), 353–363.

Repka, M. Hot Melt Extrusion. In *Encyclopedia of Pharmaceutical Technology,* 2nd ed.; Swarbrick, J., Boylan, J., Ed.; Marcel Dekker Inc.: New York, 2002; Vol. 2, pp 1488–1504.

Samy, E. M.; Shautian, L.; Daniel, G.; Udo, P. Synthesis of Nanostructured Materials Using a Laser Vaporization Condensation Technique. *Nanotechnology* (ACS Symposium Series) **1996,** *622,* 79–99.

Sangwoo, P.; Hong, Y. C.; Jeong, Y. A.; Yungwan, K.; Abiraman, S.; Jeffrey, O., et al. Photo-cross-linkable Thermoresponsive Star Polymers Designed for Control of Cell Surface Interactions. *Biomacromolecules* **2010,** *11* (10), 2647–2652.

Sao, K.; Murata, M.; Umezaki, K.; Fujisaki, Y.; Mori, T.; Niidome, T., et al. Molecular Design of Protein-based Nanocapsules for Stimulus-responsive Characteristics. *Bioorg. Med. Che.* **2009,** *17* (1), 85–93.

Sarah, A.; Ying, W. Y.; Niveen, M. K.; Fraser, J. S.; Jeffrey, I. Dual-controlled Nanoparticles Exhibiting and Logic. *J. Am. Chem. Soc.* **2009,** *131* (32), 11344–11346.

Shen, Y.; Jin, E.; Zhang, B.; Murphy, C. J.; Sui, M.; Zhao, J., et al. Prodrugs Forming High Drug Loading Multifunctional Nanocapsules for Intracellular Cancer Drug Delivery. *J. Am. Chem. Soc.* **2010,** *132* (12), 4259–4265.

Slowing, I. I.; Trewyn, B. G.; Giri, S.; Lin, V. S. Y. Mesoporous Silica Nanoparticles for Drug Delivery and Biosensing Applications. *Adv. Funct. Mater.* **2007,** *17* (8), 1225–1236.

Song, M.; Dianyu, G.; Weishan, Z.; Wei, L.; Xiuliang, M.; Zhidong, Z. Synthesis of a New Type of GdAl2 Nanocapsule with a Large Cryogenic Magnetocaloric Effect and Novel Coral-like Aggregates Self-assembled by Nanocapsules. *Nanotechnology* **2006,** *17* (21), 5406–5411.

Stenekes, R. J.; Loebis, A. E.; Fernandes, C. M.; Crommelin, D. J.; Hennink, W. E. Degrad-able Dextran Microspheres for the Controlled Release of Liposomes. *Int. J. Pharm.* **2001,** *214* (1–2), 17–20.

Stephen, A. G.; Raval, F. S.; Huynh, T.; Torres, M.; Kickhoefer, V. A.; Rome, L. H. Assembly of Vault-like Particles in Insect Cells Expressing Only the Major Vault Protein. *J. Biol. Chem.* **2001,** *276* (26), 23217–23220.

Sung, O. C.; Eun, J. L.; Hyeok, M. L.; Yue, L.; Lan, Y. H.; Dong, P. K. Hierarchical Pore Structures Fabricated by Electron Irradiation of Silicone Grease and Their Applications to Superhydrophobic and Superhydrophilic Films. *Macromol. Rapid Commun.* **2007,** *28* (3), 246–251.

Vonarbourg, A.; Passirani, C.; Desigaux, L.; Allard, E.; Saulnier, P.; Lambert, O., et al. The Encapsulation of DNA Molecules Within Biomimetic Lipid Nanocapsules. *Biomaterials* **2009,** *30* (18), 3197–3204.

Wan, L. S. C.; Heng, P. W. S.; Chia, C. G. H. Spray Drying as a Process for Microencapsulation and the Effect of Different Coating Polymers. *Drug Dev. Ind. Pharm.* **1992,** *18* (9), 997–1011.

Watnasirichaikul, S.; Davies, N. M.; Rades, T.; Tucker, I. G. Preparation of Biodegradable Insulin Nanocapsules from Biocompatible Microemulsions. *Pharm. Res.* **2000,** *17* (6), 684–689.

Xinfei, Y.; Sheng, Z.; Xiaopeng, L.; Yingfeng, T.; Shuguang, Y.; Ryan, M., et al. A Giant Surfactant of Polystyrene−(Carboxylic Acid-functionalized Polyhedral Oligomeric Silsesquioxane) Amphiphile with Highly Stretched Polystyrene Tails in Micellar Assemblies. *J. Am. Chem. Soc.* **2010,** *132* (47), 16741–16744.

Yang, H.; Mendon, S. K.; Rawlins, J. W. Ion of Blocked Isocyanates Through Aqueous Emulsion Polymerization. *eXPRESS Polym. Lett.* **2008,** *2* (5), 349–356.

Yosida, Y.; Shida, S.; Ohsuna, T.; Shiraga, N. Synthesis, Identification and Growth Mechanism of Fe, Ni and Co Crystals Encapsulated in Multivalled Carbon Nanocages. *J. Appl. Phys.* **1994,** *76,* 8–11.

Zhang, Z. D. Nanocapsules. In *Encyclopedia of Nanoscience and Nanotechnology*; Nalwa, H. S., Ed.; American Scientific Publishers; USA, 2004; Vol. 6, pp 77–160.

Zhang, Z. D.; Zheng, J. G.; Skorvanek, I.; Wen, G. H.; Kovac, J.; Wang, F. W., et al. Shell/Core Structure and Magnetic Properties of Carbon-coated Fe-Co(C) Nanocapsules. *J. Phys. Condens. Matter* **2001,** *13* (9), 1921.

Zhu, Y. Q.; Hsu, W. K.; Zhou, W. Z.; Terrones, M.; Kroto, H. W.; Walton, D. R. M. Selective Cocatalysed Growth of Novel MgO Fishbone Fractal Nanostructures. *Chem. Phys. Lett.* **2001,** *347* (4–6), 337–343.

Ziyi, Z.; Huayi, C.; Songbei, T.; Jun, D.; Jianyi, L.; Kuang, L. T. Catalytic Growth of Carbon Nanoballs WITH and Without Cobalt Encapsulation. *Chem. Phys. Lett.* **2000,** *330* (1–2), 41–47.

INDEX

E

F

X

Z

Milton Keynes UK
Ingram Content Group UK Ltd.
UKHW031139141024
449569UK00024B/1200